GCSE
Physics

The Complete Course for AQA

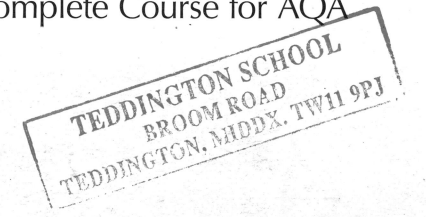

CGP

How to get your free Online Edition

Go to **cgpbooks.co.uk/extras** and enter this code...

2139 7470 8485 9363

This code will only work once. If someone has used this book before you,
they may have already claimed the Online Edition.

Contents

Published by CGP

Editors:
Robin Flello, Emily Garrett, Sharon Keeley-Holden, Duncan Lindsay, Frances Rooney,
Charlotte Whiteley and Jonathan Wray.

Contributors:
Gemma Hallam, Jason Howell.

From original material by Paddy Gannon.

ISBN: 978 1 78294 597 0

With thanks to Mark A Edwards and Glenn Rogers for the proofreading.
With thanks to Ana Pungartnik for the copyright research.

Data used to construct stopping distance diagram on page 209 from the Highway Code.
Contains public sector information licensed under the Open Government Licence v3.0.
http://www.nationalarchives.gov.uk/doc/open-government-licence/version/3/

Clipart from Corel®

How to use this book

Working Scientifically

- Working Scientifically is a big part of GCSE Physics. There's a whole section on it at the front of the book.

- Working Scientifically is also covered throughout this book wherever you see this symbol.

Learning Objectives

- These tell you exactly what you need to learn, or be able to do, for the exam.

- There's a specification reference at the bottom that links to the AQA specification.

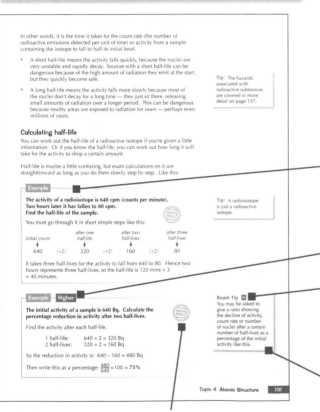

Examples

These are here to help you understand the theory.

Higher Exam Material

- Some of the material in this book will only come up in the exam if you're sitting the higher exam papers.

- This material is clearly marked with boxes that look like this:

Maths Skills

- There's a range of maths skills you could be expected to apply in your exams. The section on pages 333-342 is packed with plenty of maths that you'll need to be familiar with.

- Examples that show these maths skills in action are marked up with this symbol.

Tips and Exam Tips

- There are tips throughout this book to help you understand the theory.

- There are also exam tips to help you with answering exam questions.

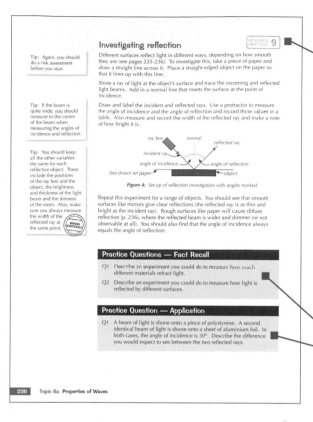

Required Practical Activities

There are some Required Practical Activities that you'll be expected to do throughout your course. You need to know all about them for the exams. They're all marked with stamps like this:

Practical Skills

There's also a whole section on pages 327-332 with extra details on practical skills you'll be expected to use in the Required Practical Activities, and apply knowledge of in the exams.

Practice Questions

- Fact recall questions test that you know the facts needed for GCSE Physics.

- Annoyingly, the examiners also expect you to be able to apply your knowledge to new situations — application questions give you plenty of practice at doing this.

- All the answers are in the back of the book.

Exam-style Questions

- Practising exam-style questions is really important — this book has some at the end of every topic to test you.

- They're the same style as the ones you'll get in the real exams.

- All the answers are in the back of the book, along with a mark scheme to show you how you get the marks.

- Higher-only questions are marked like this: **1.2**

Topic Checklist

Each topic has a checklist at the end with boxes that let you tick off what you've learnt.

Glossary

There's a glossary at the back of the book full of definitions you need to know for the exam, plus loads of other useful words.

Exam Help

There's a section at the back of the book stuffed full of things to help you with the exams.

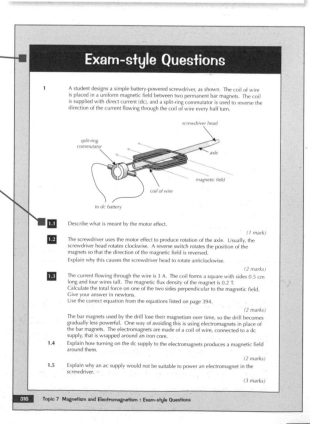

1. The Scientific Method

Science is all about finding things out and learning things about the world we live in. This section is all about the scientific process — how a scientist's initial idea turns into a theory that is accepted by the wider scientific community.

Hypotheses

Scientists try to explain things. Everything. They start by observing something they don't understand — it could be anything, e.g. planets in the sky, a person suffering from an illness, what matter is made of... anything.

Then, they come up with a **hypothesis** — a possible explanation for what they've observed. (Scientists can also sometimes form a model — a description or a representation of what's physically going on — see page 3).

The next step is to test whether the hypothesis might be right or not. This involves making a **prediction** based on the hypothesis and testing it by gathering evidence (i.e. data) from investigations. If evidence from experiments backs up a prediction, you're a step closer to figuring out if the hypothesis is true.

Tip: Investigations include lab experiments and studies.

Testing a hypothesis

Normally, scientists share their findings in peer-reviewed journals, or at conferences. **Peer-review** is where other scientists check results and scientific explanations to make sure they're 'scientific' (e.g. that experiments have been done in a sensible way) before they're published. It helps to detect false claims, but it doesn't mean that findings are correct — just that they're not wrong in any obvious way.

Once other scientists have found out about a hypothesis, they'll start basing their own predictions on it and carry out their own experiments. They'll also try to reproduce the original experiments to check the results — and if all the experiments in the world back up the hypothesis, then scientists start to think the hypothesis is true.

However, if a scientist somewhere in the world does an experiment that doesn't fit with the hypothesis (and other scientists can reproduce these results), then the hypothesis is in trouble. When this happens, scientists have to come up with a new hypothesis (maybe a modification of the old hypothesis, or maybe a completely new one).

Tip: Sometimes it can take a really long time for a hypothesis to be accepted.

Accepting a hypothesis

If pretty much every scientist in the world believes a hypothesis to be true because experiments back it up, then it usually goes in the textbooks for students to learn. Accepted hypotheses are often referred to as **theories**.

Our currently accepted theories are the ones that have survived this 'trial by evidence' — they've been tested many, many times over the years and survived (while the less good ones have been ditched). However... they never, never become hard and fast, totally indisputable fact. You can never know... it'd only take one odd, totally inexplicable result, and the hypothesising and testing would start all over again.

negatively-charged electrons

sphere of positive charge

Figure 1: *The plum pudding model of the atom.*

Example

- About 200 years ago, John Dalton proposed that atoms were tiny bits of matter which could never be split up.

- About 100 years later, the electron was discovered. This meant our hypothesis of atoms was revised, and we began to think of atoms as lumps of positive charge with negative electrons sat inside them (see Figure 1) — this was the plum pudding model (see page 120).

- After more evidence was gathered, the hypothesis was changed again and we developed the nuclear model of the atom — a tiny positive nucleus, orbited by negative electrons (see p. 122).

Models

Models are used to describe or display how an object or system behaves in reality. They're often based on evidence collected from experiments, and should be able to accurately predict what will happen in other, similar experiments. There are different types of models that scientists can use to describe the world around them. Here are just a few:

- A **descriptive model** describes what's happening in a certain situation, without explaining why. It won't necessarily include details that could be used to predict the outcome of a different scenario. For example, a graph showing the measured resistance of a device at different temperatures would be a descriptive model.

- A **representational model** is a simplified description or picture of what's going on in real life. It can be used to explain observations and make predictions. E.g. the nuclear model is a simplified way of showing the arrangement of electrons in an atom (see p. 122). It can be used to explain how electrons move between different energy levels.

- **Spatial models** are used to summarise how data is arranged within space. For example, a map showing where high levels of radiation from rocks is found (see page 136) would be a spatial model.

- **Computational models** use computers to make simulations of complex real-life processes, such as climate change. They're used when there are a lot of different variables (factors that change) to consider, and because you can easily change their design to take into account new data.

- **Mathematical models** can be used to describe the relationship between variables in numerical form (e.g. as an equation), and therefore predict outcomes of a scenario. For example, an equation can be written to predict the energy transferred to or from a material's thermal energy stores when it changes temperature. This is dependent on the change in temperature of the material, its mass and its specific heat capacity (see page 30).

Tip: Like hypotheses, models have to be tested before they're accepted by other scientists. You can test models by using them to make a prediction, and then carrying out an investigation to see whether the results match the prediction.

higher

lower

Figure 2: *A spatial model showing radiation from rocks in the United Kingdom. The scale shows how the level of radiation from rocks in different areas varies.*

Tip: Mathematical models are made using patterns found in data and also using information about known relationships between variables.

Tip: Like hypotheses, models are constantly being revised and modified, based on new data.

All models have limitations on what they can explain or predict. The Big Bang model (a model used to describe the beginning of the universe) can be used to explain why everything in the universe is moving away from us. One of its limitations is that is doesn't explain the moments before the Big Bang.

Communicating results

Some scientific discoveries show that people should change their habits, or they might provide ideas that could be developed into new technology. So scientists need to tell the world about their discoveries.

Tip: New scientific discoveries are usually communicated to the public in the news or via the internet. They might be communicated to governments and large organisations via reports or meetings.

> **Example**
>
> Radioactive materials are used widely in medicine for imaging and treatment (see p. 138). Information about these materials needs to be communicated to doctors so they can make use of them. The patients also need to be told of the risks involved so they can make informed decisions about their treatment.

Reports about scientific discoveries in the media (e.g. newspapers or television) aren't peer-reviewed. This means that, even though news stories are often based on data that has been peer-reviewed, the data might be presented in a way that is over-simplified or inaccurate, leaving it open to misinterpretation.

Tip: If you're reading an article about a new scientific discovery, always think about how the study was carried out. It may be that the sample size was very small, and so the results aren't representative (see page 11 for more on sample sizes).

It's important that the evidence isn't presented in a **biased** way. This can sometimes happen when people want to make a point, e.g. they overemphasise a relationship in the data. (Sometimes without knowing they're doing it.) There are all sorts of reasons why people might want to do this.

> **Examples**
>
> - They want to keep the organisation or company that's funding the research happy. (If the results aren't what they'd like they might not give them any more money to fund further research.)
> - Governments might want to persuade voters, other governments or journalists to agree with their policies about a certain issue.
> - Companies might want to 'big up' their products, or make impressive safety claims.
> - Environmental campaigners might want to persuade people to behave differently.

Tip: An example of bias is a newspaper article describing details of data supporting an idea without giving any of the evidence against it.

There's also a risk that if an investigation is done by a team of highly-regarded scientists it'll be taken more seriously than evidence from less well known scientists. But having experience, authority or a fancy qualification doesn't necessarily mean the evidence is good — the only way to tell is to look at the evidence scientifically (e.g. is it repeatable, valid, etc. — see page 9).

2. Scientific Applications and Issues

New scientific discoveries can lead to lots of exciting new ways of using science in our everyday lives. Unfortunately, these developments may also come with problems that need to be considered.

Using scientific developments

Lots of scientific developments go on to have useful applications.

> **Examples**
>
> - The discovery of the generator effect (which turns movement into potential difference) allowed people to develop machines to easily generate electricity. This is the basis for the widespread electricity supplies we have today.
>
> - As scientists investigated the effects of radiation, they found they had damaging effects on the body, which could be used to kill cancer cells (see pages 138-139).

Tip: **H** See page 303 for more on the generator effect.

Issues created by science

Scientific knowledge is increased by doing experiments. And this knowledge leads to scientific developments, e.g. new technologies or new advice. These developments can create issues though. For example, they could create political issues, which could lead to developments being ignored, or governments being slow to act if they think responding to the developments could affect their popularity with voters.

> **Example**
>
> Some governments were pretty slow to accept the fact that human activities are causing global warming, despite all the evidence. This is because accepting it means they've got to do something about it, which costs money and could hurt their economy. This could lose them a lot of votes.

Tip: See page 56 for more on global warming.

Scientific developments can cause a whole host of other issues too.

> **Examples**
>
> - **Economic issues**: Society can't always afford to do things scientists recommend (e.g. investing heavily in alternative energy sources) without cutting back elsewhere.
>
> - **Social issues**: Decisions based on scientific evidence affect people — e.g. should fossil fuels be taxed more highly (to encourage investment in alternative energy)? Should alcohol be banned (to prevent health problems)? Would the effect on people's lifestyles be acceptable?
>
> - **Environmental issues**: Human activity often affects the natural environment — e.g. burning fossil fuels may be cheaper than using alternative energy resources, but it has a large impact on the environment.
>
> - **Personal issues**: Some decisions will affect individuals. For example, someone might support alternative energy, but object if a wind farm is built next to their house.

Figure 1: *The ATLAS detector at CERN, part of the Large Hadron Collider (LHC). The LHC has greatly advanced our knowledge of subatomic particles, but it also cost a lot of money to build and run.*

3. Limitations of Science

Science has taught us an awful lot about the world we live in and how things work — but science doesn't have the answer for everything.

Questions science hasn't answered yet

We don't understand everything. And we never will. We'll find out more, for sure — as more hypotheses are suggested, and more experiments are done. But there'll always be stuff we don't know.

> **Examples**
>
> - Today we don't know as much as we'd like about the impacts of global warming. How much will sea levels rise? And to what extent will weather patterns change?
>
> - We also don't know anywhere near as much as we'd like about the universe. Are there other life forms out there? And what is most of the universe made of?

In order to answer scientific questions, scientists need data to provide evidence for their hypotheses. Some questions can't be answered yet because the data can't currently be collected, or because there's not enough data to support a theory. But eventually, as we get more evidence, we probably will be able to answer these questions. By then there'll be loads of new questions to answer though.

Figure 1: The night sky. We can use high powered telescopes to observe some of the universe but we still have little idea what most of it is made of or how it was formed.

Questions science can't answer

There are some questions that all the experiments in the world won't help us answer — for example, the "should we be doing this at all?" type questions.

> **Example**
>
> Take the idea of space exploration and the search for alien life — some people think it's a good idea. It increases our knowledge of the universe and the new technologies developed to explore space can be useful on Earth too. For example, a lot of developments within space exploration has led to improvements of MRI machines — these are used in hospitals to image the inside of the body. It also inspires young people to get into science.
>
> Other people say it's a bad idea. They think we should concentrate on understanding our own planet better first, or spend the vast amounts of money on solving more urgent problems here on Earth — things like providing clean drinking water and curing diseases in poor countries.

Tip: Some experiments have to be approved by ethics committees before scientists are allowed to carry them out. This stops scientists from getting wrapped up in whether they <u>can</u> do something, before anyone stops to think about whether they <u>should</u> do it.

The question of whether something is morally or ethically right or wrong can't be answered by more experiments — there is no "right" or "wrong" answer. The best we can do is get a consensus from society — a judgement that most people are more or less happy to live by. Science can provide more information to help people make this judgement, and the judgement might change over time. But in the end it's up to people and their conscience.

4. Risks and Hazards

A lot of things we do could cause us harm. But some things are more hazardous than they at first seem, whereas other things are less hazardous than they at first seem. This may sound confusing, but it'll all become clear...

What are risks and hazards?

A **hazard** is something that could potentially cause harm. All hazards have a **risk** attached to them — this is the chance that the hazard will cause harm.

The risks of some things seem pretty obvious, or we've known about them for a while, like the risk of causing acid rain by polluting the atmosphere, or of having a car accident when you're travelling in a car.

New technology arising from scientific advances can bring new risks. These risks need to be thought about alongside the potential benefits of the technology, in order to make a decision about whether it should be made available to the general public.

Example

There are now a number of imaging technologies, like CT scans, that can be used to see inside the human body. These can allow doctors to diagnose medical conditions in order to treat them.

However, these scans use ionising radiation, and so carry with them the risk of causing cancer in the patient. The risk of dying from cancer caused by the scans has to be weighed against the risk of dying from the medical condition that needs to be diagnosed. Often the risk from the medical condition far outweighs the risk from the scan, so the scan will go ahead.

Tip: A CT (computerised tomography) scan uses X-rays to form an image of the inside of a patient (see page 250).

Figure 1: *A patient being prepared for a CT scan.*

Estimating risk

You can estimate the risk based on how many times something happens in a big sample (e.g. 100 000 people) over a given period (e.g. a year). For example, you could assess the risk of a driver crashing by recording how many people in a group of 100 000 drivers crashed their cars over a year.

To make a decision about an activity that involves a hazardous event, we don't just need to take into account the chance of the event causing harm, but also how serious the consequences would be if it did.

The general rule is that, if an activity involves a hazard that's very likely to cause harm, with serious consequences if it does, that activity is considered high-risk.

Example 1

If you go for a run, you may sprain an ankle. But most small sprains recover within a few days if they're rested, so going for a run would be considered a low-risk activity.

If you go skiing, you may fall and break a bone. This would take many weeks to heal, and may cause further complications later on in life. So skiing would be considered higher risk than running.

Perceptions of risk

Not all risks have the same consequences, e.g. if you chop veg with a sharp knife you risk cutting your finger, but if you go scuba-diving you risk death. You're much more likely to cut your finger during half an hour of chopping than to die during half an hour of scuba-diving. But most people are happier to accept a higher probability of an accident if the consequences are short-lived and fairly minor.

People tend to be more willing to accept a risk if they choose to do something (e.g. go scuba diving), compared to having the risk imposed on them (e.g. having a nuclear power station built next door).

People's perception of risk (how risky they think something is) isn't always accurate. They tend to view familiar activities as low-risk and unfamiliar activities as high-risk — even if that's not the case. For example, cycling on roads is often high-risk, but many people are happy to do it because it's a familiar activity. Air travel is actually pretty safe, but a lot of people perceive it as high-risk. People may under-estimate the risk of things with long-term or invisible negative effects, e.g. using tanning beds.

Tip: Risks people choose to take are called 'voluntary risks'. Risks that people are forced to take are called 'imposed risks'.

Tip: You can find out about potential hazards by looking in textbooks, doing some internet research, or asking your teacher.

Reducing risk in investigations

Part of planning an investigation is making sure that it's safe. To make sure your experiment is safe you must identify all the hazards. Hazards include:

- Lasers: e.g. if a laser is directed into the eye, this can cause blindness.

- Gamma radiation: e.g. gamma-emitting radioactive sources can cause cancer.

- Fire: e.g. an unattended Bunsen burner is a fire hazard.

- Electricity: e.g. faulty electrical equipment could give you a shock.

Once you've identified the hazards you might encounter, you should think of ways of reducing the risks from the hazards.

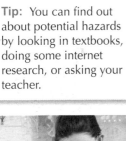

Figure 2: A scientist wearing safety goggles to protect her eyes during an experiment.

- If you're working with springs, always wear safety goggles. This will reduce the risk of the spring (or a fragment of it) hitting your eye if the spring snaps.

- If you're using a Bunsen burner, stand it on a heat proof mat. This will reduce the risk of starting a fire.

5. Designing Investigations

To be a good scientist you need to know how to design a good experiment, including how to make sure you get good quality results.

Making predictions from a hypothesis

Scientists observe things and come up with hypotheses to explain them. To decide whether a **hypothesis** might be correct you need to do an investigation to gather evidence, which will help support or disprove the hypothesis. The first step is to use the hypothesis to come up with a **prediction** — a statement about what you think will happen that you can test.

> **Example**
>
> If your hypothesis is "aluminium foil is a better thermal insulator than newspaper", then your prediction might be "a cup of hot water wrapped in newspaper will cool down to a lower temperature than a cup of hot water wrapped in aluminium foil will over a period of five minutes".

Once a scientist has come up with a prediction, they'll design an investigation to see if there are patterns or relationships between two variables. For example, to see if there's a pattern or relationship between the variables 'insulating material used' and 'change in temperature'.

Tip: A variable is just something in the experiment that can change.

Repeatable and reproducible results

Results need to be **repeatable** and **reproducible**. Repeatable means that if the same person does an experiment again using the same methods and equipment, they'll get similar results. Reproducible means that if someone else does the experiment, or a different method or piece of equipment is used, the results will be similar.

Tip: Data that's repeatable and reproducible is <u>reliable</u> and scientists are more likely to have confidence in it.

> **Example**
>
> In 1989, two scientists claimed that they'd produced 'cold fusion' (the energy source of the Sun but without the high temperatures). If it was true, it would have meant free energy for the world forever. However, other scientists just couldn't reproduce the results, so 'cold fusion' wasn't accepted as a theory.

Figure 1: *Stanley Pons and Martin Fleischmann — the scientists who allegedly discovered cold fusion.*

Ensuring the test is valid

Valid results are repeatable, reproducible and answer the original question.

> **Example**
>
> **Do power lines cause cancer?**
>
> Some studies have found that children who live near overhead power lines are more likely to develop cancer. What they'd actually found was a **correlation** (relationship) between the variables "presence of power lines" and "incidence of cancer". They found that as one changed, so did the other.
>
> But this data isn't enough to say that the power lines cause cancer, as there might be other explanations. For example, power lines are often near busy roads, so the areas tested could contain different levels of pollution. As the studies don't show a definite link they don't answer the original question.

Tip: Peer review (see page 2) is used to make sure that results are valid before they're published.

Tip: See page 17 for more on correlation.

Ensuring it's a fair test

Tip: For the results of an investigation to be <u>valid</u> the investigation must be a <u>fair test</u>.

In a lab experiment you usually change one variable and measure how it affects another variable. To make it a fair test, everything else that could affect the results should stay the same (otherwise you can't tell if the thing you're changing is causing the results or not — the data won't be valid).

> **Example**
>
> To investigate how the length of a wire affects its resistance, you must only change the wire length. You need to keep, for example, the temperature the same, otherwise you won't know if any change in the resistance was caused by the change in length, or the change in temperature.

The variable you change is called the **independent variable**. The variable you measure when you change the independent variable is called the **dependent variable**. The variables that you keep the same are called **control variables**.

> **Example**
>
> In the resistance experiment above, the length of the wire is the independent variable, the resistance is the dependent variable, and the control variables are the temperature, supplied potential difference, wire thickness, etc.

Control experiments and control groups

Tip: Control experiments let you see what happens when you don't change anything at all.

Tip: The control experiment in this example could also account for any systematic errors (see page 13) that may be affecting all of your results.

In some investigations it's useful to have a **control experiment** — an experiment that's kept under the same conditions as the rest of the investigation, but doesn't have anything done to it. This allows you to see exactly what effect changing the independent variable has in the investigation.

> **Example**
>
> You can investigate how different materials act as thermal insulators by wrapping each material round a beaker of hot water and recording how much the temperature of the water has dropped by after a given time period (see page 38). However, you would also carry out the experiment without any insulating material — the control experiment. This gives you a point of comparison, so you can evaluate by how much a thermal insulator has slowed down the cooling process. Without it, you wouldn't know if the insulators were actually slowing down the cooling rate.

It's important that a study (an investigation that doesn't take place in a lab) is a fair test, just like a lab experiment. It's a lot trickier to control the variables in a study than it is in a lab experiment though. Sometimes you can't control them all, but you can use a **control group** to help. This is a group of whatever you're studying (e.g. people) that's kept under the same conditions as the group in the experiment, but doesn't have anything done to it.

> **Example**
>
> If you were studying the link between CT scans and thyroid cancer, you'd take one group of people who have had CT scans, and another group (the control group) who haven't. Both groups should be of roughly the same age, live in the same area, have similar lifestyles, etc.
>
> The control group will help you try to account for other variables like people's diet, which could affect the results.

Sample size

Data based on small samples isn't as good as data based on large samples. A sample should be representative of the whole population (i.e. it should share as many of the various characteristics in the population as possible) — a small sample can't do that as well.

Tip: It's hard to spot anomalies if your sample size is too small.

The bigger the sample size the better, but scientists have to be realistic when choosing how big.

Example

If you were studying how exposure to sunlight affects people's risk of skin cancer it'd be great to study everyone in the UK (a huge sample), but it'd take ages and cost a bomb. Studying a thousand people with a mixture of ages, gender and race would be more realistic.

Trial runs

It's a good idea to do a **trial run** (a quick version of your experiment) before you do the proper experiment. Trial runs are used to figure out the range (the upper and lower limits) of independent variable values used in the proper experiment. If there was no change in the dependent variable between your upper and lower values in the trial run, then you might increase the range until there was an observable change. Or if there was a large change, you might want to make your higher and lower values closer together.

Tip: If you don't have time to do a trial run, you could always look at the data other people have got doing a similar experiment and use a range and interval values similar to theirs.

Example

In the experiment on p. 163, masses are added to a spring to find how the force acting on the spring is related to its extension. Enough data points need to be collected before the spring stretches beyond its limit of proportionality. Doing a trial run on the spring (or an identical one) ensures that your lower independent variable value (the smallest total mass) doesn't stretch the spring too far, but is also large enough to cause a notable extension. A trial run could also be used to check that your upper independent value (the largest mass) will tell you what the spring's limit of proportionality is.

Trial runs can be used to figure out the appropriate intervals (gaps) between the values too. The intervals can't be too small (otherwise the experiment would take ages), or too big (otherwise you might miss something).

Example

If, in the experiment in the example above, the spring extended a lot when you added a 100 g mass, you may be better using 10 g masses instead.

Tip: Consistently repeating the results is crucial for checking that your results are repeatable.

Trial runs can also help you figure out whether or not your experiment is repeatable. If you repeat it three times and the results are all similar, the experiment is repeatable.

6. Collecting Data

Once you've designed your experiment, you need to get on and do it. Here's a guide to making sure the results you collect are good.

Getting good quality results

When you do an experiment you want your results to be **repeatable**, **reproducible** and as **accurate** and **precise** as possible.

To check repeatability you need to repeat the readings and check that the results are similar — you should repeat each reading at least three times. To make sure your results are reproducible you can cross check them by taking a second set of readings with another instrument (or a different observer).

Tip: Sometimes, you can work out what result you should get at the end of an experiment (the theoretical result) by doing a bit of maths. If your experiment is accurate there shouldn't be much difference between the theoretical result and the result you actually get.

Your data also needs to be accurate. Really accurate results are those that are really close to the true answer. The accuracy of your results usually depends on your method — you need to make sure you're measuring the right thing and that you don't miss anything that should be included in the measurements. For example, finding the wavelength of water waves produced by a signal generator by measuring a single wave isn't very accurate because the wavelengths can be very small and the waves are constantly moving. It's more accurate to measure the length of multiple waves, and then divide it by the number of waves to find the wavelength. Any inaccuracy on the initial measurement is also divided, making your final result more accurate. See pages 229-230 for more on this experiment.

Your data also needs to be precise. Precise results are ones where the data is all really close to the mean (average) of your repeated results (i.e. not spread out).

Tip: For more on means see page 14.

> ### Example
>
> Look at the data in this table. Data set 1 is more precise than data set 2 because all the data in set 1 is really close to the mean, whereas the data in set 2 is more spread out.
>
> | Repeat | Data set 1 | Data set 2 |
> |--------|-----------|-----------|
> | 1 | 12 | 11 |
> | 2 | 14 | 17 |
> | 3 | 13 | 14 |
> | Mean | 13 | 14 |

Choosing the right equipment

When doing an experiment, you need to make sure you're using the right equipment for the job. The measuring equipment you use has to be sensitive enough to measure the changes you're looking for.

> ### Example
>
> If you need to measure changes of 1 cm^3 you need to use a measuring cylinder that can measure in 1 cm^3 steps — it'd be no good trying with one that only measures 10 cm^3 steps, it wouldn't be sensitive enough.

Figure 1: *Different types of measuring cylinder and glassware — make sure you choose the right one before you start an experiment.*

The smallest change a measuring instrument can detect is called its **resolution**. For example, some mass balances have a resolution of 1 g, some have a resolution of 0.1 g, and some are even more sensitive.

Also, equipment needs to be **calibrated** by measuring a known value. If there's a difference between the measured and known value, you can use this to correct the inaccuracy of the equipment.

Tip: Calibration is a way of making sure that a measuring device is measuring things accurately — you get it to measure something you know has a certain value and set the device to say that amount.

> **Example**
>
> If a known mass is put on a mass balance, but the reading is a different value, you know that the mass balance has not been calibrated properly.

Errors

Random errors

The results of an experiment will always vary a bit due to **random errors** — unpredictable differences caused by things like human errors in measuring.

> **Example**
>
> Errors made when reading from a measuring cylinder are random. You have to estimate or round the level when it's between two marks — so sometimes your figure will be a bit above the real one, and sometimes a bit below.

You can reduce the effect of random errors by taking repeat readings and finding the mean. This will make your results more precise.

Systematic errors

If a measurement is wrong by the same amount every time, it's called a **systematic error**.

Tip: If there's no systematic error, then doing repeats and calculating a mean can make your results more accurate.

> **Example**
>
> If you measured from the very end of your ruler instead of from the 0 cm mark every time, all your measurements would be a bit small.

Just to make things more complicated, if a systematic error is caused by using equipment that isn't zeroed properly it's called a **zero error**. You can compensate for some of these errors if you know about them though.

Tip: A zero error is a specific type of systematic error.

> **Example**
>
> If a mass balance always reads 1 gram before you put anything on it, all your measurements will be 1 gram too heavy. This is a zero error. You can compensate for this by subtracting 1 gram from all your results.

Tip: Repeating the experiment in the exact same way and calculating a mean won't correct a systematic error.

Anomalous results

Sometimes you get a result that doesn't seem to fit in with the rest at all. These results are called **anomalous results** (or outliers).

> **Example**
>
> Look at the data in this table. The entry that has been circled is an anomalous result because it's much larger than any of the other data values.
>
> | Experiment | A | B | C | D | E | F |
> |---|---|---|---|---|---|---|
> | Acceleration (m/s²) | 1.05 | 1.12 | 1.08 | (8.54) | 1.06 | 1.11 |

Tip: There are lots of reasons why you might get an anomalous result, but usually they're due to human error rather than anything crazy happening in the experiment.

You should investigate anomalous results and try to work out what happened. If you can work out what happened (e.g. you measured something totally wrong) you can ignore them when processing your results.

7. Processing Data

Once you've collected some data, you might need to process it.

Organising data

It's really important that your data is organised. Tables are dead useful for organising data. When you draw a table use a ruler, make sure each column has a heading (including the units) and keep it neat and tidy.

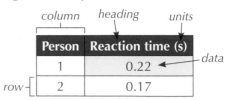

Person	Reaction time (s)
1	0.22
2	0.17

Figure 1: Table showing the time taken to react to a stimulus for two people.

Processing your data

When you've collected data from a number of repeats of an experiment, it's useful to summarise it using a few handy-to-use figures.

Mean and range

When you've done repeats of an experiment you should always calculate the **mean** (a type of average). To do this add together all the data values and divide by the total number of values in the sample.

You might also need to calculate the **range** (how spread out the data is). To do this find the largest number and subtract the smallest number from it.

> **Example**
>
> Look at the data in the table below. The mean and range of each set of data has been calculated.
>
> | Trolley | Repeat (m/s) 1 | 2 | 3 | Mean (m/s) | Range (m/s) |
> |---------|----|----|----|------------|-------------|
> | A | 3.1 | 3.6 | 3.2 | (3.1 + 3.6 + 3.2) ÷ 3 = 3.3 | 3.6 – 3.1 = 0.5 |
> | B | 4.7 | 5.1 | 5.8 | (4.7 + 5.1 + 5.8) ÷ 3 = 5.2 | 5.8 – 4.7 = 1.1 |

Median and mode

There are two more types of average, other than the mean, that you might need to calculate. These are the **median** and the **mode**.

- To calculate the median, put all your data in numerical order — the median is the middle value.

- The number that appears most often in a data set is the mode.

> **Example**
>
> **The results of a study investigating the reaction times (in seconds) of students are shown below:**
> **0.10, 0.25, 0.20, 0.15, 0.25, 0.15, 0.25, 0.20, 0.30**
>
> First put the data in numerical order:
> 0.10, 0.15, 0.15, 0.20, 0.20, 0.25, 0.25, 0.25, 0.30
>
> There are nine values, so the median is the 5th number, which is 0.20.
>
> 0.25 comes up three times. No other numbers come up more than twice. So the mode is 0.25.

Tip: If you're recording your data as decimals, make sure you give each value to the same number of decimal places.

Tip: Annoyingly, it's difficult to see any patterns or relationships in detail just from a table. You need to use some kind of graph or chart for that (see pages 16-17).

Tip: You should ignore anomalous results when calculating the mean, range, median or mode — see page 13 for more on anomalous results.

Tip: If you have an even number of values, the median is the mean of the middle two values.

Uncertainty

When you repeat a measurement, you often get a slightly different figure each time you do it due to random error. This means that each result has some **uncertainty** to it. The measurements you make will also have some uncertainty in them due to limits in the resolution of the equipment you use. This all means that the mean of a set of results will also have some uncertainty to it. Here's how to calculate the uncertainty of a mean result:

Tip: There's more about errors on page 13.

$$\text{uncertainty} = \frac{\text{range}}{2}$$

The larger the range, the less precise your results are and the more uncertainty there will be in your results. Uncertainties are shown using the '±' symbol.

Example

The table below shows the results of an experiment to determine the resistance of a piece of wire in a circuit.

Repeat	1	2	3	mean
Resistance (Ω)	4.20	3.80	3.70	3.90

1. The range is: $4.20 - 3.70 = 0.50 \ \Omega$

2. So the uncertainty of the mean is: range ÷ 2 = 0.50 ÷ 2 = 0.25 Ω. You'd write this as 3.90 ± 0.25 Ω

Tip: Since uncertainty affects precision, you'll need to think about it when you come to evaluating your results (see page 21).

Measuring a greater amount of something helps to reduce uncertainty. For example, for radioactive decay, measuring the count rate over a longer period compared to a shorter period reduces the percentage uncertainty in your results.

Rounding to significant figures

The first **significant figure** (s.f.) of a number is the first digit that isn't a zero. The second, third and fourth significant figures follow on immediately after the first (even if they're zeros). When you're processing your data you may well want to round any really long numbers to a certain number of s.f.

Exam Tip
If a question asks you to give your answer to a certain number of significant figures, make sure you do this, or you might not get all the marks.

Example

0.6874976 rounds to **0.69 to 2 s.f.** and to **0.687 to 3 s.f.**

When you're doing calculations using measurements given to a certain number of significant figures, you should give your answer to the lowest number of significant figures that was used in the calculation. If your calculation has multiple steps, only round the final answer, or it won't be as accurate.

Tip: Remember to write down how many significant figures you've rounded to after your answer.

Example

For the calculation: $1.2 \div 1.85 = 0.648648648...$

1.2 is given to 2 significant figures. 1.85 is given to 3 significant figures. So the answer should be given to 2 significant figures.

Round the final significant figure (0.6<u>4</u>8) up to 5: $1.2 \div 1.85 = 0.65$ (2 s.f.)

Tip: When rounding a number, if the next digit after the last significant figure you're using is less than 5 you should round it <u>down</u>, and if it's 5 or more you should round it <u>up</u>.

The lowest number of significant figures in the calculation is used because the fewer digits a measurement has, the less accurate it is. Your answer can only be as accurate as the least accurate measurement in the calculation.

8. Graphs and Charts

It can often be easier to see trends in data by plotting a graph or chart of your results, rather than by looking at numbers in a table.

Plotting your data on a graph or chart

One of the best ways to present your data after you've processed it is to plot your results on a graph or chart. You need to know these rules about drawing graphs and charts:

- Draw it nice and big (covering at least half of the graph paper).

- Label both axes and remember to include the units.

- If you've got more than one set of data include a key.

- Give your graph a title explaining what it is showing.

Whatever type of graph or chart you draw, make sure you follow the rules above. There are lots of different types you can use. The type you should use depends on the type of data you've collected.

Bar charts and histograms

If either the independent or dependent variable is **categoric** or **discrete**, you should use a bar chart to display the data (see Figure 1). If the independent variable is **continuous**, the frequency data should be shown on a histogram. Histograms may look like bar charts, but it's the area of the bars that represents the frequency (rather than height). The height of each bar is called the **frequency density** and is found by dividing the frequency by the class width. (The class width is just the width of the bar on the histogram, see Figure 1.)

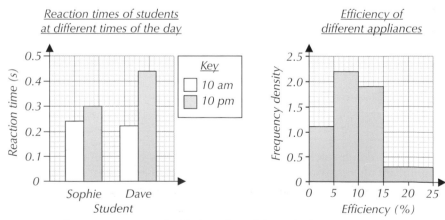

Figure 1: *An example of a bar chart (left) and a histogram (right).*

Plotting points

If the independent and the dependent variable are continuous you should plot points on a graph to display the data. Here are the golden rules specifically for plotting points on graphs:

- Put the independent variable (the thing you change) on the x-axis.

- Put the dependent variable (the thing you measure) on the y-axis.

- To plot the points, use a sharp pencil and make a neat little cross.

Tip: Categoric data is data that comes in distinct categories, e.g. 'state of matter (solid, liquid, gas)' and 'type of material (e.g. wood metal, paper)'. Discrete data can only take certain values, because there are no in-between values, e.g. 'number of people' (because you can't have half a person). Continuous data is numerical data that can have any value within a range, e.g. length, volume, temperature.

Tip: Frequency is just the number of times that something occurs. It's often shown in a frequency table.

Tip: A frequency diagram is a histogram where the width of all the bars are the same and frequency is plotted on the y-axis, rather than frequency density.

Tip: The x-axis is the horizontal axis and the y-axis is the vertical axis.

In general, you shouldn't join the crosses up. Only specific graphs, such as distance-time graphs (page 183) and velocity-time graphs (page 186), will need you to connect every point you plot. Otherwise, you'll need to draw a line of best fit (or a curve of best fit if your points make a curve). When drawing a line (or curve), try to draw the line through or as near to as many points as possible, ignoring anomalous results. When you draw a line of best fit, there should be roughly as many points above the line as underneath it.

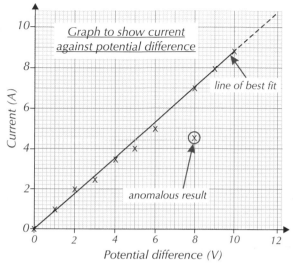

Figure 2: An example of a graph with points plotted and a line of best fit drawn.

Tip: Use the biggest data values you've got to draw a sensible scale on your axes. Here, the highest current is 8.8 A, so it makes sense to label the *y*-axis up to 10 A.

Tip: If you're not in an exam, you can use a computer to plot your graph and draw your line of best fit for you.

Correlations

Graphs are used to show the relationship between two variables. Data can show three different types of **correlation** (relationship).

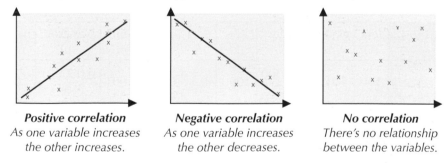

Positive correlation	**Negative correlation**	**No correlation**
As one variable increases the other increases.	As one variable increases the other decreases.	There's no relationship between the variables.

Figure 3: Examples of different types of correlations shown on a graph.

You also need to be able to describe the following types of graphs.

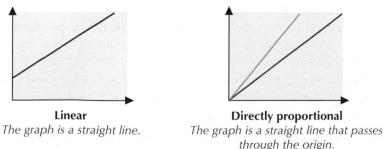

Linear	**Directly proportional**
The graph is a straight line.	The graph is a straight line that passes through the origin.

Figure 4: Examples of linear graphs. The graph on the right shows two linear graphs where the variables are directly proportional to each other.

Tip: Just because two variables are correlated doesn't mean that the change in one is causing the change in the other. There might be other factors involved, or it could be due to chance — see pages 20-21 for more.

Tip: For variables that are directly proportional, both variables increase (or decrease) in the same ratio.

Tip: You can find out a lot of useful information from the line of best fit you draw. For more on how to get the most out of your graphs, see pages 339-341.

9. Units

Using the correct units is important when you're drawing graphs or calculating values with an equation. Otherwise your numbers don't really mean anything.

S.I. units

Lots of different units can be used to describe the same quantity. For example, volume can be given in terms of cubic feet, cubic metres, litres or pints. It would be quite confusing if different scientists used different units to define quantities, as it would be hard to compare people's data. To stop this happening, scientists have come up with a set of standard units, called **S.I. units**, that all scientists use to measure their data. Here are some S.I. units you'll see in physics:

Tip: SI stands for 'Système International', which is French for 'international system'.

Quantity	S.I. Unit
mass	kilogram, kg
length	metre, m
time	second, s
electric current	ampere, A

Figure 1: *Some common S.I. units used in physics.*

Scaling prefixes

Quantities come in a huge range of sizes. For example, the volume of a swimming pool might be around 2 000 000 000 cm^3, while the volume of a cup is around 250 cm^3. To make the size of numbers more manageable, larger or smaller units are used. Figure 2 shows the prefixes which can be used in front of units (e.g. metres) to make them bigger or smaller:

prefix	tera (T)	giga (G)	mega (M)	kilo (k)	deci (d)	centi (c)	milli (m)	micro (µ)	nano (n)
multiple of unit	10^{12}	10^9	1 000 000 (10^6)	1000	0.1	0.01	0.001	0.000001 (10^{-6})	10^{-9}

Figure 2: *Scaling prefixes used with units.*

These prefixes are called **scaling prefixes** and they tell you how much bigger or smaller a unit is than the original unit. So one kilometre is one thousand metres.

Converting between units

Exam Tip
If you're going from a smaller unit to a larger unit, your number should get smaller.
If you're going from a larger unit to a smaller unit, your number should get larger.
This is a handy way to check you've done the conversion correctly.

To swap from one unit to another, all you need to know is what number you have to divide or multiply by to get from the original unit to the new unit — this is called the **conversion factor** and is equal to the number of times the smaller unit goes into the larger unit.

- To go from a bigger unit to a smaller unit, you multiply by the conversion factor.

- To go from a smaller unit to a bigger unit, you divide by the conversion factor.

There are some conversions that'll be particularly useful for GCSE Physics. Here they are...

Mass can have units of kg and g.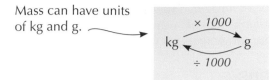

Exam Tip
Before you put values into an equation, you need to make sure they have the right units.

Length can have lots of units, including mm, μm and nm.

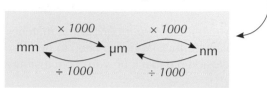

Exam Tip
Being familiar with these common conversions could save you time when it comes to doing calculations in the exam, and will help you get the right answer.

Time can have units of min and s.

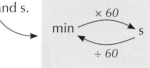

Area can have units of m², cm² and mm².

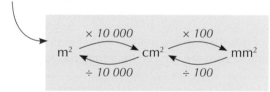

Volume can have units of m³, dm³ and cm³.

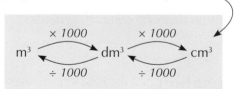

Tip: Volume is also often given in l (litres) or ml (millilitres). To convert, you'll need to remember that 1 ml = 1 cm³ and so 1 l = 1000 cm³.

Examples

- To go from dm³ to cm³, you'd multiply by 1000.

  2 dm³ is equal to 2 × 1000 = **2000 cm³**

- To go from grams to kilograms, you'd divide by 1000.

  3400 g is equal to 3400 ÷ 1000 = **3.4 kg**

10. Conclusions and Evaluations

So... you've planned and carried out an amazing experiment, got your data and have processed and presented it in a sensible way. Now it's time to figure out what your data actually tells you, and how much you can trust what it says.

How to draw conclusions

Drawing conclusions might seem pretty straightforward — you just look at your data and say what pattern or relationship you see between the dependent and independent variables.

But you've got to be really careful that your conclusion matches the data you've got and doesn't go any further. You also need to be able to use your results to justify your conclusion (i.e. back up your conclusion with some specific data).

When writing a conclusion you need to refer back to the original hypothesis and say whether the data supports it or not.

Example

This table shows the count rate detected when one of two different materials (A and B) is placed between a radioactive sample and the detector.

Material	Count rate (counts per second)
A	85
B	13
No material	98

The conclusion of this experiment would be that material B blocks more of the radiation from the sample than material A.

The justification for this conclusion is that the count rate was much lower using material B compared with using material A.

You can't conclude that material B blocks more radiation from any other radioactive sample — the results might be completely different.

Correlation and causation

Tip: Graphs are useful for seeing whether two variables are correlated (see page 17).

If two things are correlated (i.e. there's a relationship between them) it doesn't necessarily mean that a change in one variable is causing the change in the other — this is really important, don't forget it. There are three possible reasons for a correlation:

1. Chance

Tip: Causation just means one thing is causing another.

Even though it might seem a bit weird, it's possible that two things show a correlation in a study purely because of chance.

Example

One study might find a correlation between the number of people suffering from insomnia (trouble sleeping) and the distance they live from a wind farm. But other scientists don't get a correlation when they investigate it — the results of the first study are just a fluke.

2. They're linked by a third variable

A lot of the time it may look as if a change in one variable is causing a change in the other, but it isn't — a third variable links the two things.

> **Example**
>
> There's a correlation between water temperature and shark attacks. This isn't because warm water makes sharks crazy. Instead, they're linked by a third variable — the number of people swimming (more people swim when the water's hotter, and with more people in the water shark attacks increase).

3. Causation

Sometimes a change in one variable does cause a change in the other.

> **Example**
>
> There's a correlation between exposure to radiation and cancer. This is because radiation can damage body cells and cause cancer.

You can only conclude that a correlation is due to cause if you've controlled all the variables that could be affecting the result. (For the radiation example, this would include age and exposure to other things that cause cancer.)

Evaluation

An evaluation is a critical analysis of the whole investigation. Here you need to comment on the following points about your experiment and the data you gathered:

- **The method**: Was it valid? Did you control all the other variables to make it a fair test?

- **The quality of your results:** Was there enough evidence to reach a valid conclusion? Were the results repeatable, reproducible, accurate and precise?

- **Anomalous results**: Were any of the results anomalous? If there were none then say so. If there were any, try to explain them — were they caused by errors in measurement? Were there any other variables that could have affected the results? You should comment on the level of uncertainty in your results too.

Once you've thought about these points you can decide how much confidence you have in your conclusion. For example, if your results are repeatable, reproducible and valid and they back up your conclusion then you can have a high degree of confidence in your conclusion.

You can also suggest any changes to the method that would improve the quality of the results, so that you could have more confidence in your conclusion. For example, you might suggest changing the way you controlled a variable, or increasing the number of measurements you took. Taking more measurements at narrower intervals could give you a more accurate result.

You could also make more predictions based on your conclusion, then further experiments could be carried out to test them.

Tip: Lots of things are correlated without being directly related. E.g. the level of carbon dioxide (CO_2) in the atmosphere and the amount of obesity have both increased over the last 100 years, but that doesn't mean increased atmospheric CO_2 is causing people to become obese.

Tip: When suggesting improvements to the investigation, always make sure that you say why you think this would make the results better.

Learning Objectives:

- Know that an object or group of objects can be considered a system.
- Know that when a system changes, energy is transferred between stores.
- Be able to describe changes in the way energy is stored when energy is transferred in changing systems.
- Know that energy is transferred between stores when work is done by a force or by a current flowing.
- Know the principle of conservation of energy.
- Be able to calculate the amount of energy transferred to stores in a changing system.
- Know and be able to give examples to show that, whenever energy is transferred, some energy will be transferred to non-useful stores (dissipated). This energy is said to be "wasted".
- Be able to describe and give examples of the energy transfers that take place in a closed system, where there is no net change in energy.

Specification References
4.1.1.1, 4.1.2.1

1. Energy Stores and Transfers

Energy is what makes everything happen. It can be transferred between stores, and different things happen depending on which stores it is moving between.

Energy stores

When energy is transferred to an object, the energy is stored in one of the object's **energy stores**. You can think of energy stores as being like buckets that energy can be poured into or taken out of.

Here are the stores you need to know, and some examples of objects with energy in each of these stores:

Energy store	Objects with energy in this store
Kinetic	Anything moving has energy in its kinetic energy store.
Thermal	Any object. The hotter it is, the more energy it has in this store. You may also see thermal energy stores called internal energy stores.
Chemical	Anything that can release energy by a chemical reaction, e.g. food, fuels.
Gravitational Potential	Anything that has mass and is inside a gravitational field.
Elastic Potential	Anything that is stretched (or compressed) e.g. springs.
Electrostatic	Anything with electric charge that is interacting with another electric charge — e.g. two charges that attract or repel each other.
Magnetic	Anything magnetic that is interacting with another magnet — e.g. two magnets that attract or repel each other.
Nuclear	Atomic nuclei have energy in this store that can be released in nuclear reactions.

Energy transfers

A **system** is just a fancy word for a single object (e.g. the air in a piston) or a group of objects (e.g. two colliding vehicles) that you're interested in.

Closed systems are systems where neither matter nor energy can enter or leave. The net change in the total energy of a closed system is always zero.

When a system changes, energy is transferred. It can be transferred into or away from a system, between different objects in the system, or between different types of energy stores.

Energy can be transferred between stores in four main ways:

- Mechanically — an object moving due to a force acting on it, e.g. pushing, pulling, stretching or squashing.

- Electrically — a charge (current) moving through a potential difference, e.g. charges moving round a circuit.

- By heating — energy transferred from a hotter object to a colder object, e.g. heating a pan of water on a hob.

- By radiation — energy transferred by e.g. light/sound waves (for example, energy from the Sun reaching Earth by light).

Tip: See Topic 6 for more on energy transfers by radiation.

Examples

- If you're boiling water in a kettle — you can think of the water as the system. Energy is transferred to the water (from the kettle's heating element) by heating, to the water's thermal energy store (causing the temperature of the water to rise).

  You could also think of the kettle's heating element and the water together as a two-object system. Energy is transferred electrically to the thermal energy store of the kettle's heating element, which transfers energy by heating to the water's thermal energy store.

- The clown in a jack-in-the-box could also be considered as a system. When the lid is opened, energy is transferred mechanically from the elastic potential energy store of the compressed spring to the kinetic energy store and gravitational potential energy store of the clown as the spring extends.

Tip: Energy is transferred electrically by the moving charges doing work against the electrical resistance of the heating element — see page 91.

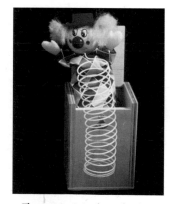

Figure 1: A jack-in-the-box toy. Energy is transferred from the extending spring to the kinetic energy store of the clown.

Work done

Work done is just another way of saying energy transferred — they're the same thing. Work can be done by a moving charge (work done against resistance in a circuit, see page 91) or by a force moving an object through a displacement (see page 155).

Here are some examples of energy transfers involving work:

Examples

- The initial force exerted by a person to throw a ball upwards does work. It causes an energy transfer from the chemical energy store of the person's arm to the kinetic energy store of the arm and the ball.

- A ball dropped from a height is accelerated by gravity. The gravitational force does work. It causes energy to be transferred from the ball's gravitational potential energy store to its kinetic energy store (see page 28).

Exam Tip
You need to be able to calculate the changes in energy involved when work is done by forces (p. 155) or when a charge flows (p. 91).

Exam Tip
In the exam, they can ask you to describe the energy transfer in any system. If you understand a few different examples, it'll be much easier to think through whatever they ask you about in the exam.

- The friction between a car's brakes and its wheels does work (see page 215). It causes an energy transfer from the wheels' kinetic energy stores to the thermal energy stores of the brakes and the surroundings, causing the car to slow down.

- In a collision between a car and a stationary object, the normal contact force between the car and the object does work. It causes energy to be transferred from the car's kinetic energy store to other energy stores, e.g. the elastic potential and thermal energy stores of the object and the car body. Some energy will also be transferred away by sound waves (p. 279).

The conservation of energy principle

Energy always obeys the **conservation of energy principle**:

> Energy can be transferred usefully, stored, or dissipated, but can never be created or destroyed.

Tip: Dissipated is a fancy way of saying spread out and lost.

You can use the conservation of energy principle to calculate how much energy is transferred to certain stores when a system is changed. You'll see lots of examples of this in GCSE Physics, and you can only do it because of conservation of energy.

Tip: Remember you can calculate energy transferred by heating (page 30) or when work is done by a force (page 155) or by moving charges (pages 91-93) using the formulas given on those pages.

Example

A car travelling at a constant speed applies the brakes and comes to a stop.

- The energy in the kinetic energy store of the car is transferred to the thermal energy stores of the brakes (see page 215).

- All the energy in the kinetic energy store is transferred, so by conservation of energy, it must be moved to other stores.

- You can calculate the energy transferred, it's just the energy that the car had in its kinetic energy store before the brakes were applied, given by $\frac{1}{2}mv^2$ — see page 26.

- If you assume all of the energy is transferred to the thermal energy stores of the brakes, then you can calculate the energy in these stores.

Tip: Another example of using conservation of energy in this way is when an object is falling (without air resistance) — see page 28.

- You'll see on page 30 that the energy in an object's thermal energy stores can be found using $mc\Delta\theta$, so you can let $\frac{1}{2}mv^2 = mc\Delta\theta$ in this transfer and work out things like $\Delta\theta$, the temperature change.

When energy is transferred between stores, not all of the energy is transferred usefully to the store that you want it to go to. Some energy is always dissipated when an energy transfer takes place. Dissipated energy is sometimes called 'wasted energy' because the energy is stored in a way which is not useful (usually energy has been transferred into thermal energy stores). See page 41 for much more on this.

> **Example**
>
> A mobile phone is a system.
>
> - When you use the phone, energy is usefully transferred from the chemical energy store of the battery in the phone.
>
> - But some of this energy is dissipated in this transfer to the thermal energy store of the phone (you may have noticed your phone feels warm if you've been using it for a while).

The conservation of energy principle explains why the total energy in a closed system is always the same. No energy can enter or leave a closed system, and since energy cannot be created or destroyed, the total energy is always the same.

> **Example**
>
> Imagine that a hot spoon is put into cold water in a perfectly insulated container (i.e. no energy can be transferred from the water to the surroundings, or vice versa). This system (the spoon and the water) is a closed system.
>
> The spoon will cool down, and the water will warm up, but the total energy of the system will not change. This is because no energy has left or entered the system — it has merely moved between stores within the system.

Practice Questions — Fact Recall

Q1 State the eight different forms of energy store.

Q2 What is meant by a closed system?

Q3 State the conservation of energy principle.

Practice Question — Application

Q1 Describe the main energy transfer between stores that occurs, including the way in which energy is transferred, when each of the following things happen:

a) An arrow is released from a bow.

b) A gas camping stove is used to heat soup.

c) A battery-powered fan is used.

- Know that all moving objects have energy in their kinetic energy store.
- Know and be able to use $E_k = \frac{1}{2}mv^2$ to calculate the energy in the kinetic energy store of an object.
- Know and be able to use $E_p = mgh$ to calculate the energy transferred to an object's g.p.e. store when it is raised above the ground.
- Be able to make calculations of energy transfers between stores, e.g. between the gravitational potential energy store and kinetic energy store of a falling object.
- Know that a stretched (or compressed) spring has energy in its elastic potential energy store.
- Be able to use $E_e = \frac{1}{2}ke^2$ to calculate the energy in the elastic potential energy store of a stretched (or compressed) spring.

Specification References 4.1.1.1, 4.1.1.2

Exam Tip
Use this formula triangle (see page 338) to rearrange the equation:

2. Kinetic and Potential Energy Stores

Some of the most common and straightforward energy transfers are between kinetic energy stores and potential energy stores. They have handy formulas so you can easily calculate how much energy is gained or lost.

Kinetic energy stores

Anything that is moving has energy in its **kinetic energy store**. Energy is transferred to this store when an object speeds up and is transferred away from this store when an object slows down.

The energy in the kinetic energy store depends on the object's mass and speed. The greater its mass and the faster it's going, the more energy there will be in its kinetic energy store. It's got a slightly tricky formula, so you'll need to concentrate a bit harder on this one:

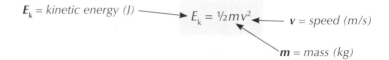

small mass, low speed — small amount of energy in kinetic energy store

large mass, high speed — large amount of energy in kinetic energy store

Figure 1: *A diagram to show how the energy in the kinetic energy store of a moving object depends on its mass (m) and speed (v).*

Example

A van of mass 2450 kg is travelling at 40.0 m/s. Calculate the energy in its kinetic energy store.

You just plug the numbers into the formula — but watch the 'v^2'.

$$E_k = \frac{1}{2}mv^2 = \frac{1}{2} \times 2450 \times 40.0^2 = 1\,960\,000\text{ J}$$

Example

A moped with 1.17×10^4 J of energy in its kinetic energy store travels at 12.0 m/s. What is the mass of the moped?

$E_k = 1.17 \times 10^4$ J, $v = 12.0$ m/s

Rearranging $E_k = \frac{1}{2}mv^2$,

$$m = (2 \times E_k) \div v^2 = (2 \times 1.17 \times 10^4) \div 12.0^2 = 162.5\text{ kg}$$

Gravitational potential energy stores

Lifting an object in a gravitational field (page 149) requires work. This causes a transfer of energy to the **gravitational potential energy (g.p.e.) store** of the raised object. The higher an object is lifted, the more energy is transferred to this store.

The amount of energy in an object's g.p.e. store depends on its mass, its height and the strength of the gravitational field the object is in. The amount of energy that is transferred to an object's gravitational potential energy store when it's raised through a certain height can be found by:

E_p = gravitational potential energy (J)

g = gravitational field strength (N/kg)

h = height (m)

m = mass (kg)

$$E_p = mgh$$

On Earth, the gravitational field strength (g) is approximately 9.8 N/kg.

No height above ground

Energy transferred to the g.p.e. store equal to $E_p = mgh$

Figure 2: *A diagram showing how the energy in the gravitational potential energy store of a mass (m) increases when it is lifted to a height (h) in a gravitational field.*

Exam Tip
If you need help with rearranging this equation, this formula triangle might help:

$\dfrac{E_p}{m \times g \times h}$

Exam Tip
You'll be given the value of gravitational field strength in the exam.

Example

A 50 kg mass is slowly raised through a height of 6.0 m. Find the energy transferred to its gravitational potential energy store. The gravitational field strength is 9.8 N/kg.

Just plug the numbers into the formula:
$$E_p = mgh = 50 \times 9.8 \times 6.0 = 2940 \text{ J}$$

Example

A flea of mass 1.0×10^{-3} g jumps vertically from the ground. The gravitational field strength is 9.8 N/kg. At the top of the jump the flea has gained 1.96×10^{-6} J of energy in its g.p.e. store. How high has the flea jumped?

$E_p = 1.96 \times 10^{-6}$ J
$m = 1.0 \times 10^{-3}$ g $= 1.0 \times 10^{-6}$ kg

Rearranging $E_p = mgh$,

$$h = E_p \div (mg) = 1.96 \times 10^{-6} \div (1.0 \times 10^{-6} \times 9.8) = 0.20 \text{ m}$$

Tip: Be careful when rearranging. Remember to do the same thing to each side of the equation, one step at a time (see page 337).

Tip: When you're doing calculations, give your final answer to the lowest number of significant figures used in any value in the calculation (see page 15 for more on this).

Energy transfer for falling objects

When something falls, energy from its gravitational potential energy store is transferred to its kinetic energy store. The further is falls, the faster it goes.

For a falling object when there's no air resistance, you can use the principle of conservation of energy to get:

Energy lost from the g.p.e. store = Energy gained in the kinetic energy store

energy in g.p.e. store at this height = mgh

energy lost from g.p.e. store = energy gained in kinetic energy store

more energy lost from g.p.e. store = more energy transferred to the kinetic energy store

Figure 3: *A diagram showing that energy is transferred from the g.p.e. store to the kinetic energy store of a falling object. The higher the object is to begin with, the more energy it'll have in its kinetic energy store when it hits the ground.*

You can use this to make calculations of energy transfers when an object falls, using the equations on the previous pages.

Example

The flea from the example on the previous page falls from the top of its jump. Assuming there is no air resistance, calculate the speed of the flea when it hits the ground. Give your answer to 2 significant figures.

All the energy the flea gained in its g.p.e. store by jumping will be transferred back to its kinetic energy store as it falls towards the ground, so E_p at the top of the jump will equal E_k when it hits the ground.

$E_p = 1.96 \times 10^{-6}$ J $= E_k$
$m = 1.0 \times 10^{-6}$ kg

$E_k = \frac{1}{2}mv^2$, so rearrange for v:

$$v = \sqrt{\frac{2E_k}{m}} = \sqrt{\frac{2 \times 1.96 \times 10^{-6}}{1.0 \times 10^{-6}}} = 1.979... = 2.0 \text{ m/s (to 2 s.f.)}$$

Elastic potential energy stores

Stretching or squashing an object can transfer energy to its **elastic potential energy store**. The energy in the elastic potential energy store of a stretched spring (provided it has not been stretched past its limit of proportionality) can be found using the formula from page 161:

E_e = elastic potential energy (J) ⟶ $E_e = \frac{1}{2}ke^2$ ⟵ e = extension (m)

k = spring constant (N/m)

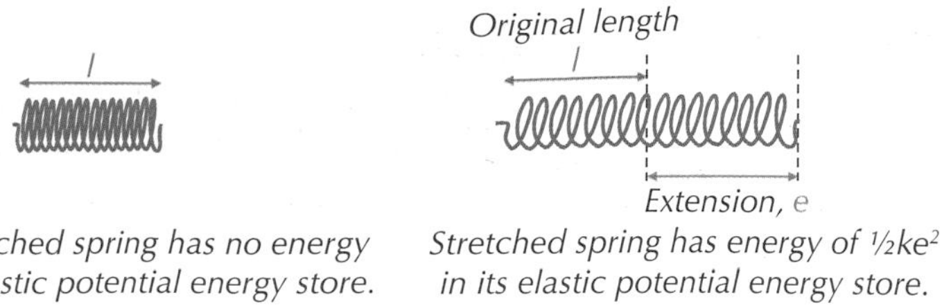

Original length

Unstretched spring has no energy
in its elastic potential energy store.

Stretched spring has energy of ½ke²
in its elastic potential energy store.

Extension, e

Figure 4: *A diagram showing how to find the energy in the
elastic potential energy store of a stretched spring.*

Example

**A spring with spring constant 40 N/m is stretched from its
normal length of 8.0 cm to a stretched length of 23.0 cm.
Calculate the energy transferred to its elastic potential energy store.**

MATHS
SKILLS

First, find *k* and *e*, and place them in the correct units.
$k = 40$ N/m
e = stretched length – normal length = $23.0 - 8.0 = 15$ cm = 0.15 m

Now simply plug the values into the equation:

$$E_e = \tfrac{1}{2}ke^2 = \tfrac{1}{2} \times 40 \times 0.15^2 = 0.45 \text{ J}$$

Practice Questions — Fact Recall

Q1 Which of these has the most energy in its kinetic energy store:
a small dog walking slowly or a large dog running fast? Explain why.

Q2 What is the formula for calculating the energy in an object's kinetic
energy store? What does each term represent and what units should
they be in?

Q3 What is the formula for working out the energy transferred to an
object's gravitational potential energy store? What does each term
represent and what units should they be in?

Q4 What is the formula for working out the energy transferred to
an object's elastic potential energy store? What does each term
represent and what units should they be in?

Practice Questions — Application

Q1 A 25 000 kg plane takes off and climbs to a height of 12 000 m
above its take off point. How much energy has been transferred to
its gravitational potential energy store?

Q2 A spring is stretched by 60 cm, which transfers 18 J of energy to its
elastic potential energy store. What is its spring constant?

Q3 A 12.5 g ball with 40 J of energy in its kinetic energy store is
travelling horizontally through the air. How fast is it moving?

Q4 A 1 kg potato falls to the ground and loses 450 J of energy from its
gravitational potential energy store. Assuming no air resistance, what
is the speed of the potato as it hits the ground?

Tip: Take gravitational
field strength (*g*) to be
9.8 N/kg.

Exam Tip
Watch out for data given
in the 'wrong' units
in the exam. Double-
check the numbers are
in the 'right' units before
you stick them into the
formula.

- Know that the specific heat capacity of a material is the energy needed to raise the temperature of 1 kg of the material by 1°C.

- Be able to use the equation $\Delta E = mc\Delta\theta$ to calculate the energy transferred to (stored by) or from (released by) the thermal energy store of a substance when its temperature changes.

- Be able to make calculations of energy transfers between energy stores, e.g. between the kinetic energy store of a car and the thermal energy store of its brakes.

- Be able to find the specific heat capacity of various materials, and to link work done with energy transferred to thermal energy stores in the experiment (Required Practical 1).

Specification References
4.1.1.1, 4.1.1.3

3. Specific Heat Capacity

Some materials are easier to heat up than others. Specific heat capacity is a measure of how much energy it takes to change the temperature of a material. Let the fun begin...

What is specific heat capacity?

More energy needs to be transferred to the thermal energy store of some materials to increase their temperature than others. For example, you need 4200 J to warm 1 kg of water by 1 °C, but only 139 J to warm 1 kg of mercury by 1 °C.

Materials that need to have a lot of energy transferred to in their thermal energy stores to warm up also transfer a lot of energy when they cool down.

How much energy needs to be transferred to the thermal energy store of a substance before its temperature increases is determined by its **specific heat capacity**.

> Specific heat capacity is the amount of energy needed to raise the temperature of 1 kg of a substance by 1 °C.

The amount of energy transferred to (i.e. stored by) or transferred from (i.e. released by) the thermal energy store of a substance for a given temperature change, $\Delta\theta$, is linked to its specific heat capacity by this equation:

$$\Delta E = mc\Delta\theta$$

ΔE = change in thermal energy (J) $\Delta\theta$ = temperature change (°C)

m = mass (kg) c = specific heat capacity (J/kg°C)

Tip: Remember, Δ just means 'change in' (see p. 337).

> **Example**
>
> **Water has a specific heat capacity of 4200 J/kg°C. How much energy is needed to heat 2.00 kg of water from 10.0 °C to 100.0 °C?**
>
> First work out the temperature difference, $\Delta\theta$, between the starting and finishing temperatures.
>
> $\Delta\theta = 100.0\ °C - 10.0\ °C = 90.0\ °C$
>
> Then plug the numbers for m, c and $\Delta\theta$ into the formula to find ΔE.
>
> $\Delta E = mc\Delta\theta = 2.00 \times 4200 \times 90.0 = 756\ 000$ J

You can use the formula above with the conservation of energy, just like on page 28, to make calculations for energy transfers in all sorts of situations. See the example on the next page.

A 3500 kg van travelling at 30.0 m/s applies its brakes and comes to a stop. Estimate the change in temperature of the brakes in this transfer, if their combined mass is 25 kg and their specific heat capacity is 420 J/kg °C.

Assume that all of the energy in the kinetic energy stores of the van is transferred to the thermal energy stores of the brakes. The energy transferred is equal to the energy in the kinetic energy stores, given by $\frac{1}{2}mv^2$ (p. 26).

$E = \frac{1}{2}mv^2 = \frac{1}{2} \times 3500 \times 30.0^2 = 1\ 575\ 000$ J

All this is transferred to the brakes' thermal energy stores, so $\Delta E = 1\ 575\ 000$ J.

Rearrange $\Delta E = mc\Delta\theta$ to give:
$\Delta\theta = \Delta E \div mc = 1\ 575\ 000 \div (25 \times 420) = 150\ °C$

Figure 1: *A thermogram showing the temperature differences around a car wheel during braking. White is hottest (where the brakes are) and blue is coolest.*

Investigating specific heat capacity

REQUIRED PRACTICAL **1**

You can do an experiment to investigate the specific heat capacity of a material. To investigate a solid material (e.g. copper), you'll need a block of the material with two holes in it (for the heater and thermometer to go into, see Figure 2).

Tip: Remember to carry out a risk assessment before you do any practical. Be careful not to burn yourself when handling the heater.

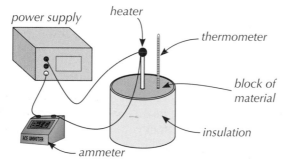

Figure 2: *A diagram of the apparatus used to investigate the specific heat capacity of a solid material.*

Measure the mass of the block, then wrap it in an insulating layer (e.g. a thick layer of newspaper) to reduce the energy transferred from the block to the surroundings (p. 37). Insert the thermometer and heater as shown in Figure 2.

Measure the initial temperature of the block and set the potential difference, V, of the power supply to be 10 V. Turn on the power supply and start a stop watch. As the block heats up, take readings of the temperature and current, I, every minute for 10 minutes. Keep an eye on the ammeter — the current through the circuit shouldn't change.

When you turn on the power, the current in the circuit (i.e. the moving charges) does work on the heater, transferring energy electrically from the power supply to the heater's thermal energy store. This energy is then transferred to the material's thermal energy store by heating, causing the material's temperature to increase.

Using your measurement of the current and the potential difference of the power supply, you can calculate the power of the heater, using $P = VI$ (p. 93). You can use this to calculate how much energy, E, has been transferred by the heater at the time of each temperature reading using the formula $E = Pt$, where t is the time in seconds since the experiment began.

Tip: If the hole's a lot bigger than the thermometer, you should put a small amount of water in with the thermometer. This lets the thermometer measure the temperature of the block better, as there is no air around it to insulate it from the block.

If you assume all the energy supplied by the heater has been transferred to the block, you can plot a graph of energy transferred to the thermal energy store of the block against temperature. You should get a graph which looks similar to Figure 3.

Tip: $\Delta\theta$ is how much the temperature has changed by since the start of the experiment.

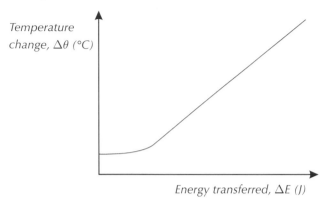

Figure 3: A graph of temperature increase against energy change for a solid block of material heated using the apparatus in Figure 2.

Since mass (m) and specific heat capacity are constant, $\Delta E = mc\Delta\theta$ shows that energy transferred is directly proportional to change in temperature. So the graph should be linear. However, as is shown in Figure 3, your graph may start off curved. You don't need to worry about why this happens, just make sure you ignore the curved part of the graph.

You can find the specific heat capacity of the block using the gradient of the linear part of your graph. The gradient is $\Delta\theta \div \Delta E$, so since $\Delta E = mc\Delta\theta$, the gradient is $1 \div mc$. So the specific heat capacity of the material of the block is: $1 \div$ (gradient $\times$ the mass of the block).

Tip: If you're just comparing the specific heat capacities, you could use identical masses of each material and identical methods for each material. If m and ΔE are constant, then c is just related to $\Delta\theta$, so a bigger change in temperature means a smaller specific heat capacity.

You can repeat this experiment with different materials to see how their specific heat capacities compare. You can also investigate the specific heat capacity of liquids with this experiment — just place the heater and thermometer in an insulated beaker filled with the liquid. Remember to measure the mass of the liquid, not the volume.

Practice Question — Fact Recall

Q1   What is the specific heat capacity of a substance?

Practice Questions — Application

Q1   A kettle heats 0.20 kg of water from a temperature of 20.0 °C to 100.0 °C. Water has a specific heat capacity of 4200 J/kg°C. How much energy is transferred to the thermal energy store of the water?

Q2   A chef heats 400 g of oil to a temperature of 113 °C. The oil is left to cool until it reaches a temperature of 25 °C. The oil transfers 70.4 kJ of energy to its surroundings. Calculate the specific heat capacity of the oil.

4. Power

Power is a really important concept that pops up all over the place because power is all about how quickly energy is transferred.

What is power?

Power is the rate of energy transfer, or the rate of doing work. Power is measured in watts, W. One watt is equivalent to one joule transferred per second (J/s).

You can calculate power using these equations:

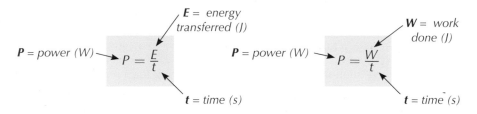

$$P = \frac{E}{t}$$

P = power (W)
E = energy transferred (J)
t = time (s)

$$P = \frac{W}{t}$$

P = power (W)
W = work done (J)
t = time (s)

Learning Objectives:

- Know that the rate of energy transfer (or work done) is power.
- Know that power is measured in watts, W, and 1 W = 1 J/s.
- Know and be able to use both $P = E \div t$ and $P = W \div t$.
- Be able to describe examples which illustate the definition of power, e.g. that the higher the power, the less time it takes to supply a given amount of energy.

Specification Reference 4.1.1.4

Example

A motor transfers 4.8 kJ of energy in 2 minutes. Find its power.

Energy transferred = 4.8 kJ = 4800 J
Time taken = 2 minutes = 2 × 60 s = 120 s

$$P = \frac{E}{t} = \frac{4800}{120} - 40 \text{ W (or 40 J/s)}$$

Tip: These formula triangles might help with rearranging the power equations:

$$\frac{E}{P \times t}$$

$$\frac{W}{P \times t}$$

Example

How long does it take for a 550 W motor to do 110 J of work?

Rearranging $P = W \div t$,
$$t = W \div P$$
$$= 110 \div 550$$
$$= 0.2 \text{ s}$$

A powerful machine is not necessarily one which can exert a strong force (although it usually ends up that way). A powerful machine is one which transfers a lot of energy in a short space of time.

Example

Consider two cars that are identical in every way apart from the power of their engines. Both cars race the same distance along a straight race track to a finish line. The car with the more powerful engine will reach the finish line faster than the other car — i.e. it will transfer the same amount of energy, but over less time.

It takes 8000 J of work to lift a stunt performer to the top of a building.
Motor A can lift the stunt performer to the correct height in 50 s.
Motor B would take 300 s to lift the performer to the same height.
Which motor is most powerful? Calculate the power of this motor.

Both motors transfer the same amount of energy, but motor A would do it
quicker than motor B. So, motor A is the more powerful motor.

Plug the time taken and work done for motor A into the equation $P = W \div t$
and find the power.
$P = W \div t = 8000 \div 50 = 160$ W

Practice Questions — Fact Recall

Q1 What is meant by the term 'power'?

Q2 State the equations for power. What does each term represent and
what units are they measured in?

Practice Questions — Application

Q1 Find the power of the following:

a) A motor that transfers 150 J of energy in 37.5 s.

b) A motor that transfers 79.8 kJ of energy in 42 s.

c) A motor that transfers 6 840 kJ of energy in 9.5 minutes.

Q2 Two lifts, A and B, operate between the ground floor and the second
floor of a building. Lift A's motor has a power of 4800 W, and lift B's
motor has a power of 5200 W. State which of the two lifts will carry
a given load up the two floors in the least time, and explain why.

Q3 How long does each of the following take?

a) A 525 W motor to do 1344 J of work.

b) A 2.86 kW toaster to transfer 1430 J of energy.

Q4 How much energy do the following transfer?

a) A machine with a power of 1240 W running for 35 s.

b) A 1500 W heater switched on for 17 minutes.

5. Conduction and Convection

Energy is transferred by heating through solids by conduction, and through liquids and gases by conduction and convection. Both types of energy transfer by heating happen because the particles in a substance move about...

Conduction

> **Conduction** is the process by which vibrating particles transfer energy to neighbouring particles.

Energy transferred to an object by heating is transferred to the thermal energy store of the object. This energy is shared across the kinetic energy stores of the particles in the object.

The particles in the part of the object being heated have more energy transferred to their kinetic energy stores, and so vibrate more and collide with each other. These collisions cause energy to be transferred between the particles' kinetic energy stores. This is conduction.

This process continues throughout the object until the energy is spread out evenly across the object. It's then usually transferred to the thermal energy store of the surroundings (or anything else touching the object).

Particles in the hotter part of a solid vibrate more. *Particles collide and pass energy between their kinetic energy stores.*

Energy is transferred in this direction through the solid.

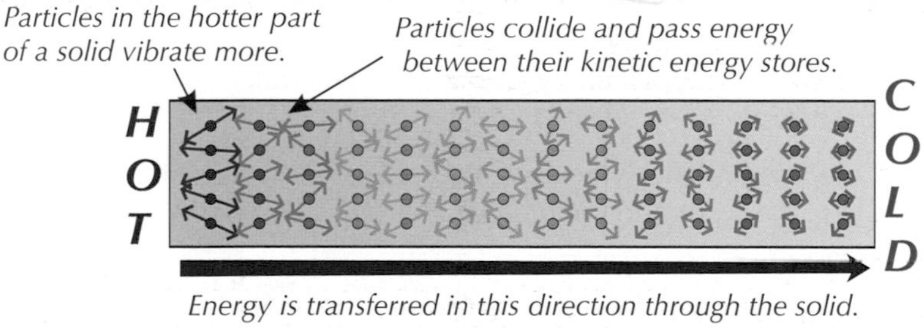

Figure 1: *Conduction of energy through a solid.*

Conduction occurs mainly in solids. Particles in liquids and gases are much more free to move around, so their particles collide less frequently, which is why they usually transfer energy by convection instead of conduction.

Thermal conductivity is a measure of how quickly energy is transferred through a material in this way. Materials with a high thermal conductivity transfer energy between their particles quickly. The higher a material's thermal conductivity, the faster energy can be transferred through it by conduction. Materials with a high thermal conductivity are known as thermal conductors, and those with a low thermal conductivity are known as thermal insulators.

> **Example**
>
> Pots and pans are usually made of metal because metals have high thermal conductivity. This means they transfer energy quickly, so the energy is transferred to the food quicker, making cooking quicker. Pan handles are usually made of a thermal insulator. They don't transfer energy by conduction very fast, so the handle stays cool and you don't end up burning your hand when touching it.

Learning Objective:
- Know that the higher a material's thermal conductivity, the faster energy can be transferred through it by conduction.

Specification Reference
4.1.2.1

Tip: You'll see on page 110 that the energy transferred to an object by heating is actually stored in the kinetic energy stores of its particles.

Tip: There's much more on convection on the next page.

Convection

Convection is where energetic particles move away from hotter to cooler regions.

Convection can happen in gases and liquids. Energy is transferred by heating to the thermal energy store of the liquid or gas. Again, this energy is shared across the kinetic energy stores of the gas or liquid's particles.

Unlike in solids, the particles in liquids and gases are able to move around, rather than vibrate in place. When you heat a region of a gas or liquid, the particles move faster and the space between individual particles increases. This causes the density (page 106) of the region being heated to decrease.

Because liquids and gases can flow, the warmer and less dense region will rise above denser, cooler regions. If there is a constant transfer of energy by heating to the substance, a convection current can be created, where the air in the room circulates as regions of it heat up and rise, then cool down and fall.

Figure 2: You can see convection currents by sticking some potassium permanganate crystals in the bottom of a beaker of cold water. Heat the beaker gently over a Bunsen flame — the potassium permanganate will start to dissolve and make a gorgeous bright purple solution that gets moved around the beaker by the convection currents as the water heats. It's real pretty.

Example

Heating a room with a radiator relies on creating convection currents in the air of the room.

Energy is transferred from the radiator to the nearby air particles by conduction (the air particles collide with the radiator surface). The air by the radiator becomes warmer and less dense (as the particles move quicker). This warm air rises and is replaced by cooler air. The cooler air is then heated by the radiator.

At the same time, the previously heated air transfers energy to the surroundings (e.g. the walls and contents of the room). It cools, becomes denser and sinks.

This cycle repeats, causing a flow of air to circulate around the room — this is a convection current.

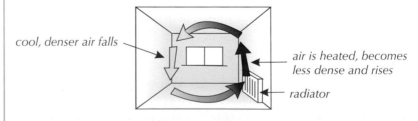

cool, denser air falls

air is heated, becomes less dense and rises

radiator

Figure 3: A radiator heating a room by convection.

Tip: The key thing to remember is that convection happens because a heated liquid or gas becomes less dense. This means it moves away, above the cooler (denser) air, taking its energy with it.

Practice Questions — Fact Recall

Q1 What is conduction? Describe how energy is transferred by conduction.

Q2 What is convection?

Q3 In which state of matter does convection not happen?

Tip: See page 107 for more on states of matter.

6. Reducing Unwanted Energy Transfers

Learning Objectives:
- Be able to describe how a building's rate of cooling is affected by the thermal conductivity and thickness of its walls.
- Be able to explain a number of methods by which unwanted energy transfers can be reduced.
- Be able to investigate the effectiveness of different materials as thermal insulators. (Required Practical 2)

Specification Reference 4.1.2.1

Some energy is always transferred to energy stores which are not useful. This is a pain, but there are a number of ways that we can reduce these unwanted transfers and maximise the energy in the useful energy stores.

Thermal insulation

The last thing you want when you've made your house nice and toasty is for that energy to escape outside. The thicker the walls and the lower their thermal conductivity, the slower the rate of energy transfer will be (so the rate of cooling will be slower). You can help reduce the amount of energy lost from a building using **thermal insulators**. Here are some examples...

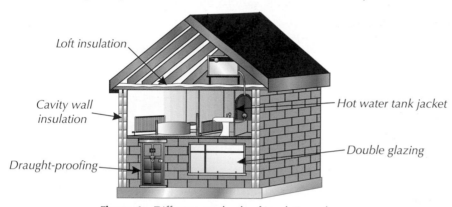

Loft insulation

Cavity wall insulation

Draught-proofing

Hot water tank jacket

Double glazing

Figure 1: *Different methods of insulating a home.*

Tip: Make sure you know how each type of insulation helps reduce the energy lost from a home.

Cavity walls and cavity wall insulation

Some houses have cavity walls, made up of an inner and an outer wall with an air gap in the middle. The air gap reduces the amount of energy transferred by conduction through the walls, because air is an insulator. Using cavity wall insulation (where the cavity wall air gap is filled with a foam) can also reduce energy transfer by convection in the cavity walls.

Loft insulation

Fibreglass is an insulating material made of thin strands of glass that trap pockets of air. A thick layer of fibreglass wool laid out across the whole loft floor reduces conduction to the attic space, as the material (and the trapped air) are insulators. Loft insulation (see Figure 3) reduces the energy transfer by convection, by preventing convection currents from forming.

Double glazing

Double-glazed windows work in the same way as cavity walls — they have an air gap between two sheets of glass to prevent energy transfer by conduction through the windows.

Draught-proofing

Draught excluders or strips of foam or plastic around doors and windows stop draughts blowing in and out, and reduce energy transfer by convection.

Figure 2: *This sleeve on a coffee cup is another example of thermal insulation. It stops your hand getting too hot. They're often made of cardboard with air pockets, as air is a good insulator.*

Figure 3: *Fibreglass loft insulation.*

Hot water tank jacket

Putting fibreglass wool around a hot water tank reduces the energy transferred by conduction from the tank's thermal energy store in the same way as loft insulation.

Thick curtains

Big bits of cloth over the window create an air gap between the room and the window, stopping hot air reaching the glass by convection. They also reduce energy transferred by conduction.

Investigating reducing energy transfers

REQUIRED PRACTICAL 2

You can investigate how effective various materials are as thermal insulators using this simple experiment.

Boil water in a kettle. Pour some of the water into a sealable container (e.g. a beaker with a lid) to a safe level. Measure the mass of the water in the container (see page 327), and use a thermometer to measure the initial temperature of the water.

Seal the container and leave it for five minutes. Measure this time using a stop watch. Then, remove the lid and measure the final temperature of the water. Pour away the water, and allow the container to cool to room temperature.

Repeat this experiment, but each time, wrap the container in a different material (e.g. foil, newspaper, cotton wool) once it has been sealed. Make sure you use the same mass of water each time and that the water starts at the same temperature.

Tip: As always, you need to do a risk assessment for this experiment.

Tip: When you measure the temperature of the water, you could use a disc of cardboard with a hole cut in it to cover the beaker, and insert the thermometer into the hole. This will prevent the water temperature decreasing too much due to evaporation (see page 110).

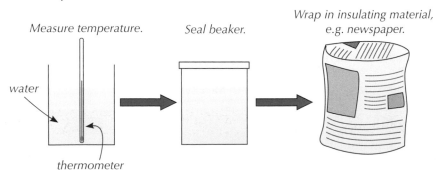

Measure temperature.   Seal beaker.   Wrap in insulating material, e.g. newspaper.

water

thermometer

Figure 4: A diagram showing steps in an investigation into the effectiveness of different materials as thermal insulators.

Figure 5: Some equipment you could use for this experiment. Examples of insulating materials include newspaper, cotton wool, foam and bubble wrap.

You should find that the temperature difference (and so the energy transferred, see page 30) is reduced by wrapping the container in thermally insulating materials like bubble wrap or cotton wool.

You can also investigate the factors affecting the effectiveness of a material as a thermal insulator, e.g. how the thickness of the material affects the temperature change of the water. Just pick a material and repeat the experiment with increasing thicknesses of that material. You should find that the thicker the insulating layer, the smaller the temperature change of the water, and so the less energy is transferred.

Lubrication

Whenever something moves, there's usually at least one frictional force acting against it. This causes some energy in the system to be dissipated, e.g. air resistance can transfer energy from a falling object's kinetic energy store to its thermal energy store and the thermal energy stores of the surroundings.

For objects that are being rubbed together, **lubricants** can be used to reduce the friction between the objects' surfaces when they move. Lubricants are usually liquids (like oil), so they can flow easily between the objects and coat them (and so reduce the friction across the whole area of contact between the objects). Streamlining can be used to reduce air resistance (see page 190).

Tip: Remember, air resistance causes a frictional force on an object moving through the air (see page 190).

Examples

▪ Lubrication is used on axles in things like cars, fans and turbines. This decreases the amount of energy lost to thermal energy stores due to friction between the axle and its supports when the axle turns.

▪ Grease can be used to lubricate door hinges and locks if they are stiff, to reduce friction.

Tip: An axle is just a rod that turns. In a car, the wheels are attached to an axle. In a fan or turbine, the blades are attached to an axle.

Practice Questions — Fact Recall

Q1 Explain how installing each of the following types of insulation helps reduce unwanted energy transfers from a home.

a) Cavity wall insulation

b) Loft insulation

c) Double glazing

Q2 Give one example of when lubrication is used to decrease unwanted energy transfers.

Practice Question — Application

Q1 Two students want to compare the effectiveness of cotton wool and newspaper as thermal insulators. They both buy a hot drink in the canteen and wrap a strip of newspaper or cotton wool around it. They measure the temperature 20 minutes later and find that there is no noticeable difference between the insulators. Explain two problems with their experiment that may have affected their results.

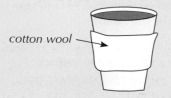

cotton wool

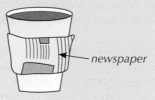

newspaper

7. Efficiency

Learning Objectives:
- Know and be able to use the equations for efficiency in terms of energy transferred and power.
- Be able to express an efficiency as either a decimal or a percentage.
- **H** Be able to describe ways in which the efficiency of an energy transfer can be improved.

Specification Reference
4.1.2.2

Sometimes it's really handy to know how much energy a device wastes compared to how much it usefully transfers — that's efficiency for you.

What is efficiency?

Useful devices are only useful because they can transfer energy from one energy store to a useful energy store. As you'll probably have gathered by now, some of the input energy is usually wasted by being transferred to a useless energy store — usually a thermal energy store.

The less energy that is 'wasted', the more efficient the device is said to be.

The **efficiency** of an energy transfer can be worked out using this equation:

$$\text{efficiency} = \frac{\text{useful output energy transfer}}{\text{total input energy transfer}}$$

This equation gives you the efficiency as a decimal, but you may be asked to express it as a percentage. To do that, simply multiply your result by 100, and stick the % symbol on the end.

Tip: Input energy is just the energy transferred to the device and useful output energy is the energy transferred to useful energy stores by the device.

Tip: Efficiency is just a number, so it doesn't matter what units your energy values are in. As long as the values for input and output energy transfer are in the same units, you'll get the right answer.

Tip: See page 335 for more on percentages.

Example 1

An electric fan is supplied with 2000 kJ of energy. 600 kJ of that is transferred to useless thermal energy stores.
What is the efficiency of the fan as a percentage?

Total input energy transfer = 2000 kJ
Useful output energy transfer = total energy in – wasted energy transfer
= 2000 – 600 = 1400 kJ

Start by working out the efficiency as a decimal:

$$\text{efficiency} = \frac{\text{useful output energy transfer}}{\text{total input energy transfer}}$$
$$= \frac{1400}{2000}$$
$$= 0.7$$

Then, multiply this by 100 to get the efficiency of the fan as a percentage:

efficiency = 0.7 × 100 = 70%

Example 2

A lamp with an efficiency of 0.740 is supplied with 350 J of energy.
How much energy is usefully transferred by the lamp?

Rearrange the equation $\text{efficiency} = \frac{\text{useful output energy transfer}}{\text{total input energy transfer}}$ to give:

useful output energy transfer = efficiency × total input energy transfer
= 0.740 × 350
= 259 J

Exam Tip
Make sure you think carefully about which value is total input energy transfer, and which is useful output energy transfer. It's easy to get them mixed up.

You might not know the energy inputs and outputs of a device, but you can still calculate its efficiency as long as you know the power input and output:

$$\text{efficiency} = \frac{\text{useful power output}}{\text{total power input}}$$

Tip: Remember, power is the rate of energy transferred, see page 33.

Example

A motor is supplied with 250 W of power and outputs 120 W of useful power. What is the efficiency of the motor? Give your answer as a decimal.

$$\text{efficiency} = \frac{\text{useful power out}}{\text{total power in}}$$
$$= \frac{120}{250}$$
$$= 0.48$$

So the motor has an efficiency of 0.48.

Wasted energy

For any given example you can talk about the input and output energy transfers, but remember this:

> No device is 100% efficient and the wasted energy is usually transferred to useless thermal energy stores.

Tip: See page 24 for more on wasted energy.

Electric heaters are the exception to this. They're usually almost 100% efficient because all the energy is transferred electrically to "useful" thermal energy stores

Ultimately, all energy ends up transferred to thermal energy stores. For example, if you use an electric drill, its transfers energy to lots of different energy stores, but it all quickly ends up in thermal energy stores.

Improving efficiency Higher

You can improve the efficiency of energy transfers by insulating objects, lubricating them or making them more streamlined (see pages 37 and 190).

Example 1 **Higher**

An electric fan usefully transfers energy electrically to the kinetic energy store of its blades, to cause the movement of air. It wastes energy in many ways, e.g. through friction between the axle and its supports. The efficiency could be improved by lubricating the axle.

Figure 1: *The internal workings of an electric fan, showing its axle.*

Tip: Improving efficiency is a useful application of science (see page 5), allowing us to maximise the energy transferred to useful energy stores and reduce the overall amount of energy used by appliances.

Example 2 **Higher**

A kettle transfers energy by heating to the thermal energy store of the water. A kettle wastes energy by heating the surroundings and by letting steam escape once the water is boiling. It could be made more efficient by insulating the kettle more and by switching off more quickly once the water starts to boil.

Figure 2: Steam escaping from the spout of a kettle as the water boils.

Practice Questions — Fact Recall

Q1 State the two equations for efficiency.

Q2 State the one device which is usually almost 100% efficient.

Q3 What eventually happens to all the energy wasted by a device?

Practice Questions — Application

Tip: A percentage efficiency must always be less than 100% and a decimal efficiency must always be less than 1.

Q1 Work out the efficiency of the following devices as a decimal.

a) A device with total power in = 90 W and useful power out = 54 W.

b) A machine supplied with 800 J of energy that wastes 280 J.

Q2 Work out the efficiency of the following devices as a percentage.

a) A device where total power in = 36 W and useful power out = 12.6 W.

b) A lamp that transfers 7.5 kJ of energy, 4.5 kJ of which is transferred usefully away by light, 2.9 kJ of which is transferred by heating to thermal energy stores of the surroundings and 100 J of which is transferred away by sound waves.

Q3 660 kJ of energy is transferred electrically to a TV. It transfers 298 kJ away by light, 197 kJ away by sound and transfers the rest by heating to thermal energy stores.

a) What is the total useful output energy transfer?

b) How much energy is wasted by the TV?

c) What is the efficiency of the TV as a decimal?

Tip: To get from a percentage to a decimal, divide by 100.
See p. 335 for more on percentages.

Q4 A machine with an efficiency of 68% transfers 816 J to useful energy stores. What is the total energy transferred to the device?

Q5 A vacuum cleaner works by transferring energy electrically to a motor which turns a fan and causes air to be drawn through the fan. Suggest two ways that a vacuum cleaner could be made more efficient.

Topic 1a Checklist — Make sure you know...

Energy Stores and Transfers

☐ That energy can be stored in: thermal energy stores, kinetic energy stores, gravitational potential energy stores, elastic potential energy stores, chemical energy stores, magnetic energy stores, electrostatic energy stores and nuclear energy stores.

☐ That a system is an object or group of objects.

☐ That energy can be transferred between stores mechanically, electrically, by heating, or by radiation.

☐ How to describe the changes in the way energy is stored when a system changes and energy is transferred.

☐ That energy is transferred when work is done by a force or by a moving charge.

☐ The conservation of energy principle — that energy can be transferred usefully, stored or dissipated, but can never be created or destroyed.

☐ How to calculate the amount of energy transferred into stores in a transfer.

☐ That during most energy transfers, some energy will be 'wasted' or 'dissipated' — i.e. transferred to energy stores which are not useful (usually thermal energy stores), and know some examples of this.

☐ That when energy transfers occur in a closed system, the net energy change of the system is zero, and know some examples of this.

Kinetic and Potential Energy Stores

☐ That any object which is moving has energy in its kinetic energy store.

☐ That the amount of energy in an object's kinetic energy store, E_k, depends on the object's mass, m, and speed, v, and can be found using the equation $E_k = \frac{1}{2}mv^2$.

☐ That an object that is lifted in a gravitational field gains energy in its gravitational potential energy store (and has energy transferred out of it when it is lowered).

☐ That the energy transferred to the gravitational potential energy store of an object, E_p, depends on its mass, m, the gravitational field strength, g, and the change in height of the object, h, and is found using the equation $E_p = mgh$.

☐ How to calculate the changes in how energy is stored for a falling object using the principle of conservation of energy.

☐ That when a spring is stretched (or compressed), energy is transferred to its elastic potential energy store.

☐ That the energy in a spring's elastic potential energy store, E_e, depends on its spring constant, k, and the extension (or compression), e, and be able to use the equation $E_e = \frac{1}{2}ke^2$.

Specific Heat Capacity

☐ That the amount of energy required to increase the temperature of a material depends on the material.

☐ That the amount of energy required to raise the temperature of 1 kg of a material by 1°C is given by the specific heat capacity.

☐ That the energy stored in (or released by) a substance when its temperature changes, ΔE, is determined by the specific heat capacity, and be able to use the equation $\Delta E = mc\Delta\theta$.

cont...

☐ How to calculate energy transfers to and from thermal energy stores using conservation of energy.

☐ How to perform an experiment to investigate the specific heat capacities of various materials.

☐ How, in the experiment above, work done by moving charges transfers energy electrically to thermal energy stores of the heater and the material.

Power

☐ That power is the rate of energy transfer (or work done), and is measured in watts (equivalent to 1 joule per second).

☐ That power, P, depends on the energy transferred (or work done) and the time taken, and is found using the equation $P = E \div t$ or $P = W \div t$.

☐ That for two otherwise identical devices with different powers, the device with the higher power will transfer the same amount of energy in less time.

Conduction and Convection

☐ That conduction is energy transfer by heating where vibrating particles transfer energy to neighbouring particles.

☐ That the higher the thermal conductivity of a material, the faster it will transfer energy by conduction.

☐ That convection is energy transfer by heating where energetic particles move away from hotter to cooler regions.

Reducing Unwanted Energy Transfers

☐ That thick walls with a low thermal conductivity can slow the rate of cooling of a house.

☐ That unwanted energy transfers can be reduced by thermal insulation, e.g. loft insulation in the home.

☐ How to investigate the effectiveness of different materials as thermal insulation.

☐ How to investigate the factors that determine the effectiveness of materials as thermal insulators.

☐ That unwanted energy transfers can be reduced by reducing friction, through methods such as lubrication (e.g. oil) and streamlining.

Efficiency

☐ That efficiency is the proportion of energy in an energy transfer that is transferred to useful energy stores by a device.

☐ That efficiency can be found by using the equation: $\text{efficiency} = \dfrac{\text{useful output energy transfer}}{\text{total input energy transfer}}$, or $\text{efficiency} = \dfrac{\text{useful power output}}{\text{total power input}}$

☐ How to express efficiency as a decimal and a percentage.

☐ H Ways of improving the efficiency of an energy transfer by reducing transfers to useless energy stores.

Exam-style Questions

1   A pot containing water is being heated on a gas hob.

   1.1   Give three energy transfers which are occurring in this scenario.

                                                          (3 marks)

   1.2   A second pan is heated at the same power, and contains the same amount of water.
                The water in each pan began at the same temperature.
                The water in the second pan reaches its boiling point faster.
                Suggest a property of material the second pan is made of which could explain this.

                                                            (1 mark)

                The hot water is poured into a flask, which is then sealed. The flask is made up of a smaller container encased within a larger one. There is a sealed gap between the walls of the two containers which is filled with air, as shown in the diagram below.

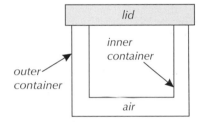

   1.3   Explain how this slows down the rate at which the water cools.

                                                       (2 marks)

2   Zaf has a mass of 85.0 kg. He runs up some stairs carrying a 10.0 kg box.
          The staircase is 10.0 m high and it takes Zaf 7.0 s to reach the top.

   2.1   How much energy has been transferred to the gravitational potential energy store of
                Zaf and the box when he reaches the top of the stairs?
                The gravitational field strength on Earth is 9.8 N/kg.
                Give your answer in joules.

                                                     (2 marks)

   2.2   What is Zaf's power output when he runs up the stairs?
                Give your answer in watts.

                                                      (2 marks)

   2.3   At the top of the stairs Zaf walks along the corridor, still carrying the box.
                Calculate Zaf's velocity if the total energy in the kinetic energy store of Zaf and
                the box is 153.9 J. Give your answer in m/s.

                                                     (3 marks)

   2.4   Zaf stops and then drops the box from a height of 1.25 m.
                Calculate the speed at which it hits the ground. State any assumptions you make.
                Give your answer in m/s and to 2 significant figures.

                                                     (4 marks)

3   A remote control car is attached to a spring, and does 4 J of work to stretch the spring by 40 cm over 0.5 s.

3.1   Write down the equation that links power, work done and time.

(1 mark)

3.2   Calculate the useful power output of the car to extend the spring.
Give your answer in W.

(2 marks)

3.3   Calculate the spring constant of the spring.
Use the correct equation from those listed on page 394. Give your answer in N/m.

(3 marks)

3.4   In the energy transfer in **3.1**, the efficiency is 62.5%.
Calculate the total input power of the car. Give your answer in W.

(2 marks)

3.5   Suggest one way that energy is transferred to a thermal energy store (and therefore wasted) by the toy car.

(1 mark)

3.6   Suggest a way to reduce the energy transfer described in **3.4**.

(1 mark)

4   A student wants to investigate the specific heat capacity of water. To do this, they are provided with a beaker, a thermometer, a heating element connected to a power supply and an ammeter.

4.1*   Describe an experiment that the student can perform with this equipment to calculate the specific heat capacity of water.

(6 marks)

4.2   Suggest one safety issue the student should consider when performing a risk assessment for this experiment.

(1 mark)

4.3   The student heats 0.50 kg of water from 10.0 °C to 100.0 °C.
189 kJ of energy was transferred to the water.
Calculate the specific heat capacity of water.
Use the correct equation from those listed on page 394. Give your answer in J/kg°C.

(3 marks)

1. Energy Resources and Their Uses

There are lots of different energy resources that we use today for all sorts of things. But first you need to know your non-renewables from your renewables.

Non-renewable energy resources

Non-renewable energy resources are the three **fossil fuels** and nuclear fuels:

- Coal
- Oil
- (Natural) gas
- Nuclear fuels (uranium and plutonium)

Fossil fuels are natural resources that form underground over millions of years. They are typically burnt to provide energy.

Non-renewable fuels are not being made at the same rate as they are being used and will all run out one day. They all do damage to the environment through emissions or because of issues with their mining, storage and disposal (see pages 56-57 for more), but they do provide most of our energy.

Renewable energy resources

Renewable energy resources are:

- Wind
- Water waves
- Tides
- Hydroelectricity
- The Sun (solar)
- Geothermal
- Bio-fuel

Renewable energy resources can be made at the same rate as they're being used and therefore will never run out. Most of them do damage to the environment, but in less nasty ways than most non-renewables. The trouble is they often don't provide as much energy as non-renewable resources and some of them are unreliable because they depend on the weather.

Use of energy resources

Energy resources, both renewable and non-renewable, are mostly used to generate electricity. There's loads more on how over the next few pages, but two other major uses are transport and heating.

Transport

Transport is one of the most obvious places where fuels are used. Traditionally, non-renewable energy resources have been the main fuel used in transportation:

- Petrol and diesel — fuel created from oil used to power many vehicles (including most cars).
- Coal — used in some old-fashioned steam trains to boil water to produce steam.

Learning Objectives:

- Know the definitions of non-renewable and renewable energy resources and be able to tell which resources are which.
- Know that non-renewable energy resources include fossil fuels (coal, oil and gas) and nuclear fuels (uranium and plutonium).
- Know that renewable energy sources include wind, water waves, tides, hydroelectricity, the Sun (solar), geothermal and bio-fuel.
- Know that energy resources are used for electricity generation, transportation and heating.
- Be able to compare how non-renewable and renewable energy resources are used in transport and heating.

Specification Reference 4.1.3

Figure 1: A bio-fuels filling station in Spain.

Recently, more methods of transportation have started using renewable energy resources, such as bio-fuels. Some vehicles run on pure bio-fuels (see p. 55) or a mix of a bio-fuel and petrol or diesel.

Heating

Energy resources are also used for heating things like your home. Non-renewable resources used for heating include:

- Natural gas — this is the most widely used fuel for heating homes in the UK. The gas is burned to heat water, which is then pumped into radiators throughout the home.

- Oil — some homes are heated by burning oil from a tank instead of gas, especially in remote places where it is difficult to connect to the gas supply.

- Coal — this is commonly burnt in fireplaces.

Renewable energy resources used for heating include:

- Geothermal power — a geothermal (or ground source) heat pump can be used to heat buildings (p. 50-51).

- Solar power — solar water heaters use electromagnetic radiation (see page 242) from the Sun to heat water which is then pumped into radiators in the building.

Electric transport and heating

Both renewable and non-renewable energy resources can also be used to generate electricity used in transport and heating. Electrically powered vehicles (e.g. trains, trams and electric cars) use electricity for transport and electric heaters (e.g. storage heaters) use electricity for heating.

Practice Questions — Fact Recall

Q1 What are the three fossil fuels?

Q2 a) Name four renewable energy sources.

    b) Give one reason why renewable energy resources aren't always used instead of non-renewable energy resources.

Q3 Give one renewable and one non-renewable resource used directly for transport (without the need for generating electricity).

Q4 How can solar power be used to heat buildings without the need for generating electricity?

2. Wind, Solar and Geothermal

Here's the first of many pages looking at using different renewable energy resources to generate electricity. First up: wind, solar and geothermal power...

Learning Objectives:
- Be able to compare the ways that wind power, solar cells and geothermal power are used for electricity generation.
- Describe the environmental issues that come from using different energy resources.
- Understand why certain energy resources are more reliable than others.

Specification Reference 4.1.3

Wind power

Generating electricity from wind involves putting up lots of wind turbines (see Figure 2) where they're exposed to the weather, like on moors or around coasts.

Each wind turbine has its own generator inside it. The electricity is generated directly from the wind turning the blades, which turns the generator.

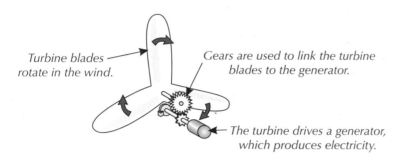

Turbine blades rotate in the wind.

Gears are used to link the turbine blades to the generator.

The turbine drives a generator, which produces electricity.

Figure 1: The structure of a wind turbine. The blades of the wind turbine rotate in the wind and directly drive a generator, which generates electricity.

Wind turbines produce no pollution (except for a little bit when they're manufactured), but according to some they do spoil the view (see Figure 2). You need about 1500 wind turbines to replace one coal-fired power station and 1500 of them cover a lot of ground — which would have a big effect on the scenery.

Wind turbines can be very noisy, which can be annoying for people living nearby. There's no permanent damage to the landscape though — if you remove the turbines, you remove the noise and the view returns to normal.

There are also problems with reliability. There's no power when the wind stops and it's impossible to increase supply when there's extra demand. Wind turbines also have to be stopped if the wind is very strong, as they could be damaged. On average, wind turbines produce electricity 70-85% of the time.

The initial costs (to build and set up the wind turbines) are quite high, but there are no fuel costs and minimal running costs.

Solar cells

Solar cells generate electric currents directly from the Sun's radiation (see Figure 3). Solar cells cause no pollution, although they do need quite a lot of energy to manufacture in the first place. Initial costs are high but after that the energy is free and running costs almost nil.

Tip: H For more on how generators produce electricity, check out page 305.

Figure 2: A wind farm in the countryside.

Tip: Be careful — solar cells are different to solar water heaters from page 48. Solar cells use some clever electronics to produce electricity from sunlight. Solar water heaters just use sunlight to heat water.

Figure 4: A solar panel generates electricity during the day, then uses it to power a lamp that lights up the road sign at night.

Figure 5: Solar cells and a wind turbine being used to power a house in Hamburg, Germany.

Exam Tip
You don't need to know exactly how electricity is generated here, but it helps to understand these details, as you'll need to be able to compare the ways different resources are used, their reliability and impact on the environment. If you're taking the Higher exam, you'll need to know how generators work — see page 305.

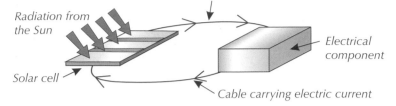

The solar cell generates electric current directly from the Sun's radiation and so can be plugged straight into electrical components, just like batteries.

Radiation from the Sun

Electrical component

Solar cell

Cable carrying electric current

Figure 3: A solar cell generating electric current from the Sun's radiation, connected directly to electrical components.

Solar cells are often the best source of energy for devices that don't use a lot of energy, or in remote places where it would be difficult to get power from other sources.

Examples

- Calculators, road signs and watches don't use much electricity, so solar cells are ideal for powering them — see Figure 4.

- Space is probably the ultimate remote location — you can't just nip up to a space satellite and top it up with fuel. Instead, satellites use large solar panels to generate the electricity they need.

In sunny countries, solar power is a very reliable source of energy, although it can only generate electricity during the daytime. Solar power can still be cost-effective even in cloudy countries like Britain. Solar cells are usually used to generate electricity on a relatively small scale, e.g. powering individual homes (see Figure 5).

Geothermal power

Geothermal power uses energy in the thermal energy stores of hot underground rocks to generate electricity. The source of much of the energy is the slow decay of various radioactive elements (see page 126), including uranium, deep inside the Earth.

Steam and hot water rise to the surface and are used to drive a turbine. The turbine turns a generator which generates electricity — see Figure 6. This is actually brilliant free energy that's reliable with very little impact on the environment.

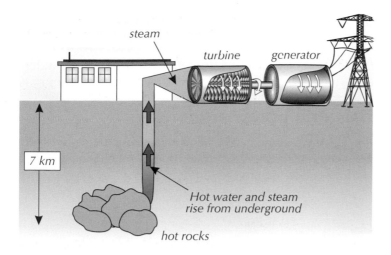

Figure 6: *The structure of a geothermal power station. Hot water and steam from underground drive a turbine, which is connected to a generator.*

In some places, geothermal power is used to heat water or buildings directly, without the need for generating electricity. The main drawback with geothermal power is that there aren't very many suitable locations for power stations. Also, the cost of building a power station is often high compared to the amount of energy we can get out of it.

Example

Iceland produces huge amounts of its energy using geothermal power. It is used to heat homes directly and to generate electricity (see Figure 7). Iceland is volcanic, so it has lots of hot rocks lying near to the surface.

Figure 7: *The Nesjavellir geothermal power station in Iceland.*

Practice Questions — Fact Recall

Q1 How is electricity generated by a wind turbine?

Q2 Give one issue associated with the reliability of using wind power.

Q3 Why are solar cells often used to power devices in remote locations?

Q4 Where does the energy transferred by geothermal power stations come from?

Practice Question — Application

Q1 For each situation below, suggest an appropriate method of generating electricity and explain your answer.

a) Street lights in a remote part of Australia.

b) A road sign on an exposed hillside in the UK.

c) Heating homes in a volcanic part of Iceland.

Learning Objectives:

- Be able to compare the ways that hydroelectric power stations, wave power and tidal power are used for electricity generation.
- Describe the environmental issues that come from using different energy resources.
- Understand why certain energy resources are more reliable than others.

Specification Reference 4.1.3

3. Hydroelectricity, Waves and Tides

Although hydroelectricity, wave power and tidal power all use water in a similar way, you need to know the details of each, and how they're different.

Hydroelectric power stations

Generating electricity using hydroelectric power usually requires the flooding of a valley by building a big dam. Water is allowed out at a controlled rate through turbines — see Figure 1.

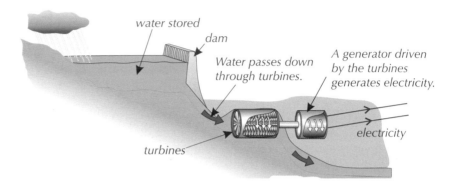

Figure 1: *A hydroelectric power station. Water is held back behind a dam. When it's released it passes through turbines, which turn a generator and generates electricity.*

There is no pollution, but there is a big impact on the environment. The flooding of the valley leads to rotting vegetation (which releases methane and carbon dioxide) and possible loss of habitat for some species (both animals and humans). Sometimes whole villages are evacuated and flooded to build a dam.

The reservoirs can also look very unsightly when they dry up. Putting hydroelectric power stations in remote valleys tends to reduce their impact on humans.

A big advantage is they can provide an immediate response to an increased demand for electricity. There's no problem with reliability except in times of drought. Initial costs are high, but there's no fuel costs and minimal running costs. It can be a useful way to generate electricity on a small scale in remote areas. However, in these cases it's often not practical or economical to connect it to the national grid.

Figure 2: *The Hoover Dam on the Colorado River in the USA — possibly the most well-known hydroelectric power station in the world.*

Tip: For more on the national grid, see page 95.

> **Example**
>
> Almost all of the electricity generated in Norway is hydroelectric. Norway is very mountainous with lots of rivers and valleys, making it ideal for hydroelectric power.

Wave power

To generate electricity using water waves, you need lots of small wave-powered turbines located around the coast. As waves come in to the shore they provide an up and down motion which can be used to drive a generator (see Figure 3).

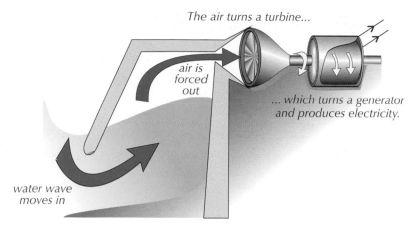

The air turns a turbine...

air is forced out

... which turns a generator and produces electricity.

water wave moves in

Figure 3: *A diagram showing how waves can be used to generate electricity. Waves force air through the turbine, which turns a generator and generates electricity.*

Tip: Electricity is also generated when the wave retreats, as air flows back through the turbine.

There is no pollution produced — the main problems are disturbing the seabed and the habitats of marine animals, spoiling the view and being a hazard to boats. They are fairly unreliable, since waves tend to die out when the wind drops. Initial costs are high, but there are no fuel costs and minimal running costs. Waves are never likely to provide energy on a large scale, but they can be very useful on small islands.

Example

- The Scottish island of Islay lies on the western coast of Scotland and is exposed to the full force of the northern Atlantic Ocean. Because of this and the fact that the island is fairly remote, it's an ideal location for a wave-powered station.

- In 2000 the Islay LIMPET (Land Installed Marine Power Energy Transmitter) was built. This was the first commercial wave power device to be connected to the UK national grid. It provides enough energy to power a few hundred homes.

- As a wave hits the structure, air is forced up through a turbine. When the wave drops, air is sucked back through the turbine. This process repeats every time a wave hits, and electricity is generated.

Figure 4: *The Islay LIMPET wave power station on the coast of Islay, a Scottish Hebridean island.*

Tidal barrages

Tides are produced by the gravitational pull of the Sun and Moon. They are used in lots of ways to generate electricity. The most common method is building a tidal barrage.

Tidal barrages are big dams built across river estuaries, with turbines in them. The turbines, as ever, are connected to electrical generators which turn and generate electricity. As the tide comes in it fills up the estuary to a height of several metres and is held back by the barrage. This water can then be allowed out through the turbines at a controlled speed when there's a height difference of water between the two sides of the barrage.

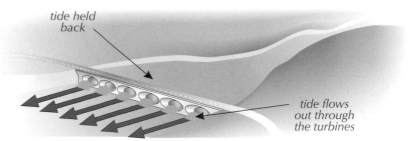

tide held back

tide flows out through the turbines

Figure 5: *The structure of a tidal barrage. Tide water is held back behind the barrage, then allowed to flow through turbines connected to generators, generating electricity.*

Figure 6: *A tidal barrage on the Rance River in France.*

This form of electricity generation causes no pollution — the main problems are preventing free access by boats, spoiling the view and altering the habitat of the wildlife, e.g. wading birds, sea creatures and beasties who live in the sand.

Tides are pretty reliable in the sense that they happen twice a day without fail, and always near to the predicted height. The only drawback is that the height of the tide is variable, so lower (neap) tides will provide significantly less energy than the bigger (spring) tides. They also don't work when the water level is the same either side of the barrage — this happens four times a day because of the tides.

Initial costs are moderately high, but there are no fuel costs and minimal running costs. Even though tidal barrages can only be used in some of the most suitable estuaries, tidal power has the potential for generating a significant amount of energy.

Practice Questions — Fact Recall

Q1   Describe how electricity is generated in a hydroelectric dam.

Q2   Give two environmental problems that can be caused by building hydroelectric dams.

Q3   Explain why hydroelectric dams are often used to generate electricity in remote areas.

Q4   Briefly explain how electricity can be generated from waves.

Q5   The total amount of electricity that can be generated by waves is fairly low. Give one other disadvantage, associated with the reliability of generating electricity using waves.

Q6   Briefly explain how a tidal barrage can be used to generate electricity.

4. Bio-fuels and Non-renewables

Bio-fuels, fossil fuels and nuclear fuels are used in much the same way to generate electricity. They're all used to heat water, which turns it into steam that drives a turbine, but they have very different issues affecting them.

Learning Objectives:

- Be able to compare the ways that bio-fuels, fossil fuels and nuclear fuels are used for electricity generation.

- Describe the environmental issues that come from using different energy resources.

- Understand why certain energy resources are more reliable than others.

Specification Reference 4.1.3

Bio-fuels

Bio-fuels are renewable energy resources — they're created from either plant products or animal dung. They can be solid, liquid or gas and can be burnt to produce electricity or run cars in the same way as fossil fuels.

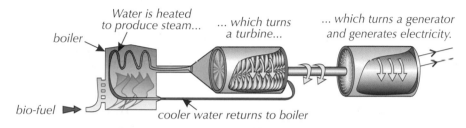

Water is heated to produce steam... ... *which turns a turbine...* ... *which turns a generator and generates electricity.*

boiler

bio-fuel ➡

cooler water returns to boiler

Figure 1: *A diagram of how bio-fuels are used to generate electricity. Bio-fuels are burned to heat water, which produces steam. The steam drives a turbine, which drives a generator and generates electricity.*

Bio-fuels are fairly reliable, as crops take a relatively short time to grow and different crops can be grown all year round. However, they cannot respond to immediate energy demands. To combat this, bio-fuels are continuously produced and stored for when they are needed.

The cost to refine bio-fuels is very high and some worry that growing crops specifically for bio-fuels will mean there isn't enough space or water to meet the demands for crops that are grown for food.

Bio-fuels made from plants are theoretically **carbon neutral**:

- The plants that grow to produce the waste absorb CO_2 from the atmosphere as they are growing.

- When the waste is burned, this CO_2 is re-released into the atmosphere. So it has a neutral effect on atmospheric CO_2 levels (although this only really works if you keep growing plants at the same rate you're burning things).

There is still debate regarding the impact of bio-fuels on the environment, once the full production is considered.

Using bio-fuels to generate electricity doesn't just produce carbon dioxide. Bio-fuel production also creates methane emissions — a lot of this comes from the animals.

In some regions, large areas of forest have been cleared to make room to grow bio-fuels (see Figure 2), resulting in lots of species losing their natural habitats. The decay and burning of this vegetation also increases CO_2 and methane emissions.

Bio-fuels have potential, but their use is limited by the amount of available farmland that can be dedicated to their production.

Figure 2: *An oil palm plantation in Indonesia. These crops can be used to make biodiesel.*

Fossil fuels

Most of the electricity we use is generated from fossil fuels (coal, oil, and gas) in big power stations. Figure 3 shows how electricity is generated in a typical fossil fuel power station.

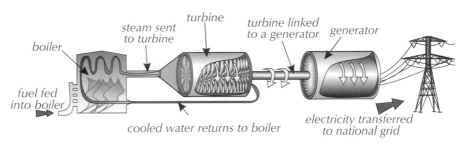

Figure 3: *Electricity generation in a fossil fuel power station.*

Exam Tip
You don't need to know exactly how electricity is generated here for the exam, but it helps to know which energy resource is which, as you'll need to compare the ways different resources are used, their reliabilities and any environmental issues they create.
You will need to know how generators work though if you're taking the Higher exam (see page 305).

Fossil fuels provide a cost-effective energy resource, that produces large amounts of energy and is readily available. While setup costs of power plants can be higher than other resources, the running costs and fuel extraction costs are fairly low.

They're also reliable — there's enough fuel to meet current demand, and they are extracted from the Earth at a fast enough rate that power plants always have fuel in stock. This means that the power plants can respond quickly to changes in demand (p. 96). However, these fuels are slowly running out. If no new resources are found, some fossil fuel stocks may run out within a hundred years.

All three fossil fuels (coal, oil and gas) release CO_2 into the atmosphere when they're burned. For the same amount of energy produced, coal releases the most CO_2, followed by oil then gas. All this CO_2 adds to the **greenhouse effect** and contributes to global warming. The greenhouse effect is where gases in the Earth's atmosphere (known as greenhouse gases) block radiation from the Sun from leaving the atmosphere. This causes the overall temperature of the atmosphere to rise (i.e. global warming).

Burning coal and oil releases sulfur dioxide, which causes acid rain. Acid rain can be harmful to trees and soils and can have far-reaching effects in ecosystems. Acid rain can be reduced by taking the sulfur out of the fuel before it is burned, or by cleaning up the emissions before they're released into the atmosphere.

Coal mining makes a mess of the landscape, especially "open-cast mining" where huge pits are dug on the surface of the Earth. Oil spillages cause serious environmental problems, affecting animals that live in and around the sea. We try to avoid them, but there's always a chance they'll happen.

Figure 4: *A pelican covered in oil from the Deepwater Horizon oil spill.*

Example

In 2010, an explosion at the oil drilling station Deepwater Horizon in the Gulf of Mexico resulted in the largest accidental oil spill in history. This resulted in thousands of miles of coastline being covered in crude oil, making the area uninhabitable for many local species of animals and plants.

Nuclear fuels

A nuclear power station is mostly the same as the one used for burning fossil fuels (see previous page), but the fuel isn't burnt. Nuclear fission of nuclear fuels such as uranium or plutonium releases the energy to heat water into steam which drives turbines. Nuclear power stations take the longest time of all the power stations to start up.

Nuclear power is 'clean', in that it doesn't cause the release of any harmful gases or other chemicals into the atmosphere. It's also reliable — there's enough fuel to meet current demand.

The biggest problem with generating electricity using nuclear fuel is that the nuclear waste produced is very dangerous and difficult to dispose of. This is because it stays highly radioactive (see page 126) for a long time and so needs to be stored safely far away from people's homes.

Nuclear fuel (e.g. uranium and plutonium) is relatively cheap but the overall cost of nuclear power is high due to the cost of the power station and final **decommissioning** — shutting down the power station so it's completely safe and poses no risk to people or the environment.

Nuclear power always carries the risk of a major catastrophe, like the Chernobyl disaster in 1986 or the Fukushima disaster in 2011.

Tip: Fission is the process where atoms split and release energy, see page 140.

Tip: The radiation produced by radioactive materials can be very dangerous to humans. See page 137 for more details.

Figure 5: *A nuclear power station at night with steam condensing into water vapour as it leaves the cooling tower.*

Practice Questions — Fact Recall

Q1 a) What is meant by a 'carbon neutral' process?

 b) Discuss how using bio-fuels to generate electricity could be a carbon neutral process.

Q2 Name two harmful gases that are released into the atmosphere by burning coal.

Q3 a) Which non-renewable resource doesn't directly release harmful gases into the atmosphere when used to generate electricity?

 b) Give two problems with using this fuel to generate electricity.

Learning Objectives:

- Understand how the ways we use energy resources have changed over time.

- Understand that although scientists have identified environmental issues with energy resources, other factors can limit our ability to deal with these issues.

Specification Reference
4.1.3

5. Trends in Energy Resource Use

Environmental issues related to energy resources have led us to change the resources we use over time. But it's not always easy to do so...

Reliance on fossil fuels

Over the 20th century, the electricity use of the UK hugely increased as the population got bigger and people began to use electricity for more things.

Since the beginning of the 21st century, electricity use in the UK has been decreasing (slowly), as we get better at making appliances more efficient (p. 40) and try to be more careful with energy use in our homes.

Most of our electricity is produced using fossil fuels (mostly coal and gas) and from nuclear power. Generating electricity isn't the only reason we burn fossil fuels — oil (diesel and petrol) is used to fuel cars, and gas is used to heat homes and cook food.

However, we are trying to increase our use of renewable energy resources (the UK aims to use renewable resources to provide 15% of its total yearly energy by 2020).

Movement towards renewable energy resources

We now know that burning fossil fuels has a lot of negative effects on the environment (p. 56). This makes many people want to use more renewable energy resources that effect the environment less.

People and governments are also becoming increasingly aware that non-renewables will run out one day. Many people think it's better to learn to get by without non-renewables before this happens.

Pressure from other countries and the public has meant that governments have begun to introduce targets for using renewable resources. This in turn puts pressure on energy providers to build new power plants that use renewable resources to make sure they don't lose business and money.

Car companies have also been affected by this change in attitude towards the environment. Electric cars and hybrids (cars powered by two fuels, e.g. petrol and electricity) are already on the market and their popularity is increasing.

Figure 1: Protest movements have applied pressure to governments to change their energy policies.

Factors limiting change

Although scientists know about the negative environmental effects of energy resources, they can only give advice and don't have the power to make people, companies or governments change their behaviour. The environmental issues aren't the only ones that matter, either. There are other political, social, ethical and economic issues to consider too (see page 5).

For example, the use of renewables is usually limited by reliability and money. Building new renewable power plants costs money, so some smaller energy providers are reluctant to do this — especially when fossil fuels are such a cost effective way of meeting demand.

Even if new power plants are built, there are a lot of arguments over where they should be. For example, many people don't want to live next to a wind farm, which can lead to them protesting.

Some energy resources like wind power are not as reliable as traditional fossil fuels and cannot increase their power output on demand. This would mean either having to use a combination of different power plants (which would be expensive) or researching ways to improve reliability.

Research into improving the reliability and cost of renewable resources takes time and money. This means that, even with funding, it might be years before improvements are made. In the meantime, dependable power stations using non-renewable resources have to be used.

Making personal changes can also be quite expensive. Hybrid cars are generally more expensive than equivalent petrol cars and things like solar panels for your home are still quite pricey. The cost of these things is slowly going down, but they are still not an option for many people.

Figure 2: *Offshore (at sea) wind turbines are relatively expensive to build and maintain but usually have fewer complaints from locals.*

Practice Questions — Fact Recall

Q1 Why has electricity usage in the UK been decreasing slightly since the start of the of the 21st century?

Q2 Give two limitations on using more renewable energy resources.

Topic 1b Checklist — Make sure you know...

Energy Resources and Their Uses

- [ ] How renewable and non-renewable energy resources are different.
- [ ] That coal, gas, oil and nuclear fuels like uranium and plutonium are non-renewable energy resources.
- [ ] That renewable energy resources include: wind, water waves, tides, hydroelectricity, the Sun (solar), geothermal and bio-fuels.
- [ ] That energy resources are mostly used for generating electricity, transport and heating.
- [ ] How certain renewable and non-renewable energy resources are used directly in transport and heating.

Wind, Solar and Geothermal

- [ ] That wind can be used to generate electricity by driving turbines.
- [ ] That solar cells generate electricity directly from radiation from the Sun.

cont...

☐ That geothermal power can be used to generate electricity by heating water to form steam which turns a turbine, or to heat buildings directly.

☐ The environmental impacts of using wind power, solar power and geothermal power.

☐ That wind power and solar power have issues with reliability due to the weather.

☐ That geothermal power is reliable once you've found a suitable location for a power station.

Hydroelectricity, Waves and Tides

☐ That in a hydroelectric power station, water flowing through a dam drives turbines directly to generate electricity.

☐ That water waves can be used to generate electricity by forcing air through a turbine.

☐ That the tides can be used to generate power. For example, tidal barrages are dams built across estuaries, where turbines are driven by tide water as it fills and leaves the estuary.

☐ The environmental impacts of using hydroelectric power, wave power and tidal power.

☐ That hydroelectricity is reliable, except in times of drought.

☐ That wave power is unreliable, as it relies on the wind causing waves.

☐ That tidal barrages are reliable as there will always be tides.

Bio-fuels and Non-renewables

☐ That bio-fuels can be burned to heat water into steam that drives a turbine to generate electricity.

☐ That fossil fuels are burned to heat water into steam that drives a turbine to generate electricity.

☐ That, in nuclear power stations, energy released from the fission of uranium or plutonium is used to heat water into steam that drives a turbine to generate electricity.

☐ The environmental impacts of using bio-fuels, fossil fuels and nuclear power to generate electricity.

☐ That fossil fuels, nuclear fuels and bio-fuels are reliable and there are enough to meet current demand, although bio-fuels can't respond to immediate energy demands.

Trends in Energy Resource Use

☐ How our use of different energy resources has changed over time.

☐ That more renewable energy resources are being used due to environmental issues with non-renewables.

☐ That although scientists have shown the environmental issues of certain energy resources, they are still used as other factors affect our ability to deal with these issues.

☐ That the cost and reliability of renewable energy resources are currently their main limitations.

Exam-style Questions

1   The table below shows the percentage of the total electricity produced from each of six energy resources in two different countries.

	Coal	Oil	Gas	Nuclear	Hydroelectric	Wind
Country 1	49.9%	2.4%	20.3%	19.6%	6.0%	1.8%
Country 2	3.9%	4.0%	4.5%	3.1%	84.2%	0.3%

1.1   What percentage of electricity in Country 1 is generated from non-renewable energy sources?

(2 marks)

1.2   Both countries generate the same amount of electricity each year.
Suggest which country will emit more pollution by generating electricity.
Explain your answer.

(2 marks)

1.3   Country 1 is considering reducing the amount of coal it burns and generating some of its electricity from bio-fuels instead. Give **one** environmental impact of using coal to generate electricity.

(1 mark)

1.4   Suggest **one** impact on reliability that switching from coal to generating electricity using bio-fuels may have.

(1 mark)

2   Coal was once the most commonly used resource for generating electricity in the UK.

2.1   Coal is a fossil fuel and a non-renewable energy resource.
Define a non-renewable energy resource.

(1 mark)

2.2   Suggest another use of coal as an energy resource, other than for generating electricity.

(1 mark)

2.3   The use of coal has decreased in recent years so that it is no longer the main fuel for UK power stations.
Suggest **one** environmental reason why the use of coal has decreased.

(1 mark)

2.4   A new coal power station is proposed for an island in Scotland. The island has a number of exposed hills, and regularly experiences windy weather.
Suggest **two** renewable energy resources which would be suitable alternatives to coal to produce electricity on the island, and explain how they are suitable.

(4 marks)

2.5   Give **one** reason why the renewable alternatives may not be able to entirely replace the coal power station as an option for generating the island's power.

(1 mark)

1. Circuits, Current and Potential Difference

Electricity flows as current around circuits. You may have seen circuit symbols before, but there are quite a few that you need to know. So first things first, let's have a bit of circuit training...

Circuit symbols

You need to know (and be able to draw) each of the following circuit symbols. You'll learn a bit more about some of them later in this section and the next.

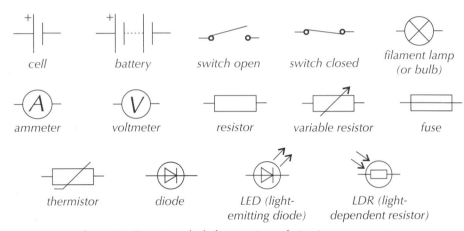

Figure 1: *Circuit symbols for a variety of circuit components.*

Circuit diagrams

Tip: Wires are just represented by straight lines.

You might be asked to draw a circuit, or to find a problem with one. One thing to make sure of is that your circuit is complete. If a component isn't connected in a circuit properly, it won't work.

A circuit is complete if you can follow a wire from one end of the battery (or other power supply), through any components to the other end of the battery (ignoring any switches) — see Figure 2.

Tip: Switches allow you to turn circuits (and so components) on and off. You can ignore them when working out if a circuit is complete, because they can always be closed to complete the circuit.

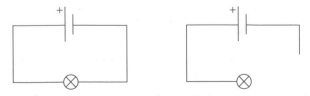

Figure 2: *(Left) A complete circuit. All the wires are joined to something at both ends and the lamp will light.
(Right) An incomplete circuit. The lamp won't light.*

Voltmeters and ammeters

Voltmeters and ammeters always have to be connected in a circuit in a certain way, as shown in Figure 3, otherwise they won't do what they are meant to.

- A **voltmeter** measures potential difference. It is always connected 'across' a component — this is known as 'in parallel'.

- An **ammeter** measures current. It is always connected 'in line' with a component — this is known as 'in series'.

Tip: Read on to find out about current and potential difference and head to pages 70-78 if you want to learn more about what 'in series' and 'in parallel' mean.

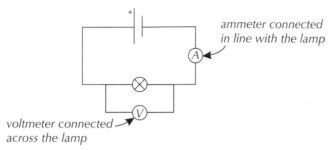

ammeter connected in line with the lamp

voltmeter connected across the lamp

Figure 3: *A circuit with a voltmeter and an ammeter connected correctly.*

Current

An electric **current** is a flow of electric charge. Cells (and other power supplies) always have a positive terminal (the longer line), and a negative terminal (the shorter line). Current flows from positive to negative around a circuit.

Tip: Electrons carry the charge. They actually flow from negative to positive (see pages 99-101). But when scientists realised this they didn't want to change the way that current was defined, so we still say it flows from positive to negative.

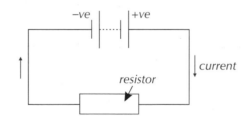

Figure 4: *Current flowing from positive to negative in an electric circuit.*

The size of the current is the rate of flow of charge. It's measured in amperes, A. When current flows past a point in a circuit for a length of time then the charge that has passed is given by this formula:

Tip: You can use a formula triangle to rearrange the equation for charge. See page 338 for how to use formula triangles.

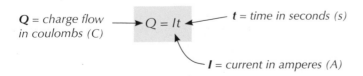

Q = charge flow in coulombs (C) $Q = It$ **t** = time in seconds (s)

I = current in amperes (A)

More charge passes around the circuit when a bigger current flows. In a single, closed loop (like the one in Figure 4) the current has the same value everywhere in the circuit (see page 71).

> ### Example
>
> **A battery charger passes a current of 2.5 A through a cell over a period of exactly 4 hours. How much charge does the charger transfer to the cell altogether?**
>
> You've got $I = 2.5$ A and $t = 4 \times 60 \times 60$ s $= 14\ 400$ s.
>
> $Q = It = 2.5 \times 14\ 400 = 36\ 000$ C

Tip: You may also see potential difference referred to as voltage, or abbreviated to pd. You'll learn a lot more about pd later on — see pages 92-93.

Potential difference

Electrical charge will only flow around a complete (closed) circuit if there is a **potential difference**, so a current can only flow is there's a source of potential difference. In a simple circuit, this is usually a cell or battery. Potential difference is the driving force that pushes the charge around. Its unit is the volt, V.

Practice Questions — Fact Recall

Q1 Draw the circuit symbol for a:

a) battery   b) resistor   c) fuse   d) lamp   e) diode

Q2 What is an incomplete circuit?

Q3 In what direction does current flow in a circuit?

Q4 What is the formula linking charge, current and time? Write down the units each quantity is measured in.

Q5 What is needed for a current to flow in a circuit?

Practice Questions — Application

Q1 Draw a complete circuit containing a cell, a lamp, and an open switch which can be used to turn the lamp on and off.

Q2 Draw a complete circuit containing a battery, a resistor and a voltmeter measuring the voltage across the resistor.

Q3 Draw a circuit containing a cell, a thermistor and an ammeter.

Q4 In which of these circuits will the lamp be lit up?

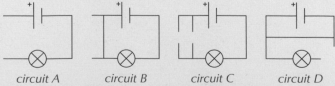

circuit A   circuit B   circuit C   circuit D

Q5 The current through a lamp is 0.2 A. Calculate the time taken for 50 C of charge to pass through the lamp.

Q6 A cell has a charge of 102 C passing through it every minute. Calculate the current flowing through the cell.

2. Resistance and *I-V* Characteristics

Resistance is how much a component in a circuit slows down the flow of current. It's different for each component, and it's related to the size of the current and potential difference.

Resistance

Resistance is anything in the circuit which reduces the flow of current. It is measured in ohms, Ω. The current flowing through a component depends on the potential difference across it, and the resistance of the component. The greater the resistance of a component, the smaller the current that flows (for a given potential difference across the component).

The resistance of a component is linked to potential difference across it and current through it by the following formula:

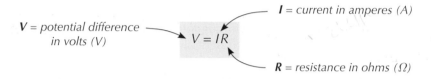

V = potential difference in volts (V)

$V = IR$

I = current in amperes (A)

R = resistance in ohms (Ω)

Example

Voltmeter *V* reads 6.0 V and resistor *R* has a resistance of 4.0 Ω.
What is the current through ammeter *A*?

MATHS SKILLS

Rearrange the formula for $V = IR$.

You need to find I, so the version you need is $I = V \div R$.

$I = V \div R = 6.0 \div 4.0 = 1.5$ A

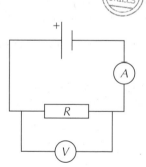

For some components, as the current through them is changed, the resistance of the component changes as well. The resistance of an **ohmic conductor** (e.g. a wire or a resistor), however, doesn't change with the current.

At a constant temperature, the current flowing through an ohmic conductor is directly proportional to the potential difference across it (i.e. R is constant in $V = IR$).

However, a lot of components aren't ohmic — their resistance changes with the current through them. These include certain types of resistor (see pages 82-83), diodes and filament lamps (p. 68).

Learning Objectives:

- Know that the resistance of and potential difference across a component determine the current flowing through it.
- Know and be able to use the equation $V = IR$.
- Know that an ohmic conductor has a constant resistance for any current through it.
- Know, and be able to explain examples of, how the resistance of some components varies with current through them.
- Be able to use a circuit to investigate the effect of wire length on the resistance of a circuit (Required Practical 3).
- Be able to use a circuit to investigate the *I-V* characteristic of a resistor at a constant temperature (ohmic conductor), filament lamp and diode (Required Practical 4).
- Be able to draw, and explain the design and use of, each of the circuits used in these practicals.
- Know the shape of these *I-V* characteristics and link it to the properties and function of the component and whether it is linear or non-linear.

Specification References
4.2.1.3, 4.2.1.4

Investigating factors affecting resistance

The resistance of a circuit can depend on a number of factors, like whether components are in series or parallel, p. 79-81, or the length of wire used in the circuit. You can investigate how the length of a wire affects resistance using the circuit shown in Figure 1.

REQUIRED PRACTICAL 3

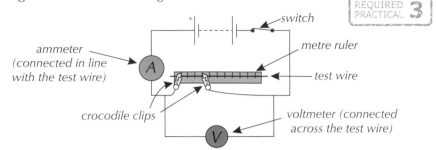

Figure 1: A test circuit for investigating how the length of a wire affects circuit resistance.

One crocodile clip is attached to the wire at the 0 cm position on the ruler. It should remain fixed at this point throughout the investigation. The second crocodile clip can be moved left and right along the test wire. The value written on the ruler at the point you attach this clip gives the length of the wire connected in the circuit.

Now you have your test circuit set up, you can begin your investigation. Attach the second crocodile clip to the wire and record the length of the wire between the two crocodile clips. Close the switch and record the current shown on the ammeter and the potential difference shown on the voltmeter.

Move the second crocodile clip, close the switch and record the new length, current and potential difference. Repeat this for a number of different lengths of wire.

Use the data you've recorded to work out the resistance for each length of wire, using $R = V \div I$ (from $V = IR$ on page 65). Plot a graph of resistance against wire length and draw a line of best fit.

The graph you obtain should look like the one shown in Figure 2. It's a straight line through the origin, which means that resistance is directly proportional to length — the longer the wire, the greater the resistance.

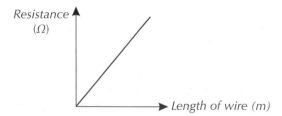

Figure 2: A graph of resistance against wire length for a test wire. It's a straight line which passes through the origin.

Your graph might not actually pass through the origin. It can be hard to attach the first crocodile clip at exactly 0 cm on the meter ruler, so all of your measurements could be slightly shorter or longer than the true values. There is also contact resistance in every circuit, e.g. between the crocodile clips and the components. These would cause systematic errors in your results.

Tip: Make sure you do a risk assessment beforehand — watch out for things like the wire becoming too hot.

Tip: Using a thin wire will give you the best results here. It should also be as straight as possible so your length measurements are accurate.

Tip: The wire may heat up, which will increase its resistance. The larger the current, the more it'll heat up, so using a small pd can help minimise this effect. It's also a good idea to open the switch between readings to let the wire cool down.

Tip: Make sure you take enough measurements to be able to spot a pattern in your graph.

Tip: There are lots of different factors you could test using this setup. Another example would be investigating how the diameter of the wire affects resistance. You'd need a selection of wires of different thicknesses to do this. A micrometer (p. 327) would be helpful for measuring their diameters.

Tip: See p. 13 for more on systematic errors.

I-V characteristics

REQUIRED PRACTICAL 4

I-V **characteristics** (current-potential difference graphs) show how the current varies as you change the potential difference across a component. Ohmic conductors (components with constant resistance — p. 65) have *I-V* characteristics that are straight lines. They're also known as linear components. Non-linear components, on the other hand, have curved *I-V* characteristics — the resistance changes depending on the current.

Since $V = IR$, you can calculate the resistance at any point of an *I-V* characteristic by reading off the values of V and I and calculating $R = V \div I$. If it's a straight-line graph through the origin, every point will give you the same value of resistance.

You can construct an *I-V* characteristic for a component yourself using the test circuit displayed in Figure 3.

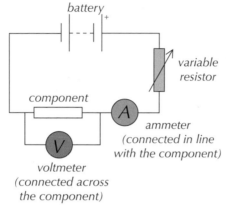

Figure 3: A circuit diagram of a test circuit used to investigate the I-V characteristic of a component.

To find a component's *I-V* characteristic, begin to vary the resistance of the variable resistor. This alters the current flowing through the circuit and the potential difference across the component.

Each time you use the variable resistor to alter the current, record the potential difference across the component for that value of current. Repeat each reading twice more to get an average pd for each measurement.

Swap over the wires connected to the battery to reverse the direction of the current. Measure negative values of current and potential difference using the same method as described above.

Now you can use your measurements to plot a graph of current against potential difference for the component. This is your *I-V* characteristic. Read on to see some examples of *I-V* characteristics for different components.

Ohmic conductors

The current through an ohmic conductor (at constant temperature) is directly proportional to potential difference, so you get a straight line *I-V* characteristic. A fixed resistor (or a length of wire) is an ohmic conductor, so you could test a few different resistors (or different wires). Different resistances result in straight lines with different slopes — see Figure 6.

Tip: Again, make sure you do a risk assessment for the practical described on this page.

Tip: This type of circuit is a dc (direct current) circuit (see p. 89). The component, ammeter and variable resistor are all connected in a line (in series). They can be in any order in the circuit. The voltmeter must be placed across (in parallel with) the component.

Figure 4: A type of variable resistor. They can be used to control the amount of current flowing in a circuit.

Tip: Once you've swapped over the wires at the battery, the readings will be negative because the current is flowing through the components in the opposite direction to before.

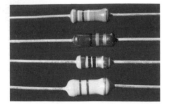

Figure 5: *Different sizes of resistors could be used in this investigation. They typically look like this.*

Tip: Remember, you can calculate the resistance at any point on an *I-V* characteristic using $R = V \div I$.

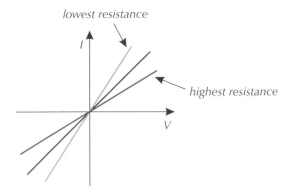

Figure 6: *I-V characteristics for ohmic conductors with different resistances.*

Filament lamps

When an electrical charge flows through a filament lamp, it transfers some energy to the thermal energy store of the filament (page 22), which is designed to heat up and glow. Resistance increases with temperature, so as more current flows through the lamp, the lamp heats up more and the resistance increases. This means less current can flow per unit potential difference, so the graph gets shallower, hence the curve of the *I-V* characteristic is shown in Figure 8.

Figure 7: *A filament lamp, with filament glowing.*

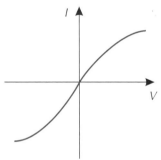

Figure 8: *An I-V characteristic for a filament lamp. It's curved because the resistance is changing.*

Diodes

A **diode** is a component that only lets current pass through it in one direction. The resistance of a diode depends on the direction of the current — it will happily let current flow through it one way, but will have a very high resistance if the current is reversed.

Figure 9: *A diagram illustrating which way current is able to flow in a diode.*

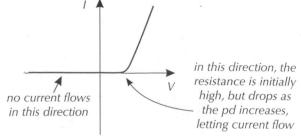

Figure 10: *An I-V characteristic for a diode. Current only flows in one direction.*

If you're investigating the *I-V* characteristic of a diode using the circuit suggested in Figure 3, you'll need to add a protective resistor to the circuit. As diodes have low resistance in one direction, the protective resistor is needed to stop the current getting too high and damaging the diode. Because you're keeping the current quite low, you should also use a milliammeter rather than a regular ammeter.

Tip: A protective resistor is just a fixed resistor connected in series with another component. It increases the overall resistance in the circuit, protecting the other component from too large a current.

Practice Questions — Fact Recall

Q1 What is resistance? What unit is resistance measured in?

Q2 Write down the equation linking resistance, pd and current.

Q3 Draw a circuit diagram of a circuit you could use to investigate how the length of a piece of wire affects the resistance of a circuit.

Q4 What is the relationship between the length of a wire and its resistance?

Q5 Draw a circuit diagram of a circuit you could use to find the *I-V* characteristic of a component.

Q6 How do you find the resistance of a fixed resistor at a constant temperature from its *I-V* characteristic?

Q7 Sketch an *I-V* graph for:

a) an ohmic conductor at a constant temperature   b) a diode

Q8 Describe the resistance of a diode as the current through it flows in:

a) the forwards direction.   b) the backwards direction.

Practice Questions — Application

Q1 A student tests component A using a suitable test circuit and plots a graph of his data, shown on the right. What would you expect component A to be?

Q2 A current of 0.015 A is flowing through a 2.0 Ω resistor. What is the potential difference across the resistor?

Q3 A current of 0.60 A flows through a motor. The potential difference across the motor is 14.4 V. Calculate the resistance of the motor.

Q4 The graph on the right shows the *I-V* characteristic of a resistor at a constant temperature. Calculate its resistance.

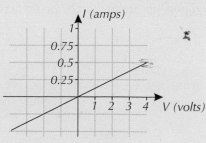

- Know that series circuits are one way of connecting electrical components.

- Be able to construct series circuits from circuit diagrams.

- Understand how series circuits can be used to measure quantities and test components.

- Know that the total potential difference from the power supply is shared between all components connected in series.

- Know that the current through each component connected in series is the same.

- Know that the total resistance of components in series is the sum of the resistances of each component.

- Understand why adding resistors in series increases the total resistance.

- Know how to calculate pd, resistance and current in series circuits.

Specification Reference 4.2.2

3. Series Circuits

So you've got the basics of circuit components covered... but how are you going to connect them? Components can either be connected in series or parallel. First up, series circuits — they're basically just big loops.

Components in series

In **series circuits**, the different components are connected in a line, end to end, between the positive and negative ends of the power supply (except for voltmeters, which are always connected across a component — see page 63).

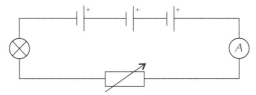

Figure 1: *A circuit in which each component is connected in series — you can draw a single line that travels along the wires and passes through every component once before returning to your starting point.*

If you remove or disconnect one component, the circuit is broken and they all stop. This is generally not very handy, and in practice very few things are connected in series. But you can design series circuits to simply measure quantities and test components (e.g. the test circuit on page 67, and the sensor circuit coming up on page 83).

Potential difference in series circuits

There is a bigger potential difference (pd) when more cells are connected in series, provided the cells are all connected the same way. For example, when two cells with a potential difference of 1.5 V are connected in series, they supply 3 V between them (see Figure 2 on the next page). You just add up all the individual cell pds to find the total power source pd.

In series circuits the total potential difference of the supply is shared between the various components. So the potential differences round a series circuit always add up to equal the source potential difference. If two or more components are the same, the pd across them will also be the same.

Tip: Ammeters are always connected in series with the components that they are measuring the current through — even if the circuit is not a series circuit (p. 63).

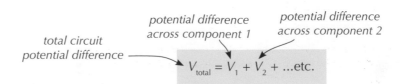

total circuit potential difference

potential difference across component 1

potential difference across component 2

$$V_{total} = V_1 + V_2 + ...\text{etc.}$$

In the diagram, two cells and two lamps are connected in series. The total potential difference across the circuit is the sum of the pds of the two batteries, so:

$$V = 1.5\,V + 1.5\,V = 3.0\,V$$

The potential differences, V_1 and V_2, of the two lamps add up to 3.0 V:

$$V = V_1 + V_2 = 3.0\,V$$

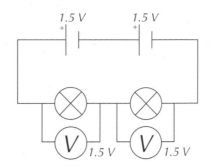

Figure 2: *A circuit diagram containing two cells and two identical lamps connected in series. The potential difference from the cells is split evenly across the lamps.*

Tip: Remember, voltmeters are always connected across the component they are measuring.

Figure 3: *Christmas tree lights are often connected in series. The bulbs can be very small because the mains voltage is shared out between them.*

Current in series circuits

In series circuits the same current flows through all parts of the circuit:

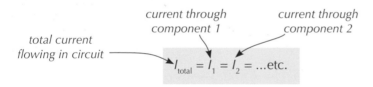

current through
component 1

current through
component 2

total current
flowing in circuit

$$I_{total} = I_1 = I_2 = \dots\text{etc.}$$

The size of the current is determined by the total potential difference of the cells and the total resistance of the circuit, using the equation $V = IR$ (from page 65).

Tip: This is why ammeters must be connected in series — so that the same current flows through the ammeter and the component that you want to measure the current through.

In the diagram, two lamps are connected in series. The current through each lamp (and in fact the whole circuit) is exactly the same.

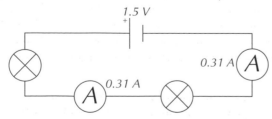

Figure 4: *A series circuit containing two lamps and two ammeters, showing that the current is the same throughout the circuit.*

Resistance in series circuits

In series circuits the total resistance is just the sum of all the resistances:

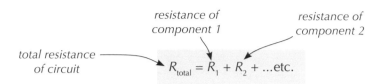

$$R_{total} = R_1 + R_2 + ...etc.$$

total resistance of circuit — *resistance of component 1* — *resistance of component 2*

This is because by adding a resistor in series, the two resistors have to share the total pd. The potential difference across each resistor is lower, so the current through each resistor is also lower. In a series circuit, the current is the same everywhere so the total current in the circuit is reduced when a resistor is added. This means that the total resistance of the circuit increases.

The bigger the resistance of a component, the bigger its share of the total potential difference.

Example

Find the total resistance of this circuit.

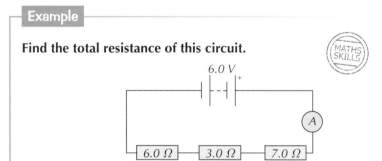

The total resistance of a series circuit is the sum of the resistances of each component:

$$R = 6.0\ \Omega + 3.0\ \Omega + 7.0\ \Omega = 16\ \Omega$$

Find the current through the circuit.

- The total potential difference in the circuit is equal to the battery potential difference: $V = 6.0\,V$.

- Rearrange the equation $V = IR$ and plug in the values of V and R to find the total current:

$$I = V \div R = 6.0 \div 16 = 0.375\ A$$

Find the potential difference across the 6.0 Ω resistor.

- The current is the same everywhere in the circuit, so the current through the 6.0 Ω resistor is $I = 0.375\ A$.

- The resistance of the resistor is clearly (I hope...) $R = 6.0\ \Omega$.

- Use the equation $V = IR$:

$$V = IR = 0.375 \times 6.0 = 2.25\ V$$

> **Tip:** The resistance of the ammeter and wires is so small that you don't need to worry about it.

Summary of series circuits

There are four simple rules to remember for series circuits:

- The pd of the cells adds up to the source pd.

- The source pd is split across the components.

- The current is the same through all the components.

- The total resistance of the circuit is the sum of all the resistances of the separate components.

You'll need to be able to use these rules in all sorts of circuit examples in the exam, so make sure you know them.

Practice Questions — Fact Recall

Q1   What does 'connected in series' mean?

Q2   What type of circuit component must always be connected in series?

Q3   How should you connect extra cells in a circuit in order to increase the total potential difference across the circuit?

Q4   True or false? Every component connected in series has the same potential difference across it.

Q5   What can you say about the current through each component connected in series?

Q6   Explain why adding a resistor in series increases the total resistance.

Practice Questions — Application

Q1   The circuit shown has a 7 Ω resistor and a filament lamp in series. Ammeter A reads a constant value of 1.5 A.

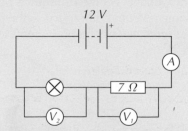

a) What is the current through the filament lamp?

b) What potential difference will voltmeter V_1 measure?

c) What potential difference will voltmeter V_2 measure?

d) What is the resistance of the filament lamp?

Tip: Remember the total V across the circuit will be equal to $V_1 + V_2$.

Learning Objectives:

- Know that parallel circuits are one way of connecting electrical components.
- Know the difference between series and parallel circuits.
- Be able to construct parallel circuits from circuit diagrams.
- Know that the pd across each component in parallel is the same.
- Know that the total current of a parallel circuit is the sum of the currents through each branch.
- Know that the total resistance of a parallel circuit is less than the resistance of the branch with the smallest resistance.
- Understand why adding resistors in parallel decreases the total resistance.
- Know that circuits can contain a combination of components wired in parallel and components wired in series.

Specification Reference 4.2.2

4. Parallel Circuits

In parallel circuits, components or groups of components are on their own separate loop or branch. They're much more useful than series circuits because you can turn off each loop separately, without turning off the rest.

Components in parallel

Unlike components in series, components connected in **parallel** each have their own branch in a circuit connected to the positive and negative terminals of the supply (except ammeters which are always connected in series). If you remove or disconnect one of them, it will hardly affect the others at all. This is because current can still flow in a complete loop (page 63) from one end of the power supply to the other through the branches that are still connected.

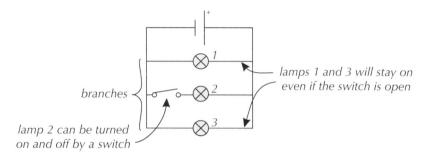

Figure 1: *A circuit diagram with three lamps connected in parallel, one of which can be switched on and off by a switch.*

This is obviously how most things must be connected, for example in cars and in household electrics. You have to be able to switch everything on and off separately. Everyday circuits often include a mixture of series and parallel parts (see page 77).

Potential difference in parallel circuits

In parallel circuits, all components get the full source pd. Each branch in a parallel circuit has the same potential difference as the power supply, so the potential difference is the same across all components connected in parallel:

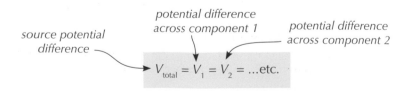

This means that identical lamps connected in parallel will all be at the same brightness.

Example

Everything electrical in a car is connected in parallel. Parallel connection is essential in a car to give these two features:

- Everything can be turned on and off separately.

- Everything always has the full pd of the battery across it. This is useful because it means that you can listen to the radio on full blast without it having much effect on the brightness of your lights.

Figure 2: A car dashboard. Everything that you can turn on and off from your dashboard will be on its own parallel circuit branch connected to the car battery.

The only slight effect is that when you turn lots of things on the lights may briefly go a bit dim because the battery can't provide full potential difference under heavy load. This is normally a very slight effect. You can spot the same thing at home when you turn a kettle on, if you watch very carefully.

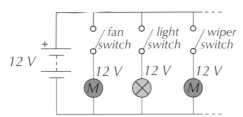

Tip: (M) is the symbol for a motor.

Figure 3: A circuit diagram to give an idea of how the electronic components in a car are connected in a parallel circuit.

Current in parallel circuits

In parallel circuits the total current flowing around the circuit is equal to the total of all the currents through the separate branches.

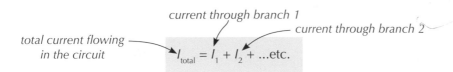

$$I_{total} = I_1 + I_2 + ...etc.$$

total current flowing in the circuit

current through branch 1

current through branch 2

This means that unlike voltage, the current going through each branch is less than the total current in the circuit. Whenever the circuit splits into one or more branches, a certain amount of the current flows through each branch.

The current through each component inside a branch is the same — it's like a mini series circuit. The same amount of current that entered the branch must then leave the branch when it rejoins the rest of the circuit.

If two identical components are connected in parallel then the same current will flow through each component.

Example

Two lamps connected in parallel will share the total current in the circuit. In Figure 4:

$$I_1 = I_2 + I_3$$

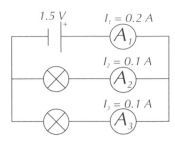

Figure 4: A circuit diagram showing identical lamps connected in parallel.

Resistance in parallel circuits

If you have two resistors in parallel, their total resistance is less than the resistance of the smallest of the two resistors. So adding a resistor in parallel reduces the total resistance.

This can be tough to get your head around, but think about it like this:

In parallel, both resistors have the same potential difference across them as the source. This means the 'pushing force' making the current flow is the same as the source pd for each resistor you add.

But by adding another additional loop, the current has more than one direction to go in. This increases the total current that can flow around the circuit. Using $V = IR$, an increase in current means a decrease in the total resistance of the circuit.

Tip: This last bit is important — adding another loop to the circuit increases the current in the circuit (because there are more paths where charge is flowing).

Tip: You might see the answer in this example written as $R_{total} < 4\ \Omega$. This just means that the total resistance is less than 4 Ω.

Example

In the circuit shown in Figure 5, there are three resistors connected in parallel.

In a parallel circuit, the total resistance is smaller than the smallest resistance on an individual branch. Therefore, the total resistance of the circuit in Figure 5 is less than 4 Ω.

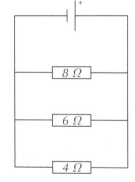

Figure 5: A circuit diagram showing three resistors connected in parallel.

Summary of parallel circuits

There are three rules for parallel circuits. Know them well:

- The pd across each branch is the same as the source pd.

- The current is split across the branches, and the total current is the sum of the current of each branch.

- The total resistance of a parallel circuit is less than the smallest resistance of an individual circuit branch.

Tip: Remember for parallel circuits:
$I_{total} = I_1 + I_2 + $ ...etc. and
$V_{total} = V_1 = V_2 = $ ...etc.

Example

Find the potential difference across, and the current through, resistor Y in the circuit shown.

The resistors are all connected in parallel, so the pd across each resistor in the circuit is the same as the supply pd. So the pd across resistor Y is 6 V.

The total current through the circuit is the same as the sum of the currents in the branches:

$$5.5 \text{ A} = 1.5 \text{ A} + I_2 + 1.0 \text{ A}$$

Rearranging to find I_2:

$$I_2 = 5.5 \text{ A} - 1.5 \text{ A} - 1.0 \text{ A} = 3.0 \text{ A}$$

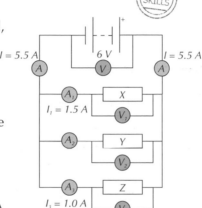

Mixed series and parallel circuits

You can have a circuit that contains components connected in series and components connected in parallel — see Figure 6. Just make sure you apply the right rules to the right bits.

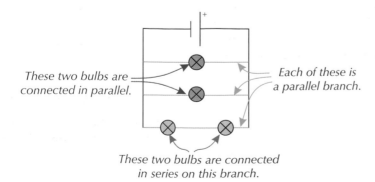

These two bulbs are connected in parallel.

Each of these is a parallel branch.

These two bulbs are connected in series on this branch.

Figure 6: *A circuit with components connected in series and parallel with each other.*

Exam Tip
You might have to deal with simple circuits that are a mixture of parallel and series like this in the exam, so make sure you understand what's going on at each component.

Example

In the circuit shown in Figure 7, the potential difference across each branch would be 1.5 V.

On the branch with two lamps connected in series, the pd is then split between the two lamps (page 70).

$$1.5\,V = V_1 + V_2 = V_3$$

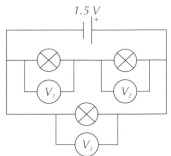

Figure 7: A circuit diagram showing lamps connected in parallel and in series.

Practice Questions — Fact Recall

Q1 What does 'connected in parallel' mean?

Q2 Why are parallel circuits more useful than series circuits in household electrics?

Q3 What can you say about the potential difference across all the components connected in parallel with a power supply? Assume each component is on a branch with no other components.

Q4 If you know the current in every branch of a parallel circuit, how can you work out the total current in the circuit?

Practice Questions — Application

Q1 A parallel circuit is shown. When the switch is closed, ammeter A reads 0.75 A.

  a) Calculate the pd shown by voltmeter V.

  b) The total current in the circuit is 2.0 A. Calculate the resistance of the lamp.

  c) Is the total resistance of the circuit more than, equal to, or less than your answer to b)?

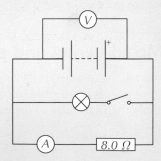

Tip: Look back at page 73 for series circuit rules.

Q2 Look at the circuit shown below.

  a) If switch S is open, ammeter A reads 0.70 A. Find the potential difference across the battery.

  b) If switch S is closed, ammeter A reads 0.50 A.

   i) Find the potential difference across lamp B.

   ii) Find the potential difference across resistor R_2.

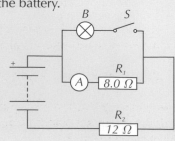

5. Investigating Resistance

As you've seen over the past few pages, the way a component is connected in a circuit determines how it contributes to the total resistance of the circuit. You can investigate this yourself by adding identical resistors in series and parallel and measuring the effect on resistance.

Resistance in series and parallel circuits

Investigating resistance for resistors in series

Before you get started, you'll need to make sure you have at least four identical resistors.

Start by constructing the circuit shown in Figure 1 using one of the resistors. Make a note of the potential difference of the battery (V).

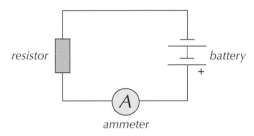

Figure 1: *A circuit diagram of the initial circuit for investigating resistance of resistors in series, with a single resistor connected in series.*

To carry out the investigation, follow these steps:

1. Measure the current through the circuit using the ammeter. Use this and the pd of the battery to calculate the resistance of the circuit using $R = V \div I$ (from $V = IR$, see page 65).

2. Add another resistor, in series with the first (as shown in Figure 2).

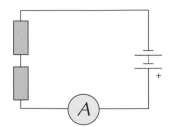

Figure 2: *A circuit diagram of the circuit in Figure 1, with a second resistor added in series with the first.*

3. Again, measure the current through the circuit and use this and the pd of the battery to calculate the overall resistance of the circuit.

4. Repeat steps 2 and 3 until you've added all of your resistors.

Now you've got your results, plot a graph of the number of identical resistors against the total resistance of the circuit. Your graph should look similar to the one shown in Figure 3.

Learning Objective:

- Be able to use a circuit to investigate the effect of connecting resistors in series and in parallel on the resistance of a circuit (Required Practical 3).

Specification Reference 4.2.1.3

Tip: As always, remember to carry out a risk assessment for this practical before you start your investigation.

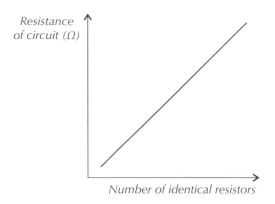

Resistance of circuit (Ω)

Number of identical resistors

Figure 3: *A graph of total resistance of a circuit against number of identical resistors connected in series.*

Tip: To refresh your memory on how current, potential difference and resistance behave in series circuits, check out pages 70-73.

You should find that adding resistors in series increases the total resistance of the circuit — i.e. adding a resistor decreases the total current through the circuit. The more resistors you add, the larger the resistance of the whole circuit. These results agree with the rules for resistance in series on page 72.

Investigating resistance for resistors in parallel

Using the same equipment as before (so the experiment is a fair test), build the same initial circuit shown in Figure 1, and carry out the following steps:

1. Measure the total current through the circuit and calculate the resistance of the circuit using $R = V \div I$ (again, V is the potential difference of the battery).

2. Next, add another resistor, in parallel with the first, as shown in Figure 4.

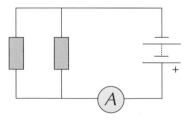

Figure 4: *A circuit diagram of the circuit in Figure 1, with a second resistor added in parallel with the first.*

Tip: Remember, ohmic conductors (like fixed resistors) only have a constant resistance at a constant temperature. Disconnecting the battery between readings will stop your circuit getting too hot, which could affect your results.

3. Measure the total current through the circuit and use this and the potential difference of the battery to calculate the overall resistance of the circuit.

4. Repeat steps 2 and 3 until you've added all of your resistors.

Just like before, use your results to plot a graph of the number of identical resistors in the circuit against the total resistance. You should get a graph which looks like the one shown in Figure 5.

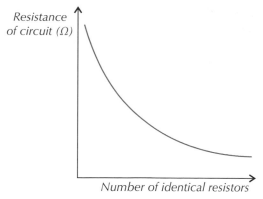

Figure 5: A graph of total resistance of a circuit against number of identical resistors connected in parallel.

Tip: You need to be able to explain why the resistance decreases when a component is added in parallel (see page 76).

When you add resistors in parallel, the total current through the circuit increases — so the total resistance of the circuit has decreased. The more resistors you add, the smaller the overall resistance becomes — as shown by the graph in Figure 4. These results agree with the rules for adding resistors in parallel from page 76.

Practice Question — Application

Q1 A student wants to investigate how the resistance of a circuit changes as they connect more resistors in parallel. They have six identical 1 Ω resistors, a battery and an ammeter.

a) Draw a circuit diagram of the circuit they should construct to find the total resistance of the circuit when three resistors are connected in parallel.

b) The student's results are displayed in the table below.

Number of 1 Ω resistors	Total resistance (Ω)
1	1.00
2	0.50
3	0.33
4	0.25
5	0.20
6	0.17

Plot a graph of their results, and draw a line of best fit.

c) Explain the trend shown by the results.

Learning Objectives:

- Know that the resistance of an LDR decreases with increasing light intensity.
- Know some applications of LDRs.
- Know that the resistance of a thermistor decreases with increasing temperature.
- Know some applications of thermistors.
- Understand how LDRs and thermistors can be used in circuits, e.g. simple sensor circuits.

Specification Reference 4.2.1.4

6. LDRs and Thermistors

You've just got two more circuit devices to learn about in this section — LDRs and thermistors...

Light-dependent resistors (LDRs)

An **LDR** is a resistor that is dependent on the intensity of light. Simple really.

- In bright light, the resistance falls (see Figure 1).

- In darkness, the resistance is highest.

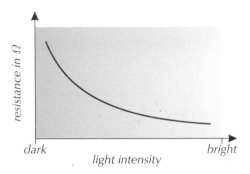

Figure 1: A graph of resistance against light intensity for a light-dependent resistor (LDR).

LDRs can be used where a function depends on light levels, e.g. when you want a component to only work in the dark.

Figure 2: An LDR (top) and its circuit symbol (bottom).

Examples

- Automatic night lights and outdoor lighting such as street lights use LDRs. When the light level falls, the resistance of the LDR increases to a level that triggers the light to turn on.

- Some burglar detectors use LDRs too. A light beam is shone at an LDR. If someone walks in front of it and breaks the light beam, the resistance of the LDR shoots up and an alarm is triggered.

Figure 3: Automatic street lighting on a motorway.

There's more on how these applications work on the next page.

Thermistors

Thermistors are another type of resistor — their resistance depends on their temperature. A thermistor's resistance decreases as the temperature increases.

- In hot conditions, the resistance drops (see Figure 4).

- In cool conditions, the resistance goes up.

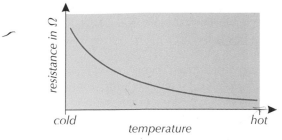

Figure 4: A graph of resistance against temperature for a thermistor.

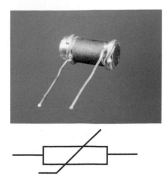

Figure 5: An example of one type of thermistor (top), and its circuit symbol (bottom).

Thermistors are useful in temperature detectors called thermostats. They are connected in a circuit where their resistance can be measured. As resistance varies with temperature, knowing the resistance means you can detect the temperature of the thermistor (and its surroundings). Thermostats can be used in car engine temperature sensors to make sure the engine isn't overheating.

Application in sensor circuits

Most applications of LDRs and thermistors involve a sensor circuit. Figure 6 shows a typical sensor circuit.

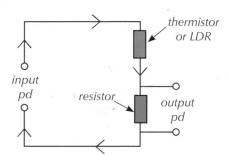

Figure 6: A circuit diagram of a typical sensor circuit using a thermistor or LDR.

If you have two circuit components in series, they share out the potential difference of the power supply relative to their resistances — the bigger the component's resistance relative to another's, the more pd it takes.

In the circuit shown, if the resistance of the thermistor or LDR increases, it takes a bigger share of the input pd from the power supply, so the pd across the resistor (the output pd) falls.

If the resistance of the thermistor or LDR decreases (i.e. the thermistor gets hotter or more light shines on the LDR) it takes a smaller share of the input pd, so the output pd rises.

You can use this effect in circuits that use the output pd across the resistor to power another component.

Tip: For a reminder on pd and resistance in series circuits, see pages 73.

Example

An air conditioning circuit can be connected to the output pd of a thermistor circuit — as the temperature rises, the resistance of the thermistor decreases and the pd across the resistor increases. This increases the pd supplied to the air conditioning circuit, helping to cool the room down.

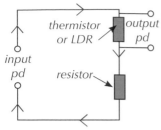

Figure 7: An alternative sensor circuit for increasing pd with decreasing temperature/light.

You could have the output pd across the thermistor/LDR instead, if you wanted the pd to increase with low temperatures or dark conditions (e.g. for an automatic night light or heater).

In most everyday applications, a more complicated switching circuit is often used to make something happen at a certain temperature or light level (e.g. a thermostat can be used to switch on the central heating when a room gets to a certain temperature).

Practice Questions — Fact Recall

Q1   How can you lower the resistance of an LDR?

Q2   Sketch a graph of resistance against temperature for a thermistor.

Q3   Give one application of:

a) LDRs   b) thermistors

Practice Question — Application

Q1   The circuit shown forms part of a burglar alarm. The ohmmeter measures the resistance of the component it is connected to. The circuit is connected to another circuit containing an alarm — if the resistance rises above a certain level, the alarm is sounded.

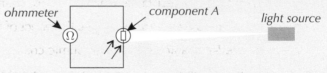

a) What is component A?

b) A cat passes through the beam of light.
Explain why this causes the alarm to sound.

Topic 2a Checklist — Make sure you know...

Circuits, Current and Potential Difference

- ☐ The circuit symbols for: a cell, a battery, a switch (open and closed), a filament lamp, an ammeter, a voltmeter, a resistor, a variable resistor, a fuse, a thermistor, a diode, an LED and an LDR.
- ☐ How to construct a basic circuit diagram.
- ☐ That voltmeters must be connected across (i.e. in parallel with) a component and ammeters must be connected in line (i.e. in series) with a component.
- ☐ That current is the flow of electric charge.
- ☐ That the size of the current (I) is the rate of flow of charge, and that the charge (Q) that flows in a given time (t) is defined by the equation $Q = It$.
- ☐ That at any point in a single closed loop, the current has the same value.
- ☐ That potential difference is the driving force that pushes current around a circuit.

Resistance and *I-V* Characteristics

- ☐ That resistance reduces the flow of current.
- ☐ That the current (I) through, potential difference (V) across, and resistance (R) of a component are related by the equation $V = IR$.
- ☐ That ohmic conductors have a resistance which does not change with a changing current.
- ☐ How to construct and use a circuit to investigate the effect of wire length on circuit resistance.
- ☐ How to investigate the *I-V* characteristics of an ohmic conductor, a filament lamp and a diode.
- ☐ The *I-V* characteristics for these three components and how to explain their shape with respect to how resistance changes with current, including whether they are linear or non-linear.

Series Circuits

- ☐ That series circuits have all components (besides voltmeters) connected in-line on a single loop.
- ☐ That multiple sources of potential difference connected in series add together to give the total pd.
- ☐ That the pd provided by a power supply is shared between all the components in the circuit.
- ☐ That the current through each component in a series circuit is the same.
- ☐ That the total resistance of a series circuit is the sum of the resistances of all components in the circuit.
- ☐ How to explain why adding resistors in series increases the total resistance of the circuit.
- ☐ How to calculate current, potential difference and resistance in a series circuit.

cont...

Parallel Circuits

☐ That parallel circuits have components which are each connected on their own branch of the circuit.

☐ That the potential difference across each parallel branch is the same.

☐ That current is split between parallel branches.

☐ That the total resistance of resistors in parallel is less than the resistance of the smallest resistor.

☐ How to explain why adding resistors in parallel decreases the total resistance of the circuit.

Investigating Resistance

☐ How to perform an experiment to investigate how adding resistors in series affects the total resistance of a circuit.

☐ How to perform an experiment to investigate how adding resistors in parallel affects to the total resistance of a circuit.

LDRs and Thermistors

☐ That the resistance of an LDR (light dependent resistor) decreases with increasing light intensity.

☐ That the resistance of a thermistor decreases with increasing temperature.

☐ Applications of LDRs (e.g. street lighting and burglar alarms) and thermistors (e.g. thermostats).

☐ How an LDR or thermistor can be used to construct simple sensor circuits.

Exam-style Questions

1   A physics student is carrying out a series of experiments to investigate the properties of a filament lamp.

1.1   The student wants to measure the current passing through the lamp using an ammeter. State how the student should connect the ammeter in the circuit, and explain why this is the case.

(2 marks)

1.2   Sketch the *I-V* characteristic of a filament lamp.

(1 mark)

1.3   Is a filament lamp an ohmic conductor? Explain your answer.

(1 mark)

The student now uses the filament lamp in a simple sensor circuit, so that the lamp gets brighter as the temperature decreases. The circuit consists of a filament lamp, a thermistor, a resistor, and a cell.

1.4   Draw a circuit diagram of a possible arrangement for this sensor circuit.

(3 marks)

1.5   Across which component should the lamp be connected if the student instead wishes to make the lamp increase in brightness as the temperature increases?

(1 mark)

2   An electric motor and filament lamp are connected in parallel to a battery, as shown.

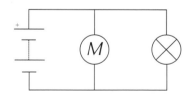

The total size of the electric current in the circuit is 1.2 A. It takes 30 seconds to move 15 C of charge through the motor. The potential difference across the motor is 14 V.

2.1   What is the size of the electric current in a circuit a measure of?

(1 mark)

2.2   What is the potential difference across the battery?

(1 mark)

2.3   Calculate the current passing through the filament lamp.
Give your answer in amperes.

(5 marks)

2.4   Calculate the resistance of the filament lamp.
Give your answer in ohms.

(4 marks)

3 The circuit shown was used to record values of the current through and the potential difference across component X at a constant temperature.

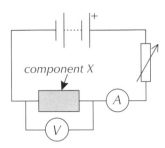

3.1 Explain how the variable resistor is used to change the current in a circuit.

(1 mark)

The data collected for component X using this circuit is displayed on the graph.

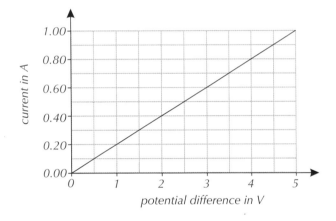

3.2 What is component X?

(1 mark)

3.3 Calculate the resistance of component X.

(3 marks)

3.4 Component Y is the same type of component as component X but has a higher resistance at the same temperature. Which of the three graphs below shows the *I-V* graphs of component X and component Y? Explain your answer.

(2 marks)

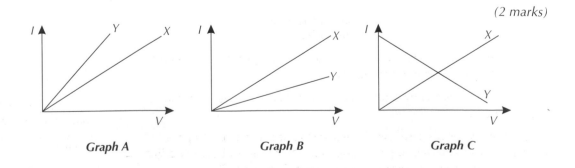

Graph A   *Graph B*   *Graph C*

1. Electricity in the Home

We use electricity all the time without thinking about it but there's actually a lot to think about. There are two-types of electricity, ac and dc, and there's a lot of clever wiring in all our appliances that keep us safe.

Ac and dc supplies

There are two types of electricity supplies — **alternating current (ac)** and **direct current (dc)**. In ac supplies the current is constantly changing direction. Alternating currents are produced by alternating voltages (page 306) in which the positive and negative ends keep alternating. The UK domestic mains supply (the electricity in your home) is an ac supply at around 230 V. The frequency of the ac mains supply (how often the current changes direction) is 50 cycles per second or 50 Hz (hertz).

By contrast, cells and batteries supply direct current (dc). Direct current is a current that is always flowing in the same direction. It's created by a direct voltage — where the positive and negative ends of the source are fixed.

Three-core cables

Most electrical appliances are connected to the mains supply by **three-core cables**. This means that they have three wires inside them, each with a core of copper and a coloured plastic coating. The colour of the insulation on each cable shows its purpose — the colours are always the same for every appliance. This is so that it is easy to tell the different wires apart.

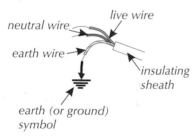

neutral wire

live wire

earth wire

insulating sheath

earth (or ground) symbol

Figure 1: *A three-core electrical cable showing the live, neutral and earth wires.*

- The brown **live wire** is what provides the alternating potential difference (at about 230 V) from the mains supply.

- The blue **neutral wire** completes the circuit and carries away current. It is around 0 V.

- The green and yellow **earth wire** is also at 0 V. It is for protecting the wiring and for safety — it stops the appliance casing from becoming live. It doesn't usually carry a current — only when there's a fault.

Tip: The potential difference between the live wire and earth is 230 V.

Figure 2: *A three-core electrical cable with a live wire, a neutral wire and an earth wire.*

Tip: A low resistance path is a path with low electrical resistance. Since $V = IR$ (page 65), a large current will flow through a low resistance path.

The live wire

Your body (just like the earth) is at 0 V. This means that if you touch the live wire, a large potential difference (pd) is produced across your body and a current flows through you. This causes a large electric shock which could injure or even kill you.

Even if a plug socket or a light switch is turned off (i.e. the switch is open) there is still a danger of an electric shock. A current isn't flowing, but there is still a pd in the live wire. If you made contact with the live wire, your body would provide a link between the supply and the earth, so a current would flow through you.

Any connection between live and the earth can be dangerous. If the link creates a low resistance path to the earth, a huge current will flow, which could result in a fire.

Practice Questions — Fact Recall

Q1   What is direct current (dc)?

Q2   What does ac stand for? How is it different from dc?

Q3   What is the colour coding of the wires in a three-core cable?

Q4   a) Describe the potential difference in the live and neutral wires in a three-core electrical cable.

b) What is the role of the earth wire in a three-core cable?

2. Power and Energy Transfer

All the useful things electricity can do are due to the transfer of energy. Energy is transferred between stores electrically (like you saw on page 23) by electrical appliances. Power tells us how quickly an appliance transfers energy.

Energy transfers in electrical appliances

You know from page 63 that electric current is the flow of electric charge. When a charge moves, it transfers energy. This is because the charge does work against the resistance of the circuit, and work done is the same as energy transferred. Electrical appliances are designed to transfer energy to components in a circuit when a current flows.

Example

- Kettles transfer energy electrically from the mains ac supply to the thermal energy store of the heating element inside the kettle.

- Energy is transferred electrically from the battery of a handheld fan to the kinetic energy store of the fan's motor.

Of course, no appliance transfers all energy completely usefully. Whenever a current flows through anything with electrical resistance (which is pretty much everything) then energy is transferred to the thermal energy stores of the components (and then the surroundings). The higher the current, the more energy is transferred to these thermal energy stores. You can calculate the efficiency of any electrical appliance (i.e. work out how much energy is transferred usefully). See page 40 for more on how to do this.

Example

Filament bulbs work by passing a current through a very thin wire, heating it up so much that it glows. Rather obviously, they waste a lot of energy through transfer to the thermal energy stores of the wire and surroundings compared to the amount of energy usefully transferred to generate light.

Power and energy transfer

The total energy transferred by an appliance depends on how long the appliance is on for and the power at which it's operating.

The **power** of an appliance is the energy that it transfers per second. So the more energy it transfers in a given time, the higher the power.

The amount of energy transferred by electrical work is given by:

$$E = Pt$$

E = energy transferred (J) t = time (s) P = power (W)

Learning Objectives:

- Know that a moving charge transfers energy, and so work is done when it flows in a circuit.
- Know that electrical appliances transfer energy electrically.
- Understand how various different appliances transfer energy from a power source to useful energy stores.
- Know that the energy transferred is determined by the power of an appliance and amount of time it's used for.
- Understand that a higher power means more energy is transferred per second.
- Know the amount of energy transferred is found by $E = Pt$.
- Know that power is measured in watts, W.
- Understand the connection between the power rating of a device and the energy transferred between stores when they are used.
- Know the amount of energy transferred by electrical work can be found by $E = QV$.
- Know that the power of a device is related to the pd across it and current through it by $P = VI$.
- Know that the power of a device is related to the current through it and its resistance by $P = I^2R$.

Specification References
4.2.4.1, 4.2.4.2

Topic 2b Domestic Electricity   91

> **Example**
>
> **If a 2.5 kW kettle is on for 5 minutes, how much energy is transferred by the kettle?**
>
> First make sure the numbers are in the right units:
>
> > power = 2.5 kW = 2500 W
> > time = 5 minutes = 300 s
>
> Then substitute into the formula:
>
> > $E = Pt = 2500 \times 300 = 750\ 000$ J

Tip: Remember:
1 kW = 1000 W.

Figure 1: *A label showing the voltage and power rating of an electrical appliance.*

Appliances are often given a power rating — they're labelled with the maximum safe power that they can operate at. You can usually take this to be their maximum operating power. The power rating tells you the maximum amount of energy transferred between stores per second when the appliance is in use. This helps customers choose between models — the lower the power rating, the less electricity an appliance uses in a given time, so the cheaper it is to run.

But a higher power doesn't necessarily mean that it transfers more energy usefully. An appliance may be more powerful than another, but less efficient, meaning that it might still only transfer the same amount of energy (or even less) to useful stores (see page 40).

Potential difference and energy transfer

When an electrical charge (Q) goes through a change in potential difference (V), then energy (E) is transferred. Energy is supplied to the charge at the power source to 'raise' it through a potential. The charge gives up this energy when it 'falls' through any potential drop in components elsewhere in the circuit, as shown in Figure 2.

Tip: Remember that energy is transferred when a moving charge does work against resistance in a circuit.

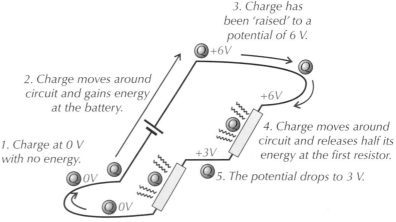

3. Charge has been 'raised' to a potential of 6 V.

2. Charge moves around circuit and gains energy at the battery.

1. Charge at 0 V with no energy.

4. Charge moves around circuit and releases half its energy at the first resistor.

5. The potential drops to 3 V.

6. The rest of the energy is released at the second resistor. The charge is at 0 V.

Figure 2: *An electrical charge passing round a circuit and transferring energy through circuit components. As it does this, its potential changes.*

The potential difference between two points is the energy transferred per unit charge passing between the two points and you can calculate it using this formula:

$$\boldsymbol{E} = energy \longrightarrow E = QV \longleftarrow \boldsymbol{V} = potential\ difference\ (V)$$
transferred (J)

$$\boldsymbol{Q} = charge\ (C)$$

So a bigger change in pd means more energy is transferred for a given charge passing through the circuit.

Tip: A battery with a bigger pd will supply more energy because the charge is raised up "higher" at the start (see Figure 2).

Example

The motor in an electric toothbrush is attached to a 3.0 V battery. The total charge that passes a point in the motor circuit during use is 140 C. Calculate the energy transferred by the motor. Explain why the energy transferred to the kinetic energy stores of the motor will be less than the energy you calculated.

To calculate the energy transferred you need to use the formula $E = QV$.

charge = 140 C, potential difference = 3.0 V

$E = QV = 140 \times 3.0 = 420$ J

The energy transferred to the kinetic energy stores of the motor will be less than 420 J because the motor won't be 100% efficient. Not all the of the energy will be transferred to the motor's kinetic energy stores — some of it will be transferred into the thermal energy store of the motor and surroundings.

Tip: There's more about efficiency on page 40.

Power, potential difference and current

If you know the potential difference across an appliance and the current flowing through it, you can work out its power using this formula:

$$\boldsymbol{P} = power\ (W) \longrightarrow P = VI \longleftarrow \boldsymbol{I} = current\ (A)$$

$$\boldsymbol{V} = potential\ difference\ (V)$$

Tip: So a higher power means a greater current flows for a given pd.

You can also find the power if you don't know the potential difference. To do this, stick $V = IR$ from page 65 into $P = VI$, giving:

$$\boldsymbol{P} = power\ (W) \longrightarrow P = I^2R \longleftarrow \boldsymbol{R} = resistance\ (\Omega)$$

$$\boldsymbol{I} = current\ (A)$$

Example 1

What's the power input of a light bulb that draws 0.20 A of current from a 230 V supply?

Just put the numbers into the equation:

$$P = VI = 230 \times 0.20$$
$$= 46\ W$$

Example 2

The motor in a toy car has a resistance of 25 Ω and an operating power of 64 W. Find the current it draws from the battery.

You don't know the potential difference, so you'll have to use $P = I^2R$. Rearrange the equation for power to make current the subject:

$$P = I^2R \text{ , so } I = \sqrt{\frac{P}{R}}$$

Then put the numbers in:

$$I = \sqrt{\frac{P}{R}} = \sqrt{\frac{64}{25}} = 1.6 \text{ A}$$

Practice Questions — Fact Recall

Q1 a) Give two examples of appliances that transfer energy electrically.

b) For each appliance given in part a), give one energy store that energy is transferred to electrically.

Q2 How is energy wasted when electric charge flows through a circuit component with electrical resistance?

Q3 What is the formula for working out the power of an appliance in terms of the energy it transfers? What does each term represent and what are their units?

Q4 What is the formula that relates power, current and potential difference? What units are each of them measured in?

Practice Questions — Application

Q1 A homeowner is choosing a cooker. Cooker A is rated at 9800 W and cooker B is rated at 10 200 W. Which cooker will use the most energy in 20 minutes?

Q2 An appliance draws 3.0 A of current from the mains supply (230 V). What's the power of the appliance?

Q3 A 2.0 kW heater is on for 30 minutes. How much energy does it transfer?

Q4 Calculate the potential difference a 4.0 C charge passes through when it transfers 22 J of energy.

Q5 A 0.2 A current flows through a 40 W filament lamp. Calculate the resistance of the filament lamp.

Q6 Microwave A is rated at 900 W and takes 4 minutes to cook a ready meal. Microwave B is rated at 650 W and takes 6 minutes to cook a ready meal. Assuming both microwaves are working at maximum power, which transfers the most energy in cooking the meal?

3. The National Grid

Now you know how electricity powers our appliances, you might be wondering how it gets to us in the first place. That's where the national grid comes in. Whoever you pay for your electricity, it's the national grid that gets it to you.

Distributing electricity

The **national grid** is the network of cables and transformers that covers the UK and connects power stations to consumers (anyone who's using electricity). The national grid transfers electrical power from power stations anywhere on the grid (the supply) to anywhere else on the grid where it's needed (the demand) — e.g. homes and industry.

To transmit the huge amount of power needed, you either need a high potential difference, or a high current (as $P = VI$, page 93). The problem with high current is that you lose loads of energy as the wires heat up and energy is transferred to the thermal energy stores of the surroundings.

It's much cheaper to boost the pd up really high (to 400 000 V) and keep the current relatively low. For a given power, increasing the pd decreases the current, which decreases the energy lost by heating the wires and the surroundings. This makes the national grid an efficient way of transferring energy.

Transformers

A **transformer** is a device used in the national grid to change the potential difference of an electrical supply. Getting the potential difference to 400 000 V for transmission requires transformers as well as big pylons and huge insulators. It's still cheaper than transmitting it through smaller wires at a high current with lots of energy loss though.

The transformers have to step the potential difference up at one end, for efficient transmission, and then bring it back down to safe, usable levels at the other end. The potential difference is increased ('stepped up') using a step-up transformer. It's then reduced again ('stepped down') for domestic use using a step-down transformer, as shown in Figure 1.

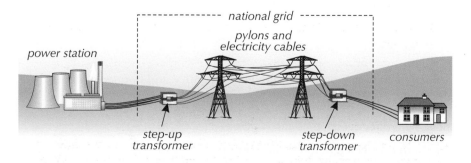

Figure 1: *A diagram of the national grid distributing electricity from a power station to consumers.*

Learning Objectives:
- Know that the national grid is a network of cables and transformers.
- Know that it is used to transfer electrical power from power stations to consumers.
- Understand why the national grid is an efficient way of distributing electrical energy.
- Know that step-up transformers are used to increase the pd before transmission, and step-down transformers are used to decrease the pd to safe levels for consumers.

Specification Reference 4.2.4.3

Tip: Remember that power is the energy transferred in a given time, so a higher power means more energy transferred.

Meeting demand

Throughout the day, electricity usage (the demand) changes. Power stations have to produce enough electricity for everyone to have it when they need it.

In order to do this, demand for electricity needs to be predicted. Demand increases when people get up in the morning, come home from school or work and when it starts to get dark or cold outside and people put the lights and heating on. Popular events like a sporting final being shown on TV could also cause a peak in demand.

Tip: See pages 47-59 for more on electricity generation and trends in energy usage.

Power stations often run at well below their maximum power output, so there's spare capacity to cope with a high demand, even if there's an unexpected shut-down of another station. Lots of smaller power stations that can start up quickly are also kept on standby just in case.

Practice Questions — Fact Recall

Q1 a) What does a transformer do?

b) Why are transformers used in the national grid?

Q2 What does each label A-E represent in this diagram?

Q3 a) State two times of the day when demand for electricity increases.

b) Suggest two ways in which electricity providers can prepare for unexpected increases in demand.

Topic 2b Checklist — Make sure you know...

Electricity in the Home

☐ That there are two types of electricity — alternating current and direct current.

☐ That alternating current (ac) is constantly changing its direction of flow.

☐ That alternating current is produced by an alternating voltage, where the positive and negative ends are always alternating.

☐ That mains electricity is a source of ac with a potential difference of 230 V and frequency of 50 Hz.

☐ That direct current (dc) flows in a fixed direction.

☐ That direct current is produced by a direct voltage.

☐ That most electrical appliances are connected to the mains by three-core cables.

☐ That three-core cables consist of a brown-coated live wire, a blue-coated neutral wire and a green-and-yellow-coated earth wire.

cont...

- [ ] That the live wire carries current to the device, and provides the alternating pd of 230 V.
- [ ] That the neutral wire carries current away from the device, and is at around 0 V.
- [ ] That the earth wire is a safety feature which does not carry current unless there is a fault, and is at 0 V.
- [ ] That your body is at 0 V.
- [ ] That touching a live wire causes an electric shock due to the potential difference between the wire and your body, which causes a current to flow through you.
- [ ] That connections between the live wire and the earth are dangerous due to the large currents that flow through them.

Power and Energy Transfer

- [ ] That work is done when charge flows in a circuit.
- [ ] That electrical appliances are designed to transfer energy electrically to a useful energy store.
- [ ] Some examples of appliances and how they transfer energy electrically to a useful energy store.
- [ ] That the energy transferred by an appliance depends on its power and the time it is switched on for.
- [ ] That the higher the power of a device, the more energy it transfers in a given time.
- [ ] That the amount of energy transferred by an appliance is given by the formula $E = Pt$.
- [ ] That power is measured in watts, W.
- [ ] How the power rating of a device informs the consumer of how much energy it will need to operate.
- [ ] That the energy transferred to or from an electrical charge across a potential difference is given by the formula $E = QV$.
- [ ] That the power of a device or component can be found using the formulas $P = VI$ and $P = I^2R$.

The National Grid

- [ ] That the national grid is a network of cables and transformers which transfers electrical power from power stations to consumers.
- [ ] That for a given power, increasing the potential difference will decrease the current.
- [ ] That low-current transmission is more efficient, as less energy is transferred to thermal energy stores which are not useful.
- [ ] That potential difference can be increased (stepped up) and decreased (stepped down) by transformers.
- [ ] That pd is increased by step-up transformers for distribution, and decreased by step-down transformers to safe levels for consumers at the other end.
- [ ] How the national grid forms an efficient system for distributing electrical power to consumers.
- [ ] How demand can change throughout the day and what actions are taken to ensure enough power can be supplied to meet unexpected high demand.

Exam-style Questions

1   A kettle with a metal casing is linked to the mains electricity using a three-core cable. The three-core cable has a live wire, an earth wire and a neutral wire inside it.

1.1   Complete the sentence by selecting the correct answer from the options given.

The live wire has a (brown / blue / green and yellow) coating.

(1 mark)

1.2   Write down the equation that links power, potential difference and current.

(1 mark)

1.3   The kettle operates at 230 V and 575 W.
Calculate the current flowing through the kettle. Give your answer in amperes.

(3 marks)

1.4   The kettle's cable is frayed, and the live wire is exposed.
Explain how this could cause a fire.

(3 marks)

2   A kitchen contains a 2.55 kW oven and a 1.15 kW dishwasher.
Both appliances are connected to the UK mains supply.

2.1   State the potential difference and frequency of the UK mains supply.

(2 marks)

The national grid is used to transmit electrical power from power stations to consumers. The electrical power is transmitted at a higher potential difference than is safe for use in homes.

2.2   Explain why the electrical power is transmitted at a high potential difference.

(3 marks)

The oven transfers 7 038 000 J of energy in the time it takes to cook a particular meal.

2.3   Write down the equation that links energy transferred, power and time.

(1 mark)

2.4   Calculate how long it takes to cook the meal. Give your answer in seconds.

(3 marks)

2.5   Write down the equation that links energy transferred, charge and potential difference.

(1 mark)

2.6   Calculate the total amount of charge that passes through the oven in the time it takes to cook the meal. Give your answer in coulombs.

(3 marks)

2.7   Calculate the operating current of the dishwasher. Give your answer in amperes.

(3 marks)

1. Static Electricity

Static electricity is the cause of all sorts of fun and games and you get the pleasure of learning all about it — joy.

Insulators and conductors

Electrical charges can move easily through some materials, and less easily through others.

- If electrical charges can easily move through a material, it is called an electrical **conductor**. Metals are known to be good conductors.

- If electrical charges cannot easily move through a material, it is called an electrical **insulator**. Plastics and rubbers are usually good insulators.

> **Example**
>
> Electrical wires and cables are usually made up of both electrical insulators and conductors. They have a core made out of an electrical conductor so that electric charge can flow through it easily, and a casing made of an electrical insulator to stop you getting an electric shock by touching the wire or cable.

Static charge

A **static charge** is an electric charge which cannot move. Static charges are often (but not always) found on electrical insulators where charge cannot flow freely. They can be positive (+ve) or negative (−ve). A static charge can build up on a conductor if it's isolated — in other words, if there's nowhere for the charge to flow to.

When certain insulating materials are rubbed together, friction causes negatively charged electrons to be scraped off one and dumped on the other.

This will leave a positive static charge on the one that loses electrons and a negative static charge on the one that gains electrons. Which way the electrons are transferred depends on the two materials involved — see the example on the next page.

Both positive and negative electrostatic charges are only ever produced by the movement of electrons. The positive charges definitely do not move.

A positive static charge is always caused by electrons moving away elsewhere. The material that loses the electrons loses some negative charge, and is left with an equal positive charge.

Learning Objectives:

- Understand how rubbing certain insulators together can lead to them becoming electrically charged.

- Know that an object which loses electrons becomes positively charged, and one which gains electrons becomes equally negatively charged.

- Know that only negative charges (electrons) move.

- Understand that a spark of charge will jump across the gap between a charged object and an earthed conductor if the potential difference is high enough.

- Know that two electrically charged materials will exert a non-contact force on each other when brought close together.

- Know that objects with opposite types of charge attract each other and objects with the same type of charge repel each other and know examples that are evidence of this.

Specification Reference 4.2.5.1

Tip: Polythene and acetate are types of plastic. They're electrical insulators.

Examples

The classic examples of static build up are polythene and acetate rods being rubbed with a cloth duster.

- With the polythene rod, electrons move from the duster to the rod. The rod becomes negatively charged and the cloth has an equal positive charge.

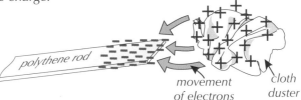

polythene rod

movement of electrons

cloth duster

Figure 1: *A polythene rod being rubbed with a cloth duster to create static charge.*

- With the acetate rod, electrons move from the rod to the duster. The rod becomes positively charged and the cloth has an equal negative charge.

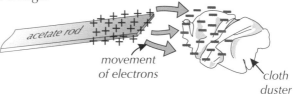

acetate rod

movement of electrons

cloth duster

Figure 2: *An acetate rod being rubbed with a cloth duster to create static charge.*

Tip: For a reminder on potential difference, see page 92.

Tip: For more on how sparks actually jump across gaps, see page 103.

Sparking

As electric charge builds up on an object, the potential difference between the object and the earth (which is at 0 V) increases. If the potential difference gets large enough, electrons can jump across the gap between the charged object and the earth — this is a **spark**. Electrons can also jump between a charged object and any earthed conductor that's nearby.

Example

You can get a static shock getting out of a car.
A charge builds up on the car's metal frame as it travels, due to friction between the car and the air. When you go to touch the car, a spark jumps across the gap and the charge travels through you to the earth.

Figure 3: *Lightning striking a TV tower. Sparks usually happen when the gap is small — but not always. Lighting is just a really big spark.*

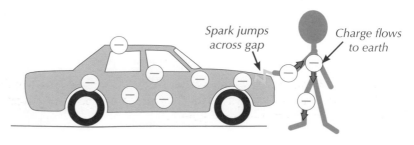

Spark jumps across gap

Charge flows to earth

Figure 4: *A spark jumping between a charged car body and a person's hand.*

Electrostatic attraction and repulsion

When two electrically charged objects are brought close together they exert a force on one another, as shown in Figure 5.

- Two things with opposite electric charges are attracted to each other.

- Two things with the same electric charge will repel each other.

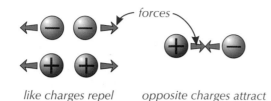

like charges repel   *opposite charges attract*

Figure 5: *A diagram showing the forces that different electric charges exert on one another.*

These forces will cause the objects to move if they are able to do so. This is known as **electrostatic attraction/repulsion** and is a **non-contact force** (the objects don't need to touch, see page 147). These forces get weaker the further apart the two things are. You need to know some examples where you can see evidence of these forces.

(see page 147)

Exam Tip
You could be asked to apply the idea of electrostatic attraction and repulsion in any situation in the exam. Just make sure you remember the two rules above and you should be fine.

Example

If a rod with a known charge is suspended from a piece of string (so it is free to move) and another rod with the same charge is placed nearby, they will repel each other. The suspended rod will swing away. If an oppositely charged rod is placed nearby instead, they will attract and the suspended rod will swing towards it.

Another charged rod — if it is negative, it will repel if it is positive, it will attract.

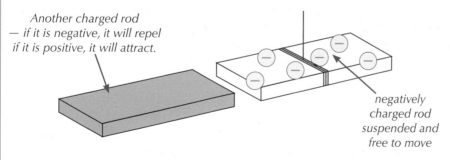

negatively charged rod suspended and free to move

Figure 6: *A diagram of a method to demonstrate electrostatic attraction and repulsion.*

Example

If you rub your hair with a balloon, electrons are transferred from your hair to the balloon. The balloon and your hair then become oppositely charged, so they will attract each other — see Figure 7.

Figure 7: *A demonstration of electrostatic attraction between a balloon and hair.*

Practice Questions — Fact Recall

Q1 How may some insulators become positively charged when they are rubbed against another insulator?

Q2 Explain why a large charge on an isolated object can result in a spark.

Q3 Say whether the following objects will attract, repel, or feel no force when brought close together:

a) Two positively charged objects.

b) Two negatively charged objects.

c) A negatively charged object and a positively charged object.

Practice Question — Application

Q1 A woman uses a plastic hair brush to brush her hair. Both the hair brush and her hair become electrically charged.

a) When the woman looks in the mirror, her hair is standing on end. Explain why.

b) The hair brush has a charge of –0.5 nC. What is the total charge on the woman's hair?

Q2 A student is walking up a carpeted staircase. As they reach out to touch the metal hand-rail, they receive a small shock. Explain how this happens.

Tip: Plastic and hair are both electrical insulators.

Tip: 1 nC = 1×10^{-9} C (it's short for nano-coulomb).

2. Electric Fields

Everything with an electric charge has an electric field. Field interactions explain the non-contact forces between charged objects, as well as sparks.

What is an electric field?

An **electric field** is created around any electrically charged object. The closer to the object you get, the stronger the field is (and the further you are from it, the weaker it is).

You can show an electric field around an object using field lines. Electric field lines point away from positive charge and towards negative charge, and they're always at right angles to a charged object's surface. You can think of field lines as showing you the path that a positively charged particle would move along in an electric field. The closer together the lines are, the stronger the field is.

You can draw the field lines for an isolated, charged sphere, as shown in Figure 1.

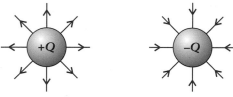

Figure 1: *The electric field pattern of a positively charged sphere (left) and a negatively charged sphere (right).*

When a charged object is placed in the electric field of another charged object, it feels a force. This force causes the attraction or repulsion you saw on page 101. It is caused by the electric fields of each charged object interacting with each other.

The size of the force between the objects is linked to the strength of the electric field at that point. You can see in Figure 1 that the further from a charge you go, the further apart the lines are and so the weaker the field is. So, as you increase the distance between the charged objects, the strength of the field decreases and the force between them gets smaller.

Example

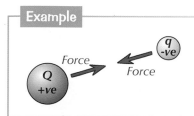

Q and q are two oppositely charged particles. The electric field of Q interacts with the electric field of q. This causes forces to act on both Q and q. These forces move q and Q closer together.

Sparking and electric fields

As you've seen on p. 100, sparks are caused when there is a high enough potential difference between a charged object and the earth (or an earthed object). A high potential difference causes a strong electric field between the charged object and the earthed object.

The strong electric field causes electrons in the air particles to be removed (known as ionisation). Air is normally an insulator, but when it is ionised it is much more conductive, so charge can flow through it. This is the spark.

Learning Objectives:

- Know that all charged objects have an electric field around them.
- Know how the strength of an electric field at a certain point is related to the distance of that point from the charged object.
- Be able to draw electric field lines for an isolated charged sphere.
- Understand that a non-contact force is experienced by a charged object placed inside the electric field of another charged object and how the size of this force depends on the distance between the two charged objects.
- Be able to explain sparking in terms of electric fields.

Specification Reference 4.2.5.2

Tip: Isolated means it is not interacting with anything else.

Q1 What direction do electric field lines point in?

Q2 Draw the field lines around an isolated negatively charged sphere.

Q3 Describe, with respect to electric fields, how a build up of static charge can lead to a spark.

Practice Questions — Application

Q1 Two particles, A and B, with equal negative charges, are located close to each other. Both are free to move.

a) Explain the motion of the particles.

b) Explain how the force on them changes as they move.

Topic 2c Checklist — Make sure you know...

Static Electricity

☐ That a static charge is a charge which cannot move.

☐ That rubbing two insulating materials together can cause negatively charged electrons to move from one to the other, giving both materials an equal but opposite charge.

☐ That a material which loses electrons is left with a positive charge and a material that gains electrons becomes negatively charged.

☐ That a greater static charge on an object leads to a greater pd between the object and earth.

☐ That if a build up of static creates a high enough pd, a spark can jump from the object to a nearby earthed conductor.

☐ That electrically charged objects exert a non-contact force on one another — objects with the same charge repel each other and objects with opposite charges attract each other.

☐ That you can see evidence of electrostatic attraction/repulsion, e.g. when you get a static shock from a car door or when you rub a balloon on your hair.

Electric Fields

☐ That all charged objects have an electric field around them.

☐ That the strength of the electric field increases the closer you are to the charged object.

☐ That electric field lines point away from positive charge and towards negative charge and the closer they are, the stronger the field.

☐ How to draw the electric field pattern of an isolated charged sphere.

☐ That a charged object within the electric field of another charged object experiences a force.

☐ That the size of the force experienced by a charged object in an electric field increases with decreasing distance from the charged object causing the field.

☐ How to explain the occurrence of sparks using the idea of electric fields.

Exam-style Questions

1   Electrostatic spray-painting is a method used to paint car bodies that uses electrostatic charges. The car body is given a negative charge, and a paint gun gives each paint droplet a positive charge, as shown in the diagram.

spray gun →

positively charged paint droplets

negatively charged car body

1.1   Use the correct words from the box below to complete the following passage.

spread out	**lost**	**clump together**
attractive	**repulsive**	**gained**

The paint droplets are positively charged because they have .... electrons.

The force between one paint droplet and another is .... This means

the droplets will ...., creating an even layer of paint on the car.

(3 marks)

1.2   Suggest why giving the paint droplets and the car opposite charges reduces the amount of paint wasted when a car is spray-painted.

(2 marks)

1.3   An engineer holds his hand close to the car and gets a small electric shock. Explain how this happens with reference to electric fields.

(5 marks)

2   A student rubs a balloon against a scarf.
The student determines that the scarf is negatively charged.

2.1   Explain how the scarf has become negatively charged.

(1 mark)

The student predicts that the balloon must be positively charged.
To test this, the student suspends the balloon from a string and brings a positively charged rod near to the balloon.

2.2   Assume the student is correct. State what they would expect to see happen to the balloon when the rod is brought close. Explain your answer.

(2 marks)

2.3   Sketch the electric field around the balloon.
You may assume the balloon is an isolated charged sphere.

(1 mark)

1. Density and States of Matter

Learning Objectives:

- Know and be able to use the equation for the density of a substance.
- Know what the three states of matter are, and be able to describe their properties using the particle model.
- Understand how the particle model can explain why substances have different densities.
- Be able to represent the particle model of the three states of matter with simple diagrams.
- Be able to measure the density of a liquid or any shape of solid using appropriate apparatus (Required Practical 5).

Specification Reference 4.3.1.1

The particle model of matter can be used to help describe the densities of substances and the different states of matter. It says that everything is made up of lots of tiny particles.

Density

Density is a measure of the 'compactness' of a substance. It relates the mass of a substance to how much space it takes up (i.e. it's a substance's mass per unit volume).

There's a formula for finding the density of a substance:

$$\rho = \text{density (kg/m}^3\text{)} \longrightarrow \rho = \frac{m}{V} \begin{array}{l} \longleftarrow m = mass\ (kg) \\ \longleftarrow V = volume\ (m^3) \end{array}$$

Density can also be measured in g/cm^3 (1 g/cm^3 = 1000 kg/m^3).

> ### Example
>
>
> **A copper cube has sides of length 5.0 cm. The density of copper is 8.96 g/cm³. Find the mass of the cube.**
>
> First find the cube's volume:
>
> $V = 5.0 \times 5.0 \times 5.0 = 125\ cm^3$
>
> Now substitute into the rearranged formula:
>
> $m = \rho V = 8.96 \times 125 = 1120\ g$

Tip: The symbol for density is the Greek letter rho (ρ) — it looks like a *p* but it isn't. The formula triangle for the density equation looks like this:

The density of an object depends on what it's made of. It can be explained using the particle model of matter.

A dense material has its particles packed tightly together. The particles in a less dense material are more spread out — if you compressed the material, its particles would move closer together, and it would become more dense. (You wouldn't be changing its mass, but you would be decreasing its volume.)

States of matter

Three **states of matter** are solid (e.g. ice), liquid (e.g. water) and gas (e.g. water vapour). The particles of a substance in each state are the same — only the arrangement and energy of the particles are different.

Solids

In solids, strong forces of attraction hold the particles close together in a fixed, regular arrangement. The particles don't have much energy so they can only vibrate about their fixed positions.

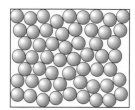

— A particle

Figure 2: *The particles in a solid.*

Liquids

There are weaker forces of attraction between the particles in liquids. The particles are close together, but can move past each other, and form irregular arrangements. They have more energy than the particles in a solid — they move in random directions at low speeds.

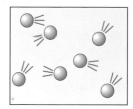

Figure 3: *The particles in a liquid.*

Gases

There are almost no forces of attraction between the particles in a gas. The particles have more energy than those in liquids and solids — they are free to move, and travel in random directions at high speeds.

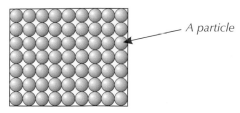

Figure 4: *The particle arrangement in a gas.*

Figure 1: *The three states of water — ice, water and water vapour.*

Exam Tip
You might be asked to use a picture or model in the exam to describe a state of matter — so make sure you know this page inside out.

Properties of solids, liquids and gases

The particle model helps to explain the properties of different states of matter.

- The density of a substance is generally highest when it is in solid form as the particles are closest together. Liquids tend to be less dense than solids. Gases are the least dense — their particles are spaced far apart.

- Gases and liquids can flow because their particles can move past each other. The particles in a solid can't move anywhere — they can only vibrate in their fixed positions, so solids can't flow.

- Gases are compressible. The particles in a gas are very spread out, which means you can squash a gas into a smaller volume — you're just reducing the distance between particles. The particles in liquids and solids can't really get much closer together, which is why only gases are easily compressible.

Measuring density

You need to be able to measure the density of a substance.
The method you should use depends on the type of substance.

Tip: As with all practicals, make sure you do a risk assessment for this practical.

Tip: The volume of a cube or cuboid is length × width × height.

Measuring the density of a solid object

First, use a balance to measure the object's mass (see p. 327). You then need to find its volume. If it's a regular solid, like a cuboid, you might be able to measure its dimensions with a ruler and calculate its volume. For an irregular solid, you can find its volume by submerging it in a eureka can of water.

A eureka can (or displacement can) is essentially a beaker with a spout — you can see this in Figure 6. To use one, fill it with water so the water level is above the spout. Let the water drain from the spout, leaving the water level just below the start of the spout. (This way, when you put your solid object in, all the water displaced will pass through the spout, giving you the correct volume.)

Place an empty measuring cylinder below the end of the spout. When you put your solid object in the eureka can, it causes the water level to rise and water to flow out of the spout.

Figure 6: An experiment to measure the density of a solid block. A string is used so the block can easily be removed.

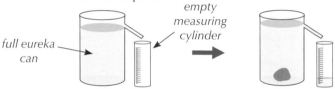

Figure 5: A eureka can and measuring cylinder being used to measure a solid's volume.

Once the spout has stopped dripping, you can measure the volume of the water in the measuring cylinder (see p. 329 for more on this). This is the volume of your solid object.

Now you can substitute the object's mass and volume into the density formula from page 106 to find its density.

Tip: You can't use this method if the object floats — the object will only displace a volume of water equal to the part of the object that's below the water line.

Measuring the density of a liquid

Place a measuring cylinder on a balance and zero the balance (see p. 327). Pour 10 ml of the liquid into the measuring cylinder and record its mass.

Pour another 10 ml into the measuring cylinder and record the total volume and mass. Repeat this process until the cylinder is full.

Tip: The measuring cylinder will give you the volume in ml. To convert to cm³, use 1 ml = 1 cm³.

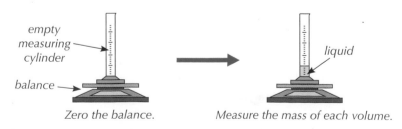

Zero the balance.   Measure the mass of each volume.

Figure 7: Using a balance to find the mass of liquid in a measuring cylinder.

Tip: To find the average, add up all the densities, then divide by the number of values. See page 14 for more.

For each set of measurements, use the formula from page 106 to find the density. Finally, take an average of your calculated densities. This will give you a more precise (p. 12) value for the density of the liquid.

Practice Questions — Fact Recall

Q1 What is the formula for density?

Q2 What are the three states of matter?

Q3 Explain the arrangement, movement and energy of the particles in a solid.

Q4 Describe how the arrangement, movement and energy of particles in a liquid is different to that in a solid.

Q5 In which state of matter do particles have the most energy?

Q6 What pieces of apparatus would you need to measure the density of an irregular solid object?

Practice Questions — Application

Q1 A block of material is in the shape of a cuboid. It has sides of length 3.0 cm, 4.5 cm and 6.0 cm, and a total mass of 0.324 kg. Find the density of the block.

Q2 The diagram below shows a box filled with small light polystyrene balls. A small fan is fitted at the bottom of the box. This apparatus can be used to model the particles in different states of matter.

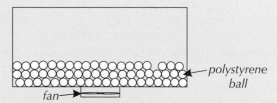

a) What state of matter is modelled by the balls at the bottom of the box when the fan is turned off? Explain your answer.

The fan is turned on, causing the small balls to fly around the inside of the box at high speeds.

b) What state of matter do the balls now model?

c) Use this model to explain the different densities of the two states of matter in parts a) and b).

- Know how energy is stored in a system by its particles.

- Know what is meant by the term 'internal energy'.

- Know that heating a system increases the energy of its particles.

- Understand that this heating will either cause an increase in temperature or a change of state.

- Know that if there is a temperature increase, its size depends on the material, the mass of the material and the energy supplied.

- Know all the different ways a substance can change state.

- Know what is meant by a physical change and how it differs from a chemical change.

- Be able to explain how a change of state is a physical change that conserves mass.

- Understand that when a substance changes state, the energy transferred changes the internal energy — but not the substance's temperature.

- Recognise and understand heating and cooling graphs for a substance undergoing changes of state.

Specification References
4.3.1.2, 4.3.2.1,
4.3.2.2, 4.3.2.3

Tip: Sublimating is the change of state from a solid directly to a gas.

2. Internal Energy and Changes of State

The state that a substance is in has a lot to do with temperature. You need to make sure you know all about specific heat capacity as part of this topic, so go back to pages 30-32 and make sure it's all fresh in your mind.

Internal energy

The particles in a system vibrate or move around — they have energy in their kinetic energy stores. They also have energy in their potential energy stores due to their positions — but you don't need to worry about this.

The energy stored in a system is stored by its particles (atoms and molecules). The **internal energy** of a system is the total energy that its particles have in their kinetic and potential energy stores.

Heating the system transfers energy to its particles (they gain energy in their kinetic energy stores and move faster), increasing the internal energy. This leads to a change in temperature (or a change in state, see below). The size of the temperature change depends on the mass of the substance, what it's made of (its specific heat capacity) and the energy input. Make sure you remember all of the stuff on specific heat capacity from p. 30 — particularly how to use the formula.

If the substance is heated enough, the particles will have enough energy in their kinetic energy stores to break the bonds holding them together. This means you get a change of state.

What is a change of state?

When you heat a liquid, it boils (or evaporates) and becomes a gas. When you heat a solid, it melts and becomes a liquid (or sublimates to become a gas). These are both changes of state.

The state can also change due to cooling. Energy is transferred from the kinetic energy stores of the particles, so they move slower and start to form bonds. They don't have enough energy to overcome the bonds.

All the changes of state are shown in Figure 1 below:

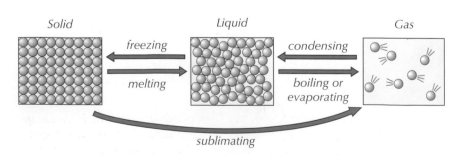

Figure 1: The different changes of state.

A change of state is a **physical change** (rather than a chemical change). This means you don't end up with a new substance — it's the same substance as you started with, just in a different form.

If you reverse a change of state (e.g. freeze a substance that has been melted), the substance will return to its original form and get back its original properties.

The number of particles doesn't change — they're just arranged differently. This means mass is conserved — none of it is lost when the substance changes state.

> **Tip:** Remember, a substance has the same particles whether it's a solid, liquid or gas — only the arrangement and energy of the particles change.

Figure 2: *Energy transferred by heating from this person's hand is enough to melt gallium metal.*

Example

As the ice in Figure 3 melts, the reading on the balance will stay the same. The mass of the beaker's contents will be conserved through the change of state.

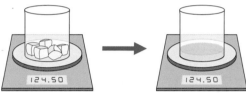

Figure 3: *A beaker of ice melting on a balance.*

Breaking bonds

When a substance is melting or boiling, you're still putting in energy and so increasing the internal energy, but the energy's used for breaking intermolecular bonds rather than raising the temperature.

There are flat spots on the heating graph in Figure 5, where energy is being transferred by heating but not being used to change the temperature.

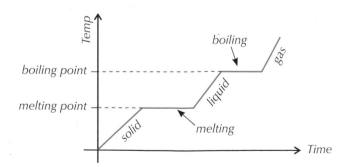

Figure 5: *Heating graph, showing temperature against time for a substance which is being heated.*

Figure 4: *When the water reaches its boiling point, the energy transferred is used to break intermolecular bonds, not to increase temperature.*

When a substance is condensing or freezing, bonds are forming between particles, which releases energy. This means the internal energy decreases, but the temperature doesn't go down until all the substance has turned to liquid (condensing) or a solid (freezing).

The flat parts of the graph in Figure 6 show this energy transfer.

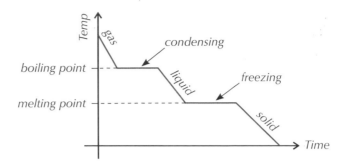

Figure 6: *Cooling graph, showing temperature against time for a substance which is being cooled.*

Practice Questions — Fact Recall

Q1   What is the definition of internal energy?

Q2   Give one way in which you can transfer energy to the particles in a system.

Q3   Name five different changes of state.

Q4   A change of state is a physical change. Explain what is meant by the term 'physical change'.

Q5   Does a change of state conserve mass? Explain your answer.

Q6   Sketch a graph of temperature against time for a solid being heated so that it undergoes two changes of state. Label the points at which the substance changes state.

Practice Questions — Application

Q1   A kettle is filled with water and switched on. Explain, in terms of energy stores, what happens to the particles in the water as it is heated to boiling point by the kettle.

Q2   Steam from a shower forms condensation on a bathroom window. Explain why this happens in terms of the steam particles.

3. Specific Latent Heat

So energy is transferred to break (or form) intermolecular bonds and change state. It's all to do with specific latent heat...

What is specific latent heat?

The energy needed to change the state of a substance is called **latent heat**. **Specific latent heat** (SLH) is the amount of energy needed to change 1 kg of a substance from one state to another without changing its temperature. For cooling, specific latent heat is the energy released by a change in state. Specific latent heat is different for different materials, and for changing between different states.

The specific latent heat for changing between a solid and a liquid (melting or freezing) is called the **specific latent heat of fusion**. The specific latent heat for changing between a liquid and a gas (evaporating, boiling or condensing) is called the **specific latent heat of vaporisation**.

You can work out the energy needed (or released) in joules when a substance of mass m changes state using this formula:

$$E = mL$$

E = energy for a change of state (J)
L = specific latent heat (J/kg)
m = mass (kg)

Example

The specific latent heat of vaporisation for water is 2 260 000 J/kg. How much energy is needed to completely boil 1.50 kg of water at 100 °C?

Just plug the numbers into the formula:

$E = mL = 1.50 \times 2\,260\,000 = 3\,390\,000$ J

Don't get confused with specific heat capacity (p. 30), which relates to a temperature rise of 1°C. Specific latent heat is about changes of state where there's no temperature change.

Learning Objectives:

- Know what is meant by the terms 'latent heat' and 'specific latent heat'.
- Be able to use the equation for specific latent heat.
- Know the difference between specific latent heat of fusion and specific latent heat of vaporisation.
- Know how specific heat capacity and specific latent heat differ.

Specification Reference 4.3.2.3

Tip: If you're finding mass or SLH, you'll need to rearrange. Here's the formula triangle:

Practice Questions — Fact Recall

Q1 State the two different types of specific latent heat for a substance, and describe the difference between them.

Q2 What are the units of specific latent heat?

Practice Questions — Application

Q1 The specific latent heat of fusion for water is 334 000 J/kg. An ice cube of mass 25.0 g is at 0 °C. How much energy is needed to melt the ice cube?

Q2 The specific latent heat of vaporisation for a liquid is 1 550 000 J/kg. What mass of the liquid (already at its boiling point) would be completely boiled by 4 960 000 J of energy?

- Know that gas particles are in constant random motion and understand how temperature is linked to the average energy in the kinetic energy stores of these particles.

- Understand how the motion of gas particles is linked to its temperature and pressure.

- Understand how a change in temperature of a gas at constant volume leads to a change in gas pressure.

- Be able to explain how a change in volume of a gas at constant temperature can lead to a change in pressure.

- Be able to use the formula linking pressure and volume for a gas at constant temperature to make calculations when pressure or volume changes.

- Know that gas pressure produces a net force at right angles to the wall of a container, so a change in gas pressure can cause a change in the volume of a gas.

- **H** Know that doing work on a gas increases the internal energy and may cause a change in temperature.

- **H** Explain how doing work on a gas in a given situation causes the gas to increase in temperature.

Specification References
4.3.3.1, 4.3.3.2,
4.3.3.3

4. Particle Motion in Gases

The temperature of a gas determines the energy in the kinetic stores of its particles. It can also affect the pressure and the volume of the gas.

Temperature of gases

The particles in a gas are constantly moving with random directions and speeds. If you increase the temperature of a gas, you transfer energy into the kinetic energy stores of its particles (you saw this on p. 110).

The temperature of a gas is related to the average energy in the kinetic energy stores of the particles in the gas. The higher the temperature, the higher the average energy.

So as you increase the temperature of a gas, the average speed of its particles increases. This is because the energy in the particles' kinetic energy stores is $\frac{1}{2}mv^2$ (see p. 26).

Gas pressure

As gas particles move about at high speeds, they bang into each other and whatever else happens to get in the way. When they collide with a surface, they exert a force on it. Since pressure is force per unit area (p. 169), this means they exert a pressure too. In a sealed container, the outward gas pressure is the total force exerted by all of the particles in the gas on a unit area of the container walls.

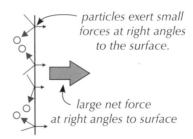

particles exert small forces at right angles to the surface.

large net force at right angles to surface

Figure 1: *Gas pressure on a surface.*

Faster particles and more collisions with the walls of the container both lead to an increase in net force, and so gas pressure. Increasing temperature will increase the speed and the number of collisions, and so the pressure (if volume is kept constant).

Alternatively, if temperature is constant, increasing the volume of a gas means the particles get more spread out and hit the walls of the container less often. The gas pressure decreases.

Pressure and volume are inversely proportional — when volume goes up, pressure goes down (and when volume goes down, pressure goes up). For a gas of fixed mass at a constant temperature, the relationship is:

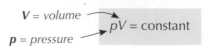

V = volume
p = pressure

$pV = \text{constant}$

Example

A gas is held in a sealed container of volume 11 m³. The pressure of the gas is 84 Pa. The volume of the container is increased and the pressure of the gas changes to 60 Pa. The temperature of the gas is unchanged. What is the new volume of the container?

Originally, $pV = 84 \times 11 = 924$
After the volume is changed, $pV = 60 \times V$

For a gas of fixed mass at a constant temperature, pV = constant, so:

$924 = 60 \times V$
$V = 924 \div 60 = 15.4$ m³

Tip: It doesn't matter what units you use for p and V, as long as you keep them the same throughout the calculation.

Tip: The pressure has decreased, so the volume must have increased. This is a good way to check your answer is along the right lines.

The pressure of a gas results in a net outwards force at right angles to the surface of its container. There is also a force on the outside of the container due to the pressure of the gas around it.

If a container can easily change its size (e.g. a balloon), then any change in these pressures will cause the gas to compress or expand, due to the overall force.

Example

If a helium balloon is released, it rises. Atmospheric pressure decreases with height (p. 171), so the pressure outside the balloon decreases. This causes the balloon to expand until the pressure inside drops to the same as the atmospheric pressure.

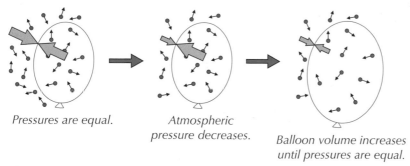

Pressures are equal.   Atmospheric pressure decreases.   Balloon volume increases until pressures are equal.

Figure 3: A change in pressure causing a change in the volume of a balloon.

Figure 2: A balloon in a vacuum chamber. Air is removed from the chamber, so the pressure outside the balloon drops. This makes the air inside the balloon expand, in order to equalise the pressure.

Doing work on a gas   Higher

If you transfer energy by applying a force, then you do work. Doing work on a gas increases its internal energy, which can increase its temperature.

Tip: There's more about doing work on p. 155.

Example

You can do work on a gas mechanically, e.g. with a bike pump. The gas applies pressure to the plunger of the pump, and so exerts a force on it. Work has to be done against this force to push down the plunger.

This transfers energy to the kinetic energy stores of the gas particles, increasing the temperature (see p. 110). If the pump is connected to a tyre, you should feel it getting warmer.

Tip: Remember, temperature of a gas is linked to the average energy in the particles' kinetic stores (see previous page).

Q1   Describe the motion of the particles in a gas.

Q2   A gas is stored inside a sealed container. How does the gas exert an outwards pressure on the walls of the container?

Q3   Explain why increasing the temperature of a fixed volume of gas will increase the pressure of the gas.

Q4   For a fixed mass of gas at a constant temperature, are pressure and volume directly proportional or inversely proportional?

Q5   The volume of a fixed mass of gas is increased. Its temperature remains constant. State what will happen to the pressure of the gas. Explain your answer by referring to the particles involved.

Q6   Work is done on a gas. What effect does this have on the internal energy of the gas?

Practice Questions — Application

Q1   A gas is compressed from a volume of 75 cm^3 to a volume of 30 cm^3. Its temperature remains the same. The pressure of the gas after it has been compressed is 110 Pa. What was the pressure of the gas before it was compressed?

Q2   A helium balloon is inflated, then placed inside a refrigerator. Explain what will happen to the volume of the balloon when it is in the refrigerator.

Topic 3 Checklist — Make sure you know...

Density and States of Matter

☐ How to calculate the density of a material.

☐ That a dense material has its particles packed closely together and a less dense material has its particles more spread out.

☐ That the three states of matter are solid, liquid and gas.

☐ How the particles are arranged in solids, liquids and gases and how to draw diagrams to show this.

☐ How the arrangement of particles in a material affects the properties of the material, including density.

☐ How to measure the density of a solid or liquid.

Internal Energy and Changes of State

☐ That energy is stored in a system by its particles.

☐ That the internal energy of a system is the total energy that its particles have in their kinetic and potential energy stores.

cont...

- [ ] That heating a system transfers energy to its particles, increasing the internal energy and causing either a change in temperature or a change of state.
- [ ] That the change in a material's temperature due to heating depends on the mass of the substance, what it's made from (its specific heat capacity) and the energy input.
- [ ] That heating or cooling a substance can lead to a change of state.
- [ ] That the different changes of state are freezing, melting, boiling/evaporating, condensing and sublimating.
- [ ] That a change of state is a physical change (not a chemical change) that conserves mass.
- [ ] That reversing a change of state will return a substance to its original form.
- [ ] That energy transferred to (or from) a substance which is changing state is used to break (or form) bonds, rather than to change the substance's temperature.
- [ ] How to sketch graphs of temperature against time for substances being heated or cooled.
- [ ] That these graphs have flat spots which show changes of state, where energy transferred does not cause a change in temperature.

Specific Latent Heat

- [ ] That the latent heat of a substance is the energy needed to change its state.
- [ ] That specific latent heat is the amount of energy needed to change (or released when changing) 1 kg of a substance from one state to another without changing its temperature.
- [ ] The difference between specific latent heat of vaporisation and specific latent heat of fusion.
- [ ] How to use the equation for specific latent heat.
- [ ] How specific heat capacity and specific latent heat differ.

Particle Motion in Gases

- [ ] That the particles in a gas are constantly moving with random directions and speeds.
- [ ] That an increase in temperature causes an increase in the average speed of gas particles.
- [ ] That particles colliding with the walls of a container exert a force, and so a pressure, on the container.
- [ ] That an increase in temperature for a gas at constant volume leads to an increase in pressure.
- [ ] That an increase in volume for a gas at constant temperature leads to a decrease in pressure.
- [ ] That the pressure of a gas in a sealed container causes a net force at right angles to the walls of the container.
- [ ] That a change in pressure can cause a change in the volume of a gas.
- [ ] **H** That doing work on a gas can increase its temperature, and how to explain why this happens in a given situation, e.g. when work is done to push down the plunger of a bike pump.

Exam-style Questions

1   A physicist is carrying out an experiment with a substance which is in solid form.
She needs 450 g of the substance for the experiment. The density of the substance
when it is a solid is 9 g/cm³.

1.1   Write down the equation that links density, mass and volume.

(1 mark)

1.2   Calculate the volume of the solid substance that the physicist needs for the experiment.
Give your answer in cm³.

(3 marks)

The physicist heats the substance for 300 seconds. The graph in **Figure 1** shows how
the substance's temperature changes over this time period.

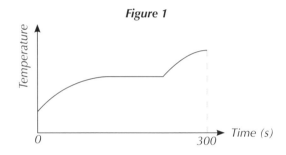

Figure 1

1.3   In what state of matter is the substance at the end of the time period shown?
Explain your answer.

(2 marks)

1.4   What is the mass of the substance at the end of the time period shown?
Explain your answer.

(2 marks)

2   A sample of gas is stored in a sealed container. The volume of the container is 0.08 m³
and the density of the gas inside the container is 0.025 kg/m³.

2.1   Calculate the mass of the gas sample. Give your answer in kg.

(3 marks)

2.2   The gas is condensed, so that it is all turned into a liquid. 5 kJ of energy is released
during the change of state. Find the specific latent heat of vaporisation of the gas.
Give the correct units. Use the correct equation from the equations listed on page 394.

(4 marks)

2.3   Will the density of the sample now be equal to, less than, or greater than 0.025 kg/m³?
Explain your answer.

(2 marks)

3   A student has an unknown liquid. He wants to know the density of the liquid.

3.1   Describe a method that the student could use to find the liquid's density.

(4 marks)

The student heats a sample of the liquid. He records the liquid's temperature every 5 seconds. The temperature of the heater is 200 °C. The student notices that the temperature of the liquid never increases past 85 °C.

3.2   Explain why this is the case.

(3 marks)

4   A substance in solid form is at its melting temperature. 67 500 J of energy is transferred to the solid, causing it to melt. The specific latent heat of fusion for the substance is 450 000 J/kg.

4.1   Find the mass of the substance. Give your answer in kg.
Use the correct equation from the equations listed on page 394.

(3 marks)

4.2   Explain why it is important that the solid was already at its melting temperature for your calculation in **4.1**.

(2 marks)

5   The gas pressure inside a sealed container is 150 Pa. The gas is compressed to a volume of 30 cm³. Following this compression, the gas pressure is 220 Pa.

5.1   What was the initial volume of the gas, before it was compressed? Give your answer in cm³. Use the correct equation from the equations listed on page 394.

(3 marks)

5.2   Explain why compressing the gas increases the gas pressure.

(2 marks)

5.3   The gas in the container has a mass of 0.132 g.
By how much has the density of the gas changed due to the compression?
Give your answer in g/cm³.

(4 marks)

6   An inflatable mattress is pumped up using a foot pump.

6.1   Explain, in terms of pressure, why pumping air into the mattress causes it to inflate.

(3 marks)

6.2   The mattress begins to feel warm as it is pumped up.
Explain why this is the case.

(3 marks)

Learning Objectives:

- Know that atoms were originally thought to be tiny balls of matter that could not be split into smaller pieces.
- Know that the discovery of the electron led to the creation of the plum pudding model.
- Know that the plum pudding model describes atoms as a sphere of positive charge with negative electrons inside them.
- Know how the results of the alpha particle scattering experiment suggested mass and positive charge was concentrated at the centre of the atom — leading to the atomic model changing.
- Know that the results of the alpha particle scattering experiment led to the creation of the nuclear model.
- Understand how new experimental evidence leads to old scientific models being changed, or new ones being created.

Specification Reference 4.4.1.3

1. The History of the Atom

You'll be learning about the 'nuclear model' of the atom shortly. But first it's time for a trip back in time to see how scientists came up with it. And it's got a bit to do with plum puddings.

The plum pudding model

The Greeks were the first to think about **atoms**. A man called Democritus in the 5th century BC thought that all matter was made up of identical lumps called "atomos". But that's about as far as the theory got until the 1800s...

In 1804 John Dalton agreed with Democritus that matter was made up of tiny spheres ("atoms") that couldn't be broken up, but he reckoned that each element was made up of a different type of "atom".

Nearly 100 years later, J J Thomson discovered particles called **electrons** that could be removed from atoms. So Dalton's theory wasn't quite right (atoms could be broken up). Thomson suggested that atoms were spheres of positive charge with tiny negative electrons stuck in them like fruit in a plum pudding — this is called the 'plum pudding model'.

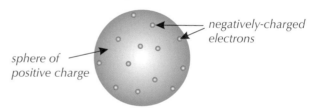

Figure 1: The plum pudding model of the atom.

That "plum pudding" theory didn't last very long though...

The alpha particle scattering experiment

In 1909, scientists in Rutherford's lab tried firing a beam of **alpha particles** (see p. 126) at thin gold foil — they used a set-up similar to Figure 2. A circular detector screen surrounds the gold foil and the alpha source, and is used to detect alpha particles deflected by any angle. This was the alpha particle scattering experiment.

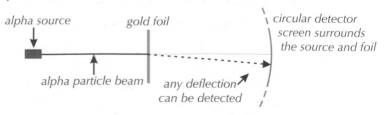

Figure 2: The experimental set-up for detecting whether alpha particles have been scattered by gold foil.

They expected that the positively-charged alpha particles would go straight through or be slightly deflected by the electrons if the plum pudding model was true.

In fact, most of the alpha particles did go straight through the foil, but some were deflected more than they had expected and the odd one came straight back at them. This was frankly a bit of a shocker. The results of the alpha particle scattering experiment showed that atoms must have small, positively-charged **nuclei** at the centre (see Figure 3).

Tip: 'Nuclei' is the plural of 'nucleus'.

Here's why:

- Most of the atom must be empty space because most of the alpha particles passed straight through the foil.

Tip: Remember, like charges repel each other (page 101).

- The nucleus must have a large positive charge as some positively-charged alpha particles were repelled and deflected by a big angle.

- The nucleus must be small as very few alpha particles were deflected back.

Tip: These results provided new evidence, causing the accepted model of the atom to be changed — see pages 2-3 for more on how new theories are accepted with new evidence.

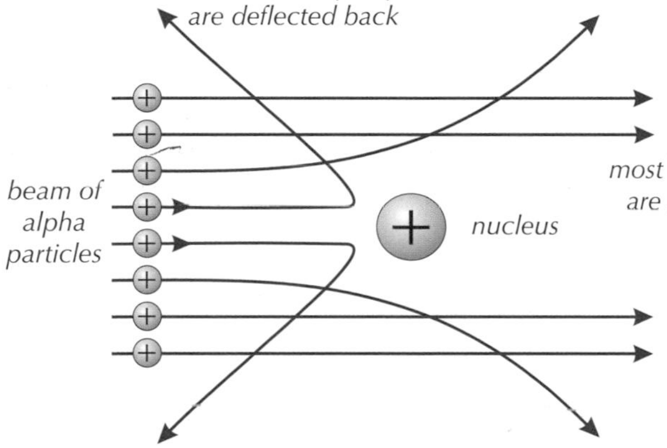

WORKING
SCIENTIFICALLY

a small number of alpha particles are deflected back

some alpha particles are deflected by a large angle due to the large positive charge of the nucleus

beam of alpha particles

nucleus

most alpha particles are not deflected

Figure 3: *A diagram showing some positively-charged alpha particles passing straight through a gold atom and some being deflected by the atom's nucleus.*

This led to the first **nuclear model** of the atom that we still use an adapted version of today (see page 122).

Figure 4: *The New Zealand physicist Ernest Rutherford.*

Practice Questions — Fact Recall

Q1 Describe the plum pudding model of the atom.

Q2 What results were expected from the alpha particle scattering experiment?

Q3 What results were seen in the alpha particle scattering experiment? How did they show that the atom has a small, positively-charged nucleus and is mostly empty space?

- Know how the nuclear model has been adapted over time.

- Understand the differences between the nuclear model and the plum pudding model.

- Know that atoms consist of a nucleus, made up of protons and neutrons, orbited by electrons.

- Know the radius of an atom is about 1×10^{-10} m.

- Know that the nucleus contains most of the mass of an atom and is 10 000 times smaller than the atom.

- Know that protons and electrons have equal and opposite charges.

- Know that atoms are not charged and contain an equal number of protons and electrons.

- Know that electrons are arranged in energy levels in an atom and move between them when they absorb and emit EM radiation.

- Know all atoms of an element have the same number of protons.

- Know what the mass number and atomic number of an element tell you.

- Be able to use the notation $^A_Z X$.

- Know that an ion is an atom with too few or too many electrons.

- Know that isotopes of an element have atoms with different numbers of neutrons.

Specification References
4.4.1.1, 4.4.1.2, 4.4.1.3

2. The Structure of the Atom

So now you know how it came to replace the plum pudding model, it's time to learn what the current nuclear model of the atom actually says.

Development of the nuclear model

The alpha particle scattering experiment led to the creation of the nuclear model, but it still needed fine-tuning. Niels Bohr adapted the initial model — he concluded that electrons orbiting the nucleus can only do so at certain distances. These distances are called energy levels. Bohr's theoretical calculations were found to agree with experimental data, so the model was accepted.

Evidence from further experiments changed the model to think of the positively charged nucleus as a group of particles (protons) which all had the same positive charge that added up to the overall charge of the nucleus. In 1932, about 20 years after the idea of the nucleus was accepted, James Chadwick proved the existence of the neutron, which explained the imbalance between atomic and mass numbers (see page 123).

All these discoveries played a part in the development of the nuclear model of the atom that we have today.

The current nuclear model

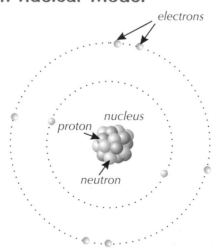

Figure 1: *The atom (not to scale).*

According to the nuclear model, the atom contains three types of particles:

- **electrons** (which are negatively charged),

- **protons** (which are positively charged) and

- **neutrons** (which are neutral — they have no charge).

The nucleus is at the centre of the atom. It is tiny but it makes up most of the mass of the atom. It contains protons and neutrons — which gives it an overall positive charge. The radius of the nucleus is about 10 000 times smaller than the radius of the atom.

The rest of the atom is mostly empty space. Negative electrons whizz round the outside of the nucleus really fast. They give the atom its overall size — the radius of an atom is about 1×10^{-10} m.

Each particle has a relative mass and a relative charge. Relative just means in relation to the other particles — it's so you can compare their masses and charges. It's useful to learn the relative charge and mass of each particle:

particle	mass	charge
proton	1	+1
neutron	1	0
electron	$\frac{1}{2000}$	−1

Figure 2: The relative masses and charges of the particles in the atom.

Atoms have no charge overall. The charge on an electron is the same size as the charge on a proton — but opposite (see Figure 2). This means the number of protons always equals the number of electrons in a neutral atom.

Electrons can move within (or sometimes leave) the energy levels of an atom. If they gain energy by absorbing EM radiation (page 243) they move to a higher energy level, further from the nucleus. If they release EM radiation, they move to a lower energy level that is closer to the nucleus.

Atomic number and mass number

You need to know how to describe the number of protons and neutrons in a nucleus:

- The number of protons in the nucleus of an atom is called the **atomic number**.

- The number of protons plus the number of neutrons in the nucleus of an atom is called the **mass number**.

An element can be described using the mass number and atomic number of its atoms. The notation looks like this:

Atoms of an element always have the same atomic number (i.e. the same number of protons and so the same charge on the nucleus), but they can have different mass numbers (these are called isotopes — see the next page).

Examples

- An atom of carbon with 6 protons and 6 neutrons would be $^{12}_{6}$C.

- An atom of oxygen with 8 protons and 9 neutrons would be $^{17}_{8}$O.

Tip: 1×10^{-10} is written in standard form. For more on standard form, see page 333.

Tip: Particles that are smaller than an atom are called 'subatomic particles'.

Tip: The mass of an electron is so small that you can often treat its relative mass as 0.

Tip: You need to know how this model differs from the plum pudding model — see page 120.

Tip: We're currently pretty happy with this model, but there's no saying it won't change. Just like for the plum pudding model, new experimental evidence sometimes means we have to change or scrap current models.

WORKING SCIENTIFICALLY

Tip: The symbol of the element is just a letter (or sometimes two letters) that tell you what element it is. For example, O is used for oxygen.

Ions

Atoms are neutral, but if some electrons are added or removed, the atom becomes a charged particle called an **ion**. The ions still have the same number of protons and neutrons as usual, but a different number of electrons.

If an atom has had electrons added or removed and has become an ion, it is said to have been ionised. This process is called ionisation.

Tip: Atoms and ions of the same element have the same number of protons in their nuclei. An element is defined by the number of protons in the nuclei of its ions or atoms, e.g. an ion or atom with eight protons in its nucleus is the element oxygen.

Tip: Most of the ionisation you'll be asked about will involve electrons being knocked off atoms, creating positive ions. This is how ionising radiation (page 126) works — it never adds electrons to an atom.

> **Example**
>
>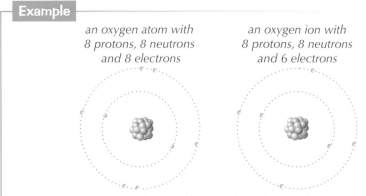
>
> an oxygen atom with 8 protons, 8 neutrons and 8 electrons
>
> an oxygen ion with 8 protons, 8 neutrons and 6 electrons
>
> **Figure 3:** A diagram showing the difference between an uncharged oxygen atom (left) and a positive oxygen ion (right).

Isotopes

Isotopes are different forms of the same element. Isotopes have atoms with the same number of protons but a different number of neutrons.

This means they have the same atomic number (and so the same charge on the nucleus), but different mass numbers.

Tip: Don't get confused between the charge of a nucleus and the charge of an atom. Nuclei are positively charged, but atoms are neutral (unless they've been ionised).

> **Example**
>
> Carbon-12 and carbon-14 are good examples of isotopes:
>
>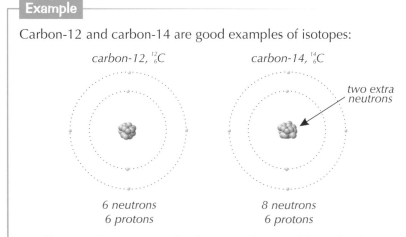
>
> carbon-12, $_{6}^{12}C$
>
> carbon-14, $_{6}^{14}C$
>
> two extra neutrons
>
> 6 neutrons
> 6 protons
>
> 8 neutrons
> 6 protons
>
> **Figure 4:** Two isotopes of carbon — carbon-12 (left) and carbon-14 (right).

All elements have different isotopes, but there are usually only one or two stable ones. The unstable isotopes are radioactive, which means they decay into other elements and give out radiation (see page 126).

Practice Questions — Fact Recall

Q1 What three types of particle make up an atom?

Q2 Which subatomic particle in an atom is not found in the nucleus?

Q3 What is the relative mass and charge of a neutron?

Q4 What can you say about the number of protons and electrons in a neutral atom? Why?

Q5 What is meant by atomic number and mass number?

Q6 What can you say about the atomic number of two atoms of the same element?

Q7 What is an isotope of an element?

Practice Question — Application

Q1 Particle A has 17 protons, 18 neutrons and 16 electrons.
Particle B has 17 protons, 20 neutrons and 17 electrons.

a) What is the overall charge of particle A?

b) Explain how you know that particle A is an ion.

c) Explain how you know that particles A and B are isotopes of the same element.

> **Tip:** You can work out the overall charge on a particle by adding up the relative charges of all its subatomic particles.

- Know that radioactive decay is where unstable nuclei emit radiation to try and become more stable.
- Know that radioactive substances may emit neutrons, alpha particles, beta particles or gamma rays from their nuclei.
- Know that an alpha particle consists of two protons and two neutrons.
- Know that a beta particle is a high-speed electron.
- Know that beta particles are emitted when a neutron changes into a proton.
- Know that a gamma ray is a high frequency electromagnetic wave.
- Know the penetration, range in air and ionising power of alpha, beta and gamma radiation.
- Understand some of the uses of alpha, beta and gamma radiation and be able to evaluate the best source for a given use.
- Know that alpha decay changes the mass and charge of a nucleus, beta decay changes only the charge and gamma radiation changes neither.
- Know the nuclear equation symbols for alpha and beta particles.
- Be able to construct and balance nuclear equations of alpha and beta decay.

Specification References
4.4.2.1, 4.4.2.2

3. Radioactivity

Radioactivity is all to do with things randomly giving out radiation. Some radiation is ionising, which means it can knock electrons off atoms to create ions. You need to know about three types of ionising radiation.

Radioactive decay

Unstable isotopes tend to decay into other elements and give out radiation as they try to become more stable. This process is called **radioactive decay**.

This process is entirely random. This means that if you have a load of unstable nuclei, you can't say when any one of them is going to decay, and neither can you do anything at all to make a decay happen. It's completely unaffected by physical conditions like temperature, or by any sort of chemical bonding, etc.

Radioactive substances emit ionising radiation. Ionising radiation is radiation that knocks electrons off atoms, creating positive ions. The ionising power of a radiation source tells you how easily it can do this.

Radioactive substances spit out one or more types of ionising radiation from their nucleus as they decay. The types you need to know about are **alpha**, **beta** and **gamma**. They can also release neutrons (n) as they decay, as they try to rebalance their atomic and mass numbers.

Alpha decay

Alpha radiation is when an **alpha particle**, α, is emitted from the nucleus. An alpha particle is two neutrons and two protons — the same as a helium nucleus. When an atom decays by emitting an alpha particle, two protons and two neutrons are lost from the nucleus.

As protons have a relative charge of +1, alpha emission decreases the charge on the nucleus (and the atomic number) by 2. The mass number decreases by 4, as protons and neutrons each have a relative mass of 1. See page 128 for an example.

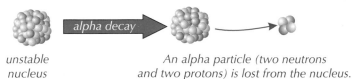

unstable nucleus

alpha decay

An alpha particle (two neutrons and two protons) is lost from the nucleus.

Figure 1: *An unstable nucleus decaying by emitting an alpha particle.*

Alpha particles are relatively big, heavy and slow-moving. This means they don't penetrate very far into materials and are stopped quickly. They only travel a few centimetres in air and are absorbed by a sheet of paper.

Because of their size they are strongly ionising — they bash into a lot of atoms and knock electrons off them before they slow down, which creates lots of ions.

Example

Alpha radiation is used in smoke detectors. It ionises air particles, causing a current to flow. If there is smoke in the air, the smoke binds to the ions, reducing the number available to carry a current. The current falls and the alarm sounds.

Beta decay

A **beta particle**, β, is just a fast-moving electron released by a nucleus. So it has virtually no mass and a relative charge of –1 (see page 123).

When a nucleus decays by beta decay, a neutron turns into a proton in the nucleus, releasing a β-particle. This increases the charge on the nucleus (and the atomic number) by 1 but leaves the mass number unchanged. See page 129 for an example.

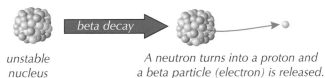

unstable
nucleus

A neutron turns into a proton and
a beta particle (electron) is released.

Figure 2: *An unstable nucleus decaying by emitting a beta particle.*

Beta particles move quite fast and they are quite small. They are moderately ionising and penetrate moderately far into materials before colliding. They have a range in air of a few metres and can be absorbed by a sheet of aluminium (around 5 mm thick).

> **Example**
>
> Beta emitters are used to test the thickness of thin sheets of metal, as the particles are not immediately absorbed by the material like alpha radiation would be, and do not penetrate as far as gamma rays.

Gamma decay

Gamma rays, γ, are very short wavelength electromagnetic (EM) waves (see page 242) released by the nucleus. Gamma rays have no mass and no charge.

They penetrate far into materials without being stopped and pass straight through air. This means they are weakly ionising because they tend to pass through rather than collide with atoms. Eventually they hit something and do damage. They can be absorbed by thick sheets of lead or metres of concrete.

> **Example**
>
> Gamma radiation is used in situations where a source needs to be detected through a thick material — e.g. for detecting cracks or blockages in underground pipes, or flaws in thick sheets of metal.

Nuclear equations

You can write alpha and beta decays as nuclear equations. They are just equations that show what atoms you start with, what radiation is emitted and what atoms you're left with. The mass and atomic numbers have to balance on both sides of the equation (before and after decay).

You don't write nuclear equations for gamma decays because they do not change the atomic mass or atomic number of the atom. Gamma emission is just a way of getting rid of excess energy, and can happen after an alpha or beta decay.

Tip: You don't need to know why a neutron turning into a proton releases a beta particle, but make sure you know that it comes from the nucleus. It's not just one of the electrons that are whizzing around outside the nucleus jumping off.

Tip: In both alpha and beta decay, a new element will be formed, as the number of protons in the nucleus (the atomic number) changes.

Exam Tip
You may be asked to talk about how the properties of different types of radiation make them suitable for different uses.

Tip: Gamma radiation is slightly different to alpha and beta radiation because it's an EM wave instead of a particle.

Tip: Some more uses of gamma radiation are covered on pages 138-139.

Tip: Gamma rays (symbol: γ) have no protons or neutrons and no charge so we just write them as $^0_0\gamma$ if we need to.

You'll need to be familiar with the notation on page 123, and how alpha and beta particles can be written in this notation:

- Alpha particles are helium nuclei (symbol: He or α) with 2 protons and 2 neutrons, so they are written $^4_2$He.

- Beta particles are electrons (symbol: e or β) so they have no protons or neutrons, and the mass number is 0. The number of protons is 0, but we write –1 where the atomic number goes because a beta particle has a charge of –1. This helps us balance the charges on each side. So a beta particle is written $^{0}_{-1}$e.

Example 1 — alpha decay

Uranium-238 can decay into thorium-234 by emitting an alpha particle. Uranium has 92 protons and thorium has 90 protons.

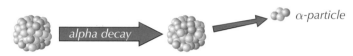

uranium-238   thorium-234 α-particle

The nuclear equation for this decay looks like this:

$$^{238}_{92}\text{U} \longrightarrow {}^{234}_{90}\text{Th} + {}^4_2\text{He}$$

On the left-hand side:

- The mass number is 238.

- The atomic number is 92.

On the right-hand side:

- The total of the mass numbers is: 234 + 4 = 238.

- The total of the atomic numbers is: 90 + 2 = 92.

So both sides of the equation balance.

Exam Tip
You won't have to work out what the elements are (e.g. Pu, U) in nuclear equations. You only need to be able to balance the atomic and mass numbers and identify the type of decay (alpha or beta).

Example 2 — alpha decay

Balance the following equation: $^{238}_{94}\text{Pu} \longrightarrow {}^{234}_{}\text{U} + {}^{}_2\text{He}$

Make these equations balance: 238 $\longrightarrow$ 234 + ...

94 $\longrightarrow$ ..... + 2

Balancing mass numbers, 238 = 234 + **4**, so the mass number of He is 4.

Balancing atomic numbers, 94 = **92** + 2, so the atomic number of U is 92.

The full equation is:

$$^{238}_{94}\text{Pu} \longrightarrow {}^{234}_{92}\text{U} + {}^4_2\text{He}$$

Example 3 — beta decay

Carbon-14 can decay into nitrogen-14 by emitting a beta particle (when a neutron turns into a proton).

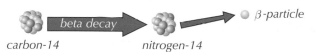

carbon-14   nitrogen-14 β-particle

The equation is: $^{14}_{6}C \rightarrow ^{14}_{7}N + ^{0}_{-1}e$

and the mass and atomic numbers balance on each side. Brill.

Practice Questions — Fact Recall

Q1   What does it mean if a substance is radioactive?

Q2   How can you change the rate of radioactive decay?

Q3   Put alpha, beta and gamma radiation in order of their ionising strength.

Q4   Describe what each type of radiation named in Q3 is made of.

Q5   What 2 numbers must be equal on both sides of a nuclear equation?

Practice Questions — Application

Q1   This diagram shows the paths of three types of radiation, A, B and C, being directed towards a human hand and a thick metal sheet.

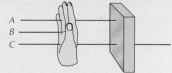

    a) Which radiation (A or B) is more penetrating? How can you tell?

    b) Which radiation (A or B) is likely to be alpha radiation? Why?

    c) Explain what radiation C is likely to be and how you know.

Q2   Americium-241 decays into neptunium-237.
The nuclear equation for this decay is:

$$^{241}_{95}Am \rightarrow ^{237}_{93}Np + ^{4}_{2}He$$

What type of decay is this — alpha or beta?

Q3   Radium-228 decays into actinium-228 by emitting an electron.

    a) What is the name of this decay?

    b) Complete the following nuclear equation for this decay:

$$^{.....}_{88}Ra \rightarrow ^{228}_{.....}Ac + ^{0}_{-1}e$$

Q4   Radioactivity can be used to measure the thickness of paper during manufacture. A beta source is placed on one side of the paper, and a detector detects how much beta radiation gets through it. Explain why this process couldn't use:

    a) an alpha source.   b) a gamma source.

4. Activity and Half-life

Radioactive samples give out less and less radiation over time, but they never stop giving out radiation altogether.

Activity

Radioactive substances will give out radiation from the nuclei of their atoms — no matter what. This radiation can be measured with a **Geiger-Muller tube** and counter, which records the count-rate — the number of radiation counts reaching it per second.

Radioactive decay is entirely random. So you can't predict exactly which nucleus in a sample will decay next, or when any one of them will decay. But you can find out the time it takes for the amount of radiation emitted by a source to halve, this is known as the **half-life**. It can be used to make predictions about radioactive sources, even though their decays are random.

Half-life can be used to find the rate at which a source decays — its **activity**. Activity is measured in becquerels, Bq (where 1 Bq is 1 decay per second).

Half-life

The radioactivity of a sample always decreases over time. Each time a radioactive nucleus decays to become a stable nucleus, the activity as a whole will decrease — so older sources emit less radiation.

How quickly the activity drops off varies a lot. For some isotopes, it takes just a few hours before nearly all of the unstable nuclei have decayed, whilst others last for millions of years.

The problem with trying to measure this is that the activity never reaches zero, which is why we have to use the idea of half-life to measure how quickly the activity drops off.

Learn the definition of half-life:

> Half-life is the time it takes for the number of nuclei of a radioactive isotope in a sample to halve.

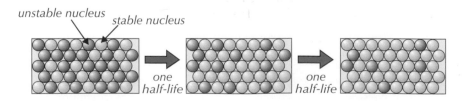

Figure 1: A diagram showing how the number of unstable nuclei of a radioactive isotope in a sample decreases over two half-lives.

In other words, it is the time it takes for the count rate (the number of radioactive emissions detected per unit of time) or activity from a sample containing the isotope to fall to half its initial level.

- A short half-life means the activity falls quickly, because the nuclei are very unstable and rapidly decay. Sources with a short half-life can be dangerous because of the high amount of radiation they emit at the start, but they quickly become safe.

- A long half-life means the activity falls more slowly because most of the nuclei don't decay for a long time — they just sit there, releasing small amounts of radiation over a longer period. This can be dangerous because nearby areas are exposed to radiation for years — perhaps even millions of years.

Tip: The hazards associated with radioactive substances are covered in more detail on page 137.

Calculating half-life

You can work out the half-life of a radioactive isotope if you're given a little information. Or if you know the half-life, you can work out how long it will take for the activity to drop a certain amount.

Half-life is maybe a little confusing, but exam calculations on it are straightforward as long as you do them slowly, step by step. Like this:

Example

The activity of a radioisotope is 640 cpm (counts per minute).
Two hours later it has fallen to 80 cpm.
Find the half-life of the sample.

You must go through it in short simple steps like this:

initial count		after one half-life		after two half-lives		after three half-lives
↓		↓		↓		↓
640	(÷2)	320	(÷2)	160	(÷2)	80

It takes three half-lives for the activity to fall from 640 to 80. Hence two hours represents three half-lives, so the half-life is 120 mins ÷ 3 = 40 minutes.

Tip: A radioisotope is just a radioactive isotope.

Example — **Higher**

The initial activity of a sample is 640 Bq. Calculate the percentage reduction in activity after two half-lives.

Find the activity after each half-life.

1 half-life:	$640 \div 2 = 320$ Bq
2 half-lives:	$320 \div 2 = 160$ Bq

So the reduction in activity is: $640 - 160 = 480$ Bq

Then write this as a percentage: $\frac{480}{640} \times 100 = \mathbf{75\%}$

Exam Tip H
You may be asked to give a ratio showing the decline of activity, count rate or number of nuclei after a certain number of half-lives as a percentage of the initial activity like this.

Using graphs

You can plot or use a graph of radioactive activity against time to work out the half-life of a radioactive isotope.

The data for the graph will usually be several readings of activity which may have been taken with a Geiger-Müller tube with a counter.

The graph will always be shaped like the one shown in Figure 3.

The half-life is found from the graph by finding the time interval on the bottom axis corresponding to a halving of the activity on the vertical axis.

Figure 2: You can do an experiment to simulate half-life using cubes with one black face. The cubes represent unstable nuclei and if they land black-side up, they have 'decayed'. Take note of how many cubes you start with and throw them all, removing any that 'decay'. Count the number remaining, then throw those remaining cubes again. Plotting your results on a graph will allow you to calculate the 'half-life' in 'number of throws'.

Tip: Remember — activity never drops to zero, so the graph will never touch the horizontal axis.

Example

Figure 3 shows how the activity of a radioactive sample decreases with time. The half-life can be found form the graph as follows:

- The initial activity is 80, so after one half-life it will be 40 (and after two it will be 20 and after three it will be 10).

- To find the half-life of the sample, draw a line from 40 on the activity axis across to the curve and down to the time axis (green dotted line). This tells you that the half-life is 4 hours.

- You can check you were right by doing the same for an activity of 20 and checking that you get a time of 8, and so on...

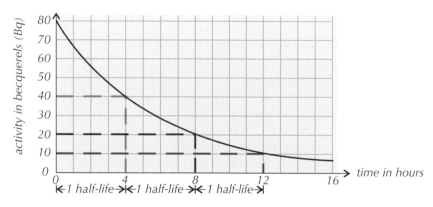

Figure 3: A graph of activity against time for a radioisotope, taking into account background radiation (see page 135).

Practice Questions — Fact Recall

Q1 Why can we not measure the time it takes for a radioactive sample to decay completely?

Q2 Write down two definitions of half-life.

Practice Questions — Application

Q1 A radioactive source with a half-life of 15 minutes has an initial count rate of 240 cpm. What will the count rate be after 1 hour?

Q2 A radioactive source has an initial count rate of 16 cpm and after 2 hours it has decreased to 4 cpm. What is the half-life of this source?

Q3 The activity of a source is 32 Bq. After how many half-lives will the activity have dropped to 4 Bq?

Q4 Find the half-life of the radioisotope from this graph of the activity of the radioisotope against time.

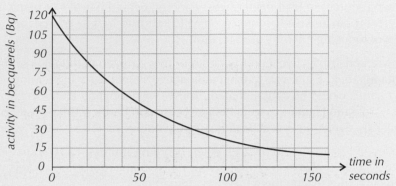

Tip: Make sure you read the axes carefully — you don't want to be saying the half-life is 'so many' hours when it's actually 'so many' seconds.

Q5 The initial activity of a radioactive source is 9600 Bq. Calculate the activity after 3 half-lives as a percentage of the initial activity.

- Know that exposure to radiation from a radioactive source is known as irradiation.

- Know that the unwanted presence of radioactive atoms on or in another material is known as contamination.

- Know some precautions which can be taken to minimise the risk of irradiation and contamination.

- Understand how the hazards of irradiation and contamination vary depending on the type of radiation.

- Be able to compare the dangers associated with contamination and irradiation.

- Understand the importance of publishing research into the effects of radiation on human health.

Specification Reference 4.4.2.4

Tip: See page 137 for more on the dangers to being exposed to radioactive decay.

5. Irradiation and Contamination

Radiation can cause harm to your body. How dangerous a radiation source is depends on which type of radiation it emits and how you're using it.

What are irradiation and contamination?

Objects near a radioactive source are **irradiated** by it. This simply means they're exposed to the radiation. Irradiating something does not make it radioactive, but exposure to radiation can be harmful to living things.

Keeping sources in lead-lined boxes, standing behind barriers or being in a different room and using remote-controlled arms to handle sources are all ways of reducing the risks of irradiation.

If unwanted radioactive atoms get onto or into a material, then it is said to be **contaminated**. E.g. if you touch a radioactive source without wearing gloves, your hands would be contaminated. These contaminating atoms might then decay, releasing radiation which could cause you harm. Contamination is especially dangerous because radioactive particles could get inside your body.

Gloves and tongs should be used when handling sources, to avoid particles getting stuck to your skin or under your nails. Some industrial workers wear protective suits to stop them breathing in particles.

Contamination or irradiation can cause different amounts of harm based on the radiation type:

- Outside the body, beta and gamma sources are the most dangerous. This is because beta and gamma can penetrate the body and get to the delicate organs. Alpha is less dangerous because it can't penetrate the skin and is easily blocked by a small air gap. High levels of irradiation from all sources are dangerous, but especially from ones that emit beta and gamma.

- Inside the body, alpha sources are the most dangerous. They do all their damage in a very localised area. Beta and gamma sources are less dangerous inside the body because they mostly pass straight out without doing much damage (they have a lower ionising power). So contamination, rather than irradiation, is the major concern when working with alpha sources.

The more we understand how different types of radiation affect our bodies, the better we can protect ourselves when using them. This is one of the reasons why it's so important that research about this is published. The data is peer-reviewed and can quickly become accepted, leading to many improvements in our use of radioactive sources.

(WORKING SCIENTIFICALLY)

Practice Questions — Fact Recall

Q1   State what is meant by the terms irradiation and contamination.

Q2   Which type(s) of ionising radiation is most dangerous outside the body? Why?

Q3   Which type(s) of ionising radiation is most dangerous inside the body? Why?

6. Background Radiation

Where you live and what you do as a job can affect how much radiation you are exposed to, so you need to be aware of where it comes from.

Radiation dose

You'll see later that radiation can cause damage to humans, including cancer or even death (see page 137). How likely you are to suffer damage if you're exposed to nuclear radiation depends on the radiation dose.

Radiation dose is a measure of the risk of harm to your body due to exposure to radiation. It depends on the type and amount of radiation you've been exposed to. The higher the radiation dose, the more at risk you are of developing cancer. Radiation dose is measured in sieverts (Sv). The radiation dose due to background radiation (see below) is small, so you'll often see it given in millisieverts instead (1 Sv = 1000 mSv).

Background radiation

Background radiation is low-level radiation that is present at all times, all around us, wherever you go. The background radiation we receive comes from many sources, including:

- Radioactivity of naturally occurring unstable isotopes which are all around us — in the air, in food, in building materials and in the rocks under our feet.

- Radiation from space, which is known as **cosmic rays**. These come mostly from the Sun. Luckily, the Earth's atmosphere absorbs a lot of the radiation from cosmic rays, but at very high altitudes a lot more of them can get through.

- Radiation due to man-made sources, e.g. fallout from nuclear weapons tests, nuclear accidents (such as Chernobyl — see Figure 1) or dumped nuclear waste.

Learning Objectives:
- Know what is meant by radiation dose (in sieverts, Sv).
- Know that 1 Sv = 1000 mSv.
- Know that we are always surrounded by background radiation.
- Know that background radiation comes from natural sources (rocks, cosmic rays, etc.) and man-made sources (e.g. the nuclear industry and fallout from nuclear weapons).
- Know that the amount of background radiation you are exposed to depends on factors such as your location and occupation.

Specification Reference 4.4.3.1

Tip: To avoid systematic errors (p. 13), you should always measure and subtract the background radiation from your results when you're investigating the activity of a source.

> ### Example
>
> Figure 2 shows that more than half of the background radiation that a typical person in the UK is exposed to comes from radon gas produced by rocks and only 1% comes from the nuclear industry.
>
>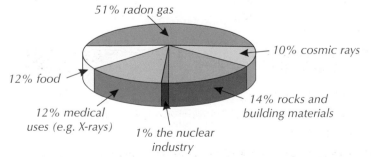
>
> 51% radon gas
> 10% cosmic rays
> 12% food
> 14% rocks and building materials
> 12% medical uses (e.g. X-rays)
> 1% the nuclear industry
>
> **Figure 2:** *A pie chart showing the relative proportions of background radiation that a typical person is exposed to in the UK from different sources.*

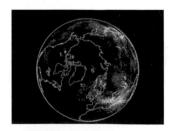

Figure 1: *A simulation of the radiation (pink) in the Northern hemisphere 10 days after the Chernobyl disaster in which a nuclear power plant in Ukraine exploded in 1986 and released lots of radiation.*

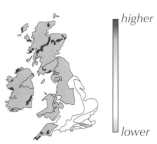

higher

lower

Figure 3: A map of the United Kingdom showing radiation from rocks. The scale shows how the level of radiation from rocks in different areas varies.

Figure 4: A nuclear power station worker wearing a radiation suit.

Tip: Uranium is mined to use as a fuel for nuclear power stations — see page 57.

Effect of location and occupation

The amount of radiation you're exposed to (and hence your radiation dose) can be affected by your location and occupation.

Location

- Certain underground rocks (e.g. granite) can cause higher levels of radiation at the surface, especially if they release radioactive radon gas, which tends to get trapped inside people's houses. A radon detector can tell you if your house has a dangerous level of radon and a radon outlet pipe can be used to keep the level down.

- People who live at high altitudes are exposed to more background radiation in the form of cosmic rays than people who live at sea level.

Occupation

- Nuclear industry workers and uranium miners are typically exposed to 10 times the normal amount of radiation. They wear protective clothing and face masks to stop them from touching or inhaling the radioactive material, and monitor their radiation doses with special radiation badges and regular check-ups.

- Radiographers work in hospitals using ionising radiation and so have a higher risk of radiation exposure. They wear lead aprons and stand behind lead screens to protect them from prolonged exposure to radiation.

- Underground (e.g. in mines, etc.) the radiation dose increases because of the rocks all around, posing a risk to miners.

Practice Questions — Fact Recall

Q1 Give three main sources of background radiation.

Q2 Why do people living in locations with certain underground rocks get a higher radiation dose than people living in other areas?

Practice Question — Application

Q1 An astronaut working aboard the International Space Station (ISS), which orbits the Earth, is worried that his job puts him at risk of a higher dose of background radiation.

   a) What source of background radiation is likely to be higher on board the ISS than on Earth?

   b) Suggest how this influences the decision that the crew of the ISS should be regularly changed.

7. Risks and Uses of Radiation

Ionising radiation gets loads of bad press, but it is pretty essential in all sorts of everyday situations, especially medicine. While it can cause cancer, it can also help to treat it.

Effect of radiation on living cells

Ionising radiation can be very harmful to living cells. Alpha, beta and gamma radiation will enter living cells and collide with molecules. These collisions cause ionisation, which damages or destroys the molecules.

Lower doses tend to cause minor damage without killing the cell. This can give rise to mutant cells which divide uncontrollably — see Figure 1. The cells keep dividing, making more cells and forming a tumour — this uncontrolled cell division is cancer.

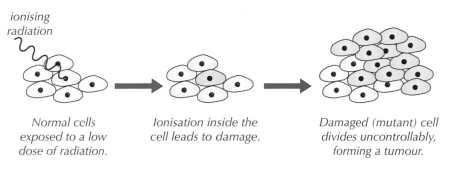

ionising radiation

Normal cells exposed to a low dose of radiation.	*Ionisation inside the cell leads to damage.*	*Damaged (mutant) cell divides uncontrollably, forming a tumour.*

Figure 1: *A cell being damaged by a low dose of radiation, leading to it multiplying uncontrollably.*

Higher doses tend to kill cells completely, which causes radiation sickness if a lot of body cells are killed at once.

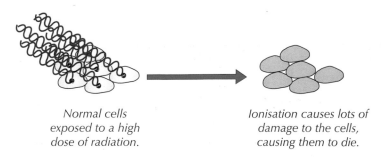

Normal cells exposed to a high dose of radiation.	*Ionisation causes lots of damage to the cells, causing them to die.*

Figure 2: *Cells being killed by a high dose of radiation.*

The extent of the harmful effects of radiation depends mainly on two things:

- How much exposure you have to the radiation.

- The energy and penetration of the radiation, since some types are more hazardous than others (see page 134).

Learning Objectives:

- Know how radiation can damage the human body.

- Know that ionising radiation has applications in medicine.

- Understand how gamma and beta emitters are used as tracers to investigate the function of internal organs.

- Understand how radiation is used to control or kill cancer cells.

- Be able to assess the risks of using radiation and compare them to the benefits of the treatment.

Specification Reference 4.4.3.3

Medical uses of radiation

Although ionising radiation can be harmful to living cells, it can also have its benefits if used correctly.

Medical tracers

Certain radioactive isotopes can be injected into people (or they can just swallow them) and their progress around the body can be followed using an external detector. These isotopes are known as **medical tracers**. A computer converts the readings from the external detector to a display showing where the strongest readings are coming from. This can help doctors to investigate whether the patient's internal organs are functioning as they should be.

Example

A well-known example is the use of iodine-123 or iodine-131. These are absorbed by the thyroid gland in the neck just like normal iodine-127, but give out gamma radiation. The radiation can be detected to indicate whether the thyroid gland is taking in iodine as it should.

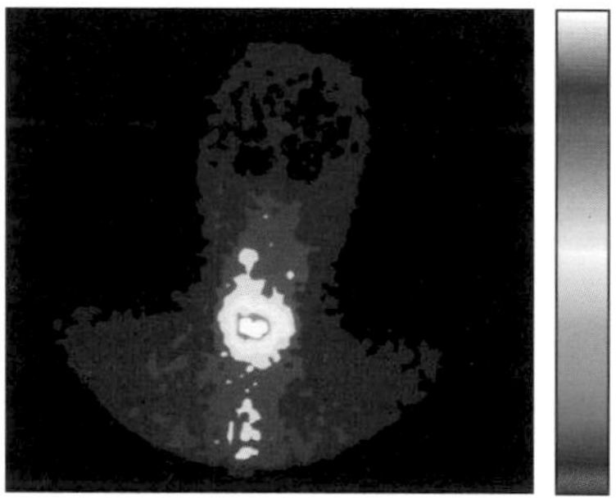

high

intensity of gamma radiation detected

low

Figure 3: *An image of the gamma radiation detected from a person who has been injected with gamma-emitter iodine-131. The image shows that most gamma radiation is coming from the thyroid, indicating that the iodine-131 has collected there.*

Tip: The intensity of gamma radiation means the amount detected per unit time.

All isotopes which are taken into the body must be gamma or beta emitters, so that the radiation passes out of the body. Alpha sources should never be used as they are highly ionising and do their damage in a localised area — see page 134. The source should only last a few hours too, so that the radioactivity inside the patient quickly disappears (i.e. they should have a short half-life).

Tip: There's more on half-life on page 130.

Radiotherapy

Radiotherapy is the treatment of cancer using ionising radiation, e.g. gamma rays. It can be used to control or destroy cancer cells. High doses of radiation will kill all living cells, including cancer cells.

The radiation has to be directed carefully and at just the right dosage so as to kill the cancer cells without damaging too many normal cells. Radioactive implants (usually beta-emitters) can also be put next to or inside tumours.

A fair bit of damage is done to normal cells, which makes the patient feel very ill. But if the cancer is successfully killed off in the end, then it's worth it.

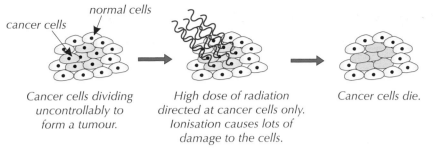

Cancer cells dividing uncontrollably to form a tumour.

High dose of radiation directed at cancer cells only. Ionisation causes lots of damage to the cells.

Cancer cells die.

Figure 5: Radiation being used to kill cancer cells.

Figure 4: *A male undergoing radiotherapy for brain cancer. The laser crosshair marks the point where the radiation should be focused, and a head brace is worn to keep the head perfectly still.*

Weighing up the risks

For every situation, it's worth considering both the benefits and risks of using radioactive materials. For example, tracers can be used to diagnose life-threatening conditions, while the risk of cancer from one use of a tracer is very small.

Whilst prolonged exposure to radiation poses risks and causes many side effects, many people with cancer choose to have radiotherapy as it may get rid of their cancer entirely. For them, the potential benefits outweigh the risks.

Perceived risk is how risky a person thinks something is. It's not the same as the actual risk of a procedure and the perceived risk can vary from person to person. See page 8 for more on this.

Tip: See page 251 for more on comparing the risks associated with different medical procedures.

Practice Questions — Fact Recall

Q1 Explain how radiation can cause cancer.

Q2 How can radiation cause cell death?

Q3 Explain how medical tracers work. Which types of ionising radiation sources can be used and why?

Q4 What is radiotherapy?

Exam Tip
In the exam you could be asked to explain why certain radioactive sources are chosen for certain tasks. Just think about the properties of each one (p. 126-127 and p. 134), and how that would make them useful.

Learning Objectives:

- Know that nuclear fission is a reaction where a large unstable nucleus splits into two smaller ones.

- Know that fission usually has to be started by the nucleus absorbing a neutron.

- Know that fission produces two smaller nuclei and two or three neutrons with energy in their kinetic energy stores, as well as energy in the form of gamma rays.

- Know that the neutrons produced can lead to further fission, and may cause a chain reaction.

- Be able to draw and understand diagrams of nuclear fission and chain reactions.

- Know that chain reactions are controlled in nuclear power plants and used to release energy.

- Know that a nuclear weapon uses an uncontrolled chain reaction to cause an explosion.

- Know that in nuclear fusion, two small nuclei combine to form a heavier nucleus.

- Know that some of the mass of the lighter nuclei is converted to energy and is emitted as radiation.

Specification References 4.4.4.1, 4.4.4.2

8. Nuclear Fission and Fusion

Nuclear fusion and nuclear fission are two different nuclear reactions which release large amounts of energy. Harnessing this energy isn't so easy though...

Nuclear fission

Nuclear fission is a type of nuclear reaction that is used to release energy from large and unstable atoms (e.g. uranium or plutonium) by splitting them into smaller atoms.

Spontaneous (unforced) fission rarely happens. Usually, the nucleus has to absorb a neutron before it will split. When the atom splits, it forms two new lighter elements that are roughly the same size (and that have some energy in their kinetic energy stores).

Two or three neutrons are also released when an atom splits. If any of these neutrons are moving slowly enough to be absorbed by another nucleus, they can cause more fission to occur. This is a **chain reaction**. It is this process which is used to generate power in a nuclear power plant.

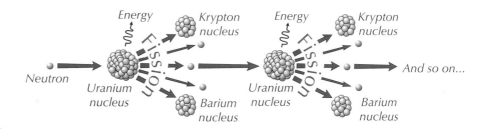

Figure 1: *A chain reaction that can occur when uranium undergoes nuclear fission.*

The energy not transferred to the kinetic energy stores of the products is carried away by gamma rays. The energy carried away by the gamma rays, and in the kinetic energy stores of the remaining free neutrons and the other decay products, can be used to heat water, making steam to turn turbines and generators (p. 55 and p. 305).

The amount of energy released by fission in a nuclear reactor is controlled by changing how quickly the chain reaction can occur. This is done using control rods, which are lowered and raised inside the nuclear reactors to absorb neutrons, slow down the chain reaction and control the amount of energy released.

Uncontrolled chain reactions quickly lead to lots of energy being released as a nuclear explosion — this is how atomic bombs work.

Nuclear fusion

Nuclear fusion is the opposite of nuclear fission. In nuclear fusion, two light nuclei collide at high speed and join (fuse) to create a larger, heavier nucleus. For example, hydrogen nuclei can fuse to produce a helium nucleus.

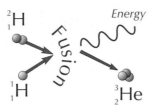

Figure 2: A diagram to show the formation of a helium nucleus from the nuclear fusion of two hydrogen nuclei. Energy is released in this process.

Tip: Nuclear fusion is the reaction which generates energy in stars, keeping them hot and bright. See page 318.

This heavier nucleus does not have as much mass as the two separate, light nuclei did. Some of the mass of the lighter nuclei is converted into energy (don't panic, you don't need to know how). This energy is then released as radiation.

So far, scientists haven't found a way of using fusion to generate energy for us to use. The temperatures and pressures needed for fusion are so high that fusion reactors are really hard and expensive to build.

Practice Questions — Fact Recall

Q1 What is nuclear fission?

Q2 What must usually happen before a nucleus will undergo nuclear fission?

Q3 Explain how a nuclear fission chain reaction can happen.

Q4 What is nuclear fusion?

Topic 4 Checklist — Make sure you know...

The History of the Atom

☐ How the concept of the atom has developed over time.

☐ That the plum pudding model describes an atom as a sphere of positive charge studded with negatively charged electrons.

☐ How the results of the alpha particle scattering experiment showed that the atom must contain a small, positively charged nucleus at the centre and how this led to the nuclear model of the atom.

The Structure of the Atom

☐ How the nuclear model was adapted and altered to give its current form.

☐ That the nuclear model of the atom describes the atom as a small, central nucleus containing protons and neutrons, with electrons moving around outside of the nucleus.

☐ The radius of an atom is approximately 1×10^{-10} m and the nucleus is more than 10 000 times smaller.

☐ The relative charges and masses of electrons, protons and neutrons.

☐ That an atom has no overall charge because the number of protons and electrons is equal and they have equal and opposite charges.

☐ That the atomic number of an atom is the number of protons in its nucleus and the mass number of an atom is the number of protons and neutrons in its nucleus.

☐ That every atom of an element has the same number of protons (atomic number).

☐ That an ion is an atom with fewer or more electrons than protons, giving it an overall charge.

☐ That isotopes are forms of an element that have atoms with the same atomic number but different mass numbers.

Radioactivity

☐ That a radioactive substance will undergo radioactive decay, where it gives out radiation from the nuclei of its atoms, and know that this is a random process.

☐ That when a substance undergoes radioactive decay it may emit alpha particles, beta particles, gamma rays or neutrons from its nucleus.

☐ That alpha decay is the process of a nucleus giving out an alpha particle, which is made up of two protons and two neutrons. It is strongly ionising and weakly penetrating.

☐ That beta decay is the process of a nucleus giving out a beta particle, which is an electron. It is moderately ionising and moderately penetrating.

☐ That gamma decay is where a nucleus gives out gamma rays, which are short-wavelength EM waves. Gamma rays are weakly ionising and strongly penetrating, passing straight through air.

☐ How alpha, beta and gamma decay affect the mass and atomic number of the nucleus.

☐ Examples of applications of radioactive sources in everyday life, e.g. alpha sources in smoke detectors.

☐ How to balance nuclear equations for alpha and beta decay.

cont...

Activity and Half-life

- [ ] The definitions of activity and count rate.
- [ ] That although radioactive decay random, the time taken for the activity to half can be predicted.
- [ ] That radioactive half-life is the time it takes for the number of nuclei in a radioactive isotope sample to halve, or the time it takes for the count rate (or activity) to reach half of its initial level.
- [ ] How to find the half-life of a radioactive substance, including from activity-time graphs.

Irradiation and Contamination

- [ ] That exposure to radiation is known as irradiation.
- [ ] That the unwanted presence of radioactive material on (or in) an object is known as contamination.
- [ ] The relative hazards of contamination and irradiation for different sources.
- [ ] The precautions that should be taken when handling radioactive substances.

Background Radiation

- [ ] What is meant by the term 'radiation dose' in sieverts.
- [ ] That background radiation is radiation that is all around us. Its sources include naturally occurring isotopes on Earth, radiation from space and man-made radiation from the nuclear industry.
- [ ] How your occupation and location can affect the radiation dose you receive.

Risks and Uses of Radiation

- [ ] How ionising radiation can damage or kill cells, and that the use of radiation in medicine comes with risks.
- [ ] How gamma and beta sources can be used as medical tracers.
- [ ] How gamma and beta sources can be used in radiotherapy to treat cancer.

Nuclear Fission and Fusion

- [ ] That nuclear fission is when an atomic nucleus splits up to form two smaller nuclei — this process releases lots of energy and so can be used to generate electricity.
- [ ] That nuclear fission usually happens when a (slow-moving) neutron is absorbed into the nucleus of an atom, causing it to become unstable and split in two.
- [ ] How nuclear fission can initiate a chain reaction through the emission of neutrons.
- [ ] That nuclear fission chain reactions can be controlled and used to release energy in power plants.
- [ ] How to sketch and label a diagram showing how a chain reaction of nuclear fission can happen.
- [ ] That nuclear fusion is when two nuclei (e.g. hydrogen nuclei) join to create a larger nucleus, and that this process releases lots of energy.

Exam-style Questions

1   The diagram shows two atoms.

1.1   Put these three words into the following sentences.
You may only use each word once.

protons	neutrons	electrons

The nucleus contains _____ and _____.

The numbers of protons and _____ in a neutral atom are equal.

(2 marks)

1.2   The two atoms shown are isotopes of each other. Describe the similarities and differences in the nuclei of two different isotopes of the same element.

(1 mark)

2   This incomplete table gives some information about three types of ionising radiation.

Radiation type:	Made up of:	Stopped by:
Alpha particles		Thin paper
	Electrons	Thin aluminium
Gamma rays	Short-wavelength EM waves	Thick lead

2.1   Complete the table.

(2 marks)

2.2   Explain what is meant by the term 'ionising'.

(1 mark)

2.3   Alpha sources are the most dangerous radioactive sources when inside the body. Explain why.

(1 mark)

2.4   With respect to the atomic number and mass number, describe how the nucleus changes when it emits an alpha particle.

(2 marks)

2.5   State what happens in the nucleus when it emits a beta particle.

(1 mark)

3   In March 2011, the Fukushima nuclear power plant in Japan was damaged by a tsunami and leaked some nuclear radiation into the air. One of the radioactive isotopes that was leaked in this incident was caesium-137, which has a half-life of 30 years.

3.1   Explain what it means for caesium-137 to have a half-life of 30 years.

(1 mark)

3.2   A sample containing caesium-137 is found to have an activity of 24 Bq.
Calculate what the activity of the sample will be in 90 years' time.

(3 marks)

Shortly after the disaster, the Japanese government decided to evacuate all people from their homes within a 12 mile radius of the power plant, because the background radiation was significantly higher than average due to the disaster.

3.3   Name **two** sources of background radiation, other than man-made nuclear sources.

(2 marks)

3.4   No one was allowed to move back into any of these homes until at least one year later. Suggest why it takes such a long time for evacuated areas to be considered safe after a nuclear disaster.

(1 mark)

Caesium-137 decays into barium-137.
The following incomplete equation shows this decay:

$$^{137}_{55}\text{Cs} \longrightarrow {}^{137}_{\,\,\text{......}}\text{Ba} + {}^{0}_{-1}\text{e}$$

3.5   What type of radiation is being given out in this decay?

(1 mark)

3.6   Complete the nuclear equation for this decay.

(1 mark)

4*   An underground pipe is thought to be cracked, so that the substance carried by the pipe is leaking out. An engineer intends to use a radioactive isotope to locate the cracks. The isotope is put into the substance carried by the pipe. A radiation detector on the ground is moved along above the pipe, as shown.

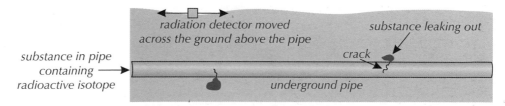

Describe how the engineer would be able to tell where there was a crack in the pipe. Explain what type of radiation the isotope in the pipe should emit and what the half-life of the source should be.

(6 marks)

5   In radiotherapy, beta-emitting implants can be used to treat cancer.
They are placed inside the body next to or inside the tumour being treated.

5.1   Describe how ionising radiation kills cancer cells.

(3 marks)

5.2   The patient may feel very ill during this treatment. Explain why this is the case.

(1 mark)

The radioactive material used is stored inside a casing, which is then implanted into the patient's body. The casing must be securely sealed to prevent contamination.

5.3   Explain why there would be a risk of contamination if the casing wasn't sealed.

(1 mark)

5.4   Suggest **one** problem with using an alpha-emitting implant sealed in a similar casing.

(1 mark)

6   Nuclear workers wear radiation dose badges containing photographic film.
Developing the film can tell the workers how much they have been irradiated in the time they've been wearing it, and what sort of radiation they've been exposed to.

6.1   Why do nuclear workers need to monitor their radiation dose?

(1 mark)

6.2   Give **two** other precautions the workers should take to control their radiation dose.

(2 marks)

7   A nuclear power station uses the energy released when uranium-235 nuclei split in two to generate electricity.

7.1   Give the term used to describe the splitting of an atomic nucleus.

(1 mark)

7.2   Give the name of the particle that can cause a nucleus of a uranium-235 atom to split when it is absorbed by that nucleus.

(1 mark)

7.3   The reaction that occurs in a nuclear reactor is a chain reaction.
Complete the diagram below to illustrate how a chain reaction occurs.

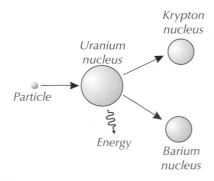

(2 marks)

1. Contact and Non-Contact Forces

Before you learn about forces, you need to understand the difference between vector and scalar quantities.

Vectors and scalars

Force is a **vector** quantity — vector quantities have a magnitude (size) and a direction. Lots of physical quantities are vector quantities — some examples are force, velocity, displacement, acceleration and momentum.

Some physical quantities have magnitude but no direction. These are called **scalar** quantities, and some examples are speed, distance, mass, temperature and time.

Vectors are usually represented by an arrow — the length of the arrow shows the magnitude, and the direction of the arrow shows the direction of the quantity.

> **Example**
>
> Velocity is a vector, but speed is a scalar quantity. The bikes in Figure 1 are travelling at the same speed, v (the length of each arrow is the same). They have different velocities because they are travelling in different directions.
>
>
>
> **Figure 1:** *Two motorcycles travelling in opposite directions have the same speed but a different velocity.*

Contact and non-contact forces

A **force** is a push or a pull on an object that is caused by it interacting with something. All forces are either contact or non-contact forces.

When two objects have to be touching for a force to act, that force is called a contact force. For example, friction, air resistance, tension in ropes and the normal contact force (page 148) are all contact forces. If the objects do not need to be touching for the force to act, the force is a non-contact force. Magnetic forces, gravitational forces and electrostatic forces are all non-contact forces.

When two objects interact, there is a force produced on both objects. An interaction pair is a pair of forces that are equal and opposite and act on two interacting objects. (This is basically Newton's Third Law — see p. 201.)

(page 148) ... see p. 201.

Learning Objectives:

- Know that vector quantities have both magnitude and direction.
- Know that scalar quantities only have a magnitude.
- Know that force is a vector.
- Know how arrows can be used to represent a vector quantity.
- Be able to define a force as the push or pull acting on an object due to an interaction with another object.
- Know the difference between contact forces and non-contact forces, including examples of both.
- Be able to describe the forces between two objects interacting with each other.
- Be able to represent the forces between two interacting objects as vectors.

Specification References 4.5.1.1, 4.5.1.2

Figure 2: *A sledge being pulled. The tension in the rope is an example of a contact force.*

Tip: There's more on gravitational forces on the next page.

Tip: The attraction between the Sun and the Earth is what causes the Earth to move around (orbit) the Sun. See pages 320-321 for more.

Tip: Any object resting on the ground will have a normal contact force acting on it.

Examples

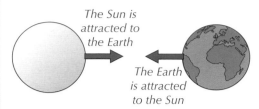

The Sun is attracted to the Earth

The Earth is attracted to the Sun

The Sun and the Earth are attracted to each other by a gravitational force. This is a non-contact force. An equal but opposite force of attraction is felt by both the Sun and the Earth.

A chair exerts a force on the ground, whilst the ground pushes back at the chair with the same force (the normal contact force). Equal but opposite forces are felt by both the chair and the ground.

Ground pushes on chair

Chair pushes on ground

Practice Questions — Fact Recall

Q1 What is the difference between a vector quantity and a scalar quantity?

Q2 State whether the following are vector quantities or scalar quantities:

   a) force

   b) acceleration

   c) speed

   d) velocity

   e) temperature

Q3 What is meant by a contact force?

Q4 What is meant by a non-contact force?

Q5 What is an interaction pair?

Practice Question — Application

Q1 The diagram shows a boat travelling at a constant velocity. Draw another diagram of the boat, this time with an arrow showing what the velocity of the boat would be if it was travelling at a higher speed in the opposite direction.

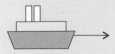

2. Weight, Mass and Gravity

Gravity is pretty important. It not only gives everything a weight, it also keeps us glued to Earth.

Gravitational force

Gravity attracts all masses, but you only notice it when one or more of the masses are really big, e.g. a planet. Anything near a planet or star is attracted to it very strongly.

This has two important effects:

1. On the surface of a planet, it makes all things accelerate (see page 181) towards the ground.

2. It gives everything a weight (see below).

Weight and mass

Weight and mass are not the same — mass is just the amount of 'stuff' in an object. For any given object this will have the same value anywhere in the universe.

Weight is the force acting on an object due to gravity (the pull of the gravitational force on the object). This force is caused by gravitational fields, and the size of the force depends on the gravitational field strength. Close to Earth, this force is caused by the gravitational field around the Earth.

Gravitational field strength varies with location. It's stronger the closer you are to the mass causing the field, and stronger for larger masses. The weight of an object depends on the strength of the gravitational field at the location of the object. This means that the weight of an object changes with its location.

An object has the same mass whether it's on Earth or on the Moon — but its weight will be different. A 1 kg mass will weigh less on the Moon than it does on Earth (about 1.6 N on the Moon compared to about 9.8 N on Earth). This is because the gravitational field strength on the surface of the Moon is less.

Measuring weight

Weight is a force measured in newtons. You can think of the force as acting from a single point on the object, called its centre of mass (a point at which you assume the whole mass is concentrated). For a uniform object (one that's the same density throughout, p. 106, and a regular shape), this will be at the centre of the object.

Figure 2: Any object with a mass will have a point at which its weight appears to act, known as the centre of mass.

Weight is measured using a calibrated spring balance (newtonmeter — see Figure 1). Mass isn't a force. It's measured in kilograms with a mass balance.

Learning Objectives:

- Be able to define weight as the force acting on an object due to gravity.
- Know that the gravitational force acting on an object close to Earth is due to the Earth's gravitational field.
- Know that the weight of an object is dependent on the gravitational field strength at the object's location.
- Know that weight is measured using a calibrated spring balance.
- Understand what is meant by the centre of mass of an object.
- Know the equation that relates weight, mass and gravitational field strength.
- Understand that weight is directly proportional to mass.

Specification Reference 4.5.1.3

Figure 1: Measuring the weight of an apple using a newtonmeter.

Calculating weight

Tip: When you see two letters written next to each other in an equation, it means they're multiplied together. So *mg* just means *m* × *g*.

You can calculate the weight of an object if you know its mass (*m*) and the strength of the gravitational field at that point (*g*):

$$W = mg$$

W = weight in N

m = mass in kg

g = gravitational field strength in N/kg

Tip: The ≈ sign means 'approximately equal to'.

For Earth, $g \approx 9.8$ N/kg and for the Moon it's around 1.6 N/kg. Increasing the mass of an object increases its weight. If you double the mass, the weight doubles too, so you can say that weight and mass are directly proportional.

Exam Tip
You'll always be given a value of *g* to use in the exam.

You can write this, using the direct proportionality symbol, as $W \propto m$.

Example 1

What is the weight, in newtons, of a 5 kg mass, both on Earth (g = 9.8 N/kg) and on the Moon (g = 1.6 N/kg)?

Just use the formula $W = mg$ in each case:

On Earth:

$$W = mg = 5 \times 9.8$$
$$= 49 \text{ N}$$

On the Moon:

$$W = mg = 5 \times 1.6$$
$$= 8 \text{ N}$$

Tip: The value of *g* is greater for Earth than the Moon because Earth has a stronger gravitational field.

Example 2

The value of g on Mars is 3.71 N/kg. What is the mass of a buggy if its weight on Mars is 4452 N?

Just rearrange the formula to make mass the subject, then plug in the correct numbers:

$$W = mg, \text{ so } m = W \div g = 4452 \div 3.71 = 1200 \text{ kg}$$

Tip: Remember — the mass of an object is always the same, no matter where it is.

Tip: You can use a formula triangle to help rearrange this equation:

There's more on formula triangles on page 338.

Practice Questions — Fact Recall

Q1   Define weight.

Q2   What is meant by the centre of mass of an object?

Q3   What apparatus is used to measure the weight of an object on Earth?

Practice Questions — Application

Q1   A rock on Earth has a mass of 15 kg. *g* on Earth = 9.8 N/kg.

     a) Calculate the weight of the rock on Earth.

     b) What would happen to the mass of the rock if it was moved to the surface of the Moon?

3. Resultant Forces

Resultant forces can help make it easier to work out what's going on in a complicated situation with lots of forces involved.

What is a resultant force?

The notion of a **resultant force** is a really important one for you to get your head round. In most real situations there are at least two forces acting on an object, but thinking about the resultant force can simplify all that.

If you have a number of forces acting at a single point, you can replace them with a single force which has the same effect as all the original forces acting all together. This single, overall force you get is called the resultant force.

Determining the resultant force

If the forces on an object all act along the same line (in the same or opposite directions), the resultant force is found by finding the sum of all the forces.

To do this, set a direction as being positive. Then add all the forces acting in that direction and subtract all the forces acting in the opposite direction.

Example

A vintage sports car is driving along with a driving force of 1000 N. Air resistance of 600 N is acting in the opposite direction. What is the resultant horizontal force on the car?

Do a sketch to visualise the problem. The driving force and the air resistance can be drawn as shown on the right. Weight and the normal contact force are also acting vertically on the car, but we can ignore them as we're only interested in horizontal forces.

air resistance
600 N

driving force
1000 N

Set the forwards direction as the positive, then add any forces in this direction and subtract any forces in the opposite direction.

So the resultant force = 1000 − 600 = 400 N (forwards).

You can use another diagram to show the resultant force acting on the car:

resultant force
400 N

Free body diagrams   Higher

You need to be able to describe all the forces acting on an isolated object or a system (p. 22) — i.e. every force acting on the object or system but none of the forces the object or system exerts on the rest of the world.

This can be shown on a **free body diagram**. Each force is represented by an arrow — the lengths of the arrows show the relative magnitudes of the forces and the directions of the arrows show the directions of the forces. You can use a free body diagram to help you figure out the resultant force on an object.

Learning Objectives:

- Know that the resultant force acting on an object has the same effect as all the individual forces acting on that object.

- Be able to calculate the resultant of two forces acting on an object along the same line.

- **H** Be able to give examples of the forces acting on a single object.

- **H** Be able to use free body diagrams to show the forces acting on an object and the resultant force.

- **H** Be able to use scale drawings to find the resultant force acting on an object and to determine whether an object is in equilibrium.

- **H** Be able to resolve a force into two components that have the same combined effect as the force.

Specification Reference 4.5.1.4

Figure 1: *A tug of war is an example of two forces acting in opposite directions. If the two forces have the same magnitude, the rope will not move.*

This free body diagram shows the forces acting on a person who's running.

Air resistance acts in the opposite direction to the forwards thrust.

The person's weight acts downwards and the normal contact force acts upwards.

The weight and normal contact arrows are the same length, so these two forces balance and the resultant vertical force is zero.

The thrust arrow is longer than the air resistance arrow, so there is a resultant force in the direction of motion (forwards).

normal contact force from ground

thrust

air resistance

weight

Tip: If the air resistance arrow was the same length as the thrust arrow, then they would balance too. There would be no resultant force and the person would be in equilibrium (see the next page for more on this).

Scale drawings Higher

You can also use scale drawings to find the resultant force acting on an object. First draw all the forces acting on an object 'tip-to-tail', making sure they're to scale and in the correct directions.

Then draw a straight line from the start of the first force to the end of the last force — this is the resultant force. Measure the length of the resultant force on the diagram to find the magnitude and measure the angle to find the direction of the force.

Tip: **H** Scale drawings are particularly useful when working with forces that are acting at different angles.

A man is on an electric bicycle that has a driving force of 4 N north. However, the wind produces a force of 3 N east. Find the magnitude and direction of the resultant force.

Start by doing a scale drawing of the forces acting, tip-to-tail. Make sure you choose a sensible scale (e.g. 1 cm = 1 N).

Draw the resultant from the tail of the first arrow to the tip of the last arrow. Measure the length of the resultant with a ruler and use the scale to find the force in N. Then use a protractor to measure the direction as a bearing.

Tip: **H** Drawing forces 'to scale' means each unit of length represents the same quantity. So in this example, each cm drawn represents 1 N of force.

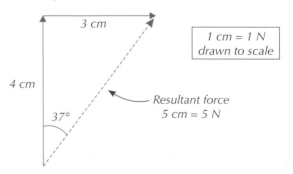

3 cm

4 cm

37°

1 cm = 1 N drawn to scale

Resultant force 5 cm = 5 N

The resultant force is 5 N on a bearing of 037°.

Tip: **H** A bearing is an angle measured clockwise from north, given as a 3 digit number, e.g. 10° = 010°.

Balanced forces **Higher**

If all of the forces acting on an object combine to give a resultant force of zero, the forces are balanced and the object is in **equilibrium**. On a scale diagram, this means that the tip of the last force you draw should end where the tail of the first force you drew begins. E.g. for three forces, the scale diagram will form a triangle.

Tip: H For an object in equilibrium, it can be useful to think of the forces 'cancelling each other out'.

Example — Higher

The free body diagram for an object is shown on the right.

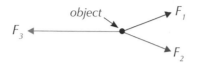

If all the forces acting on the object are drawn tip-to-tail, a complete loop is formed.
This shows that the object is in equilibrium.

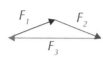

You might be given the forces acting on an object and told to find a missing force, given that the object is in equilibrium. To do this, draw out the forces you do know (to scale and tip-to-tail), then join the end of the last force to the start of the first force. This line is the missing force, so you can measure its size and direction.

Tip: H If the object wasn't in equilibrium, the forces wouldn't join up at the end to form a complete loop.

Tip: H When trying to find a missing force for an object in equilibrium, make sure you draw it in the correct direction. All the arrows should point in the same direction around the loop — unlike when you're trying to find the resultant.

Resolving forces **Higher**

Not all forces act horizontally or vertically — some act at awkward angles. To make these easier to deal with, they can be split into two "components" at right angles to each other (usually horizontal and vertical). Acting together, these components have the same effect as the single force.

You can resolve a force (split it into components) by drawing it on a square grid. Draw the force to scale, and then add the horizontal and vertical components using the grid lines. Then you can just measure them.

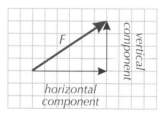

Figure 2: A force, F, split into horizontal and vertical components, drawn on a square grid.

Example — Higher

The scale diagram shows a toy car being pulled along horizontally by a string. The tension in the string has a magnitude of 2.5 N. Resolve the tension to find the magnitude of this force acting in the direction of the car's motion.

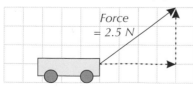

Measure the length of the black arrow — it's 2.5 cm.
The resultant force = 2.5 N, so the scale is 1 cm = 1 N.

The dotted arrows show the force resolved into two components. The car is moving horizontally, and the length of the horizontal component = 2 cm.
So the magnitude of the force acting in the direction of the car's motion = 2 N.

Tip: H Resolving forces is a bit like the reverse of finding the resultant. They sound similar, so make sure you don't get them mixed up.

Tip: H You need to find the scale of the diagram to work out the magnitude of the horizontal component.

Practice Questions — Fact Recall

Q1   What is meant by the resultant force acting on an object?

Q2   What is a free body diagram?

Q3   Describe how a scale diagram of all the forces acting on an object can be used to determine whether the object is in equilibrium.

Q4   What is meant by 'resolving a force'?

Practice Questions — Application

Q1   A bike is being pushed. The magnitude of the force pushing the bike forwards is equal to 87 N. A resistive force with a magnitude of 24 N is acting in the opposite direction. Find the magnitude and direction of the resultant of these forces.

Q2   A boat is being pulled by three tugboats. The boat also experiences a frictional force from the water. These forces are shown to scale on the diagram below. Using a scale diagram, work out whether the boat is in equilibrium.

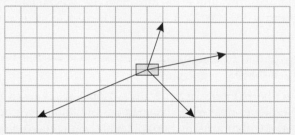

Tip: **H** Don't forget to include the weight and normal contact force in your free body diagram.

Q3   A box is being dragged along a table. The direction of the force that is pulling the box is parallel to the surface of the table. Draw a free body diagram to show all the forces acting on the box. (You can assume there is no friction between the box and the table.)

Q4   A train is being pulled along a track (shown in blue) with a force of 20 N. A frictional force and a reaction force from the side of the track act on the train. A scale diagram of the train, as viewed from above, is shown below. By resolving forces, find the resultant force acting on the train in its direction of motion.

Tip: **H** When finding the resultant force in the direction of motion, you don't need to take into account any components that are at right angles to the direction of motion.

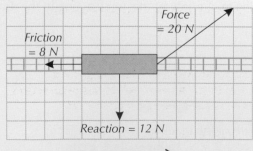

4. Work Done

If a force moves an object from one position to another, energy must be transferred from one energy store to another.

What is work done?

In physics '**work done**' has a specific meaning, and you need to know what that is.

> Work done is the energy transferred when a force moves an object through a distance.

Luckily, it's not as complicated as that statement makes it sound.

Try this:

1. Whenever something moves, something else might be providing some sort of 'effort' to move it (see Figure 1 on the next page).

2. The thing putting the effort in needs a source of energy (like fuel or food etc.).

3. It then does 'work' by moving the object — energy is transferred from one store to another.

4. Whether this energy is transferred 'usefully' (e.g. by lifting a load) or is 'wasted' (e.g. transferred to the thermal energy stores of the surroundings), you can still say that 'work is done'.

So work done is simply the energy transferred when a force acts on an object. You can work out how much work is done using the following formula:

W = work done (J) ⟶ $W = Fs$ ⟵ F = force (N)
s = distance (m)

One joule of work is done when a force of one newton causes an object to move a distance of one metre. So one joule is equal to one newton metre: $1\text{ J} = 1\text{ Nm}$. You need to be able to convert between J and Nm.

Example

Some kids drag a tractor tyre 5 m over the ground. They pull with a resultant force of 340 N in the direction of motion. Find the work done to move the tyre.

Force = 340 N   Distance = 5 m

Then put the numbers into the formula for work done:

$W = Fs = 340 \times 5 = 1700\text{ J}$

Learning Objectives:

- Know that if a force is applied to an object and displaces it, the force does work on the object.
- Know the equation that relates work done, force and distance moved.
- Understand that 1 J of work is done when a force of 1 N moves an object 1 m and be able to convert between J and Nm.
- Be able to describe the energy transfers that are involved when work is done on an object.
- Understand that work done against friction can cause a rise in the temperature of an object and its surroundings.

Specification Reference 4.5.2

Tip: There's more about energy being transferred and work done on pages 22-24.

Tip: The distance, *s*, in the equation is the distance moved in the direction of the force, *F*.

Tip: The formula triangle for this equation is:

Energy transfers

Remember, 'work done' and 'energy transferred' are one and the same. (And they're both given in joules.)

Figure 2: A wheelchair user pushing the wheels of her chair. She does work against friction by transferring energy from her chemical energy store to the kinetic energy store of the chair.

Energy is transferred from the chemical energy store of the person to their kinetic energy store.

Energy is transferred to thermal energy stores and to the kinetic energy store of the broom, causing it to move through a distance, so work is done.

Figure 1: *A diagram showing how work is done when a person supplies energy to a broom and sweeps.*

When you push something along a rough surface (like a carpet) you are doing work against frictional forces. Energy is transferred to the kinetic energy store of the object because it starts moving, but some is also transferred to the thermal energy stores of the object, the surface and the surroundings due to friction (p. 190). This causes the overall temperature of the object and surface to increase (like rubbing your hands together to warm them up).

Example

A brick is pushed along rough ground with a total force of 45 N. The brick moves a distance of 1.4 m. Find the total energy transferred.

Force = 45 N   Distance = 1.4 m

Energy transferred is the same as work done, so just use the equation:

$$W = Fs = 45 \times 1.4 = 63 \text{ J}$$

So the energy transferred is 63 J

Tip: Here, you're finding the total energy transferred. Some of this is transferred usefully to the brick's kinetic energy stores. The rest is transferred to thermal energy stores due to work done against friction.

Practice Questions — Fact Recall

Q1   Define work done.

Q2   What is 1 joule in newton metres?

Q3   Explain why doing work against friction can cause a rise in temperature.

Practice Questions — Application

Q1   A child pulls a sledge across snow for 14 m. The resultant force applied over this distance is 24 N. Calculate the work done to move the sledge.

Q2   a) A bike is pushed 20 m using a steady force of 250 N in the direction of motion. How much energy is transferred?

   b) If the bike continues to be pushed with the same force, calculate how far the bike will move if 750 J of work is done.

Tip: Remember — work done and energy transferred are the same thing.

Topic 5a Checklist — Make sure you know...

Contact and Non-Contact Forces

- ☐ That scalar quantities only have magnitude but vector quantities have magnitude and direction.
- ☐ That when an arrow is used to represent a vector, the size of the arrow indicates the magnitude of the vector and the direction of the arrow shows the direction of the vector.
- ☐ That a force on an object is defined as the push or pull acting on the object and is a vector.
- ☐ That a contact force between two interacting objects is a result of them touching and a non-contact force between two interacting objects can occur when they are not touching.

Weight, Mass and Gravity

- ☐ That gravity is responsible for giving objects weight.
- ☐ That the force due to gravity at the Earth's surface is a result of the Earth's gravitational field.
- ☐ That gravitational field strength at a point is responsible for an object's weight at that point.
- ☐ That a calibrated spring balance is used for measuring weight.
- ☐ That the centre of mass of an object is the point at which the object's weight appears to act.
- ☐ The equation weight = mass × gravitational field strength.

Resultant Forces

- ☐ That the resultant force acting on an object is the single force that has the same effect as the original forces all acting together.
- ☐ How to calculate the resultant of two forces acting on an object along the same line.
- ☐ **H** How to describe all of the forces acting on a single object.
- ☐ **H** That a free body diagram of an object shows all the forces acting on the object.
- ☐ **H** How scale drawings can be used to find the resultant force acting on an object.
- ☐ **H** How free body diagrams and scale drawings can be used to show that all the forces acting on an object are balanced and so the object is in equilibrium.
- ☐ **H** How to resolve a force into two components that are at right angles to each other.

Work Done

- ☐ That work is done on an object when a force displaces it and that it can be calculated using work done = force × distance (moved along the line of action of the force), where work done is measured in joules or newton metres.
- ☐ That if a force of 1 N is applied to an object over a distance of 1 m, then 1 J of work is done.
- ☐ How to convert between J (joules) and Nm (newton metres).
- ☐ How to describe the energy transfers for a situation in which work is done on an object.
- ☐ That if work is done against friction then energy can be transferred to thermal energy stores, causing an increase in temperature of the object and surroundings.

Exam-style Questions

1   A 1.8 kg book is resting on a table. The gravitational field strength = 9.8 N/kg.

1.1   Calculate the weight of the book. Give your answer to 2 significant figures.

(2 marks)

1.2   A 5 N force is applied to the book so that the book slides across the table.
A 1.4 N frictional force acts on the book in the opposite direction.
These two forces are shown in the diagram below. The diagram is not to scale.

Calculate the magnitude of the resultant force acting on the book.

(1 mark)

1.3   The work done by the resultant force to move the book across the table is 19.8 J.
Calculate the distance travelled by the book.

(3 marks)

2   A canal boat is towed along by two horses on the bank. Each horse is pulling the
canal boat with a different force. The diagram shown is to scale.

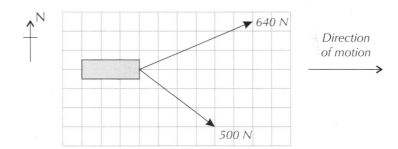

2.1   Are the horses pulling the canal boat with a contact force or a non-contact force?

(1 mark)

2.2   Calculate the total force provided by the horses in the direction of the canal boat's
motion.

(4 marks)

2.3   There is a drag force on the boat acting in the opposite direction to its motion.
If the resultant force acting on the boat in the direction of its motion is 800 N,
calculate the magnitude of the drag force.

(1 mark)

1. Forces and Elasticity

Applying forces to some objects can cause them to stretch. Stretching an object can either be a permanent change or a temporary change.

Elastic and inelastic deformation

When you apply a force to an object you may cause it to stretch, compress or bend. To change the shape of an object in this way, you need more than one force acting on the object, in different directions (otherwise the object would simply move in the direction of the applied force, instead of changing shape).

An object has been **elastically deformed** if it can go back to its original shape and length after the force has been removed. An object has been **inelastically deformed** if it doesn't return to its original shape and length after the force has been removed (i.e. it's been permanently deformed). Objects which can be elastically deformed are called **elastic objects** (e.g. a spring).

Examples

If a spring is supported at the top and a weight is attached to the bottom, it stretches.

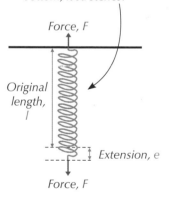

A spring can also be compressed or bent by applying forces at different points.

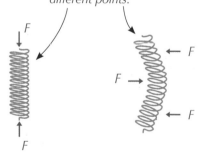

Figure 1: *A spring can be stretched, compressed or bent when more than one force is acting on it. The extension, e, of a stretched spring is the difference between its stretched length and its original length (i.e. with no force applied).*

Force and extension

The extension of a stretched spring (or other elastic object) is directly proportional to the load or force applied — so $F \propto e$ (see p. 337). This is true as long as you don't stretch it too far — more on this shortly.

Learning Objectives:

- Know that in order to stretch, bend or compress an object, at least two forces must be applied to it.
- Know what is meant by elastically deformed and inelastically deformed.
- Know that the force applied to an elastic object is directly proportional to the extension of the object, up to the limit of proportionality.
- Know and be able to use the equation $F = ke$, where k is the spring constant and e is the extension or compression of the object.
- Understand that work is done on a spring when it is stretched or compressed.
- Understand that for elastic deformations, all work done on a spring to stretch or compress it is transferred to its elastic potential energy store.
- Be able to calculate the work done in stretching (or compressing) a spring using the equation for the energy in its elastic potential energy store.

Specification Reference 4.5.3

Figure 2: *A bungee jumper leaping from a bridge. The bungee cord is an elastic object which stretches due to the person's weight. Look back at page 149 for a reminder on weight.*

This is the equation:

$$F = ke$$

F = force (N) ⟶ ⟵ *k* = spring constant (N/m)

⟵ *e* = extension (m)

The spring constant depends on the material that you are stretching — a stiffer spring has a greater spring constant.

The equation also works for compression (where *e* is just the difference between the natural and compressed lengths — the compression).

Example 1

When no force is applied, a spring has a length of 23.2 cm. When a lead ball is suspended from it, the spring extends to a length of 25.1 cm. If the spring constant *k* = 60 N/m, calculate the weight of the lead ball.

Start by finding the extension of the spring after the lead ball is attached:

$$e = 25.1 - 23.2 = 1.9 \text{ cm}$$

Be careful with units here — the formula uses extension in metres, so make sure you convert any numbers first:

$$1.9 \text{ cm} = 1.9 \div 100 \text{ m} = 0.019 \text{ m}$$

Then put the numbers into the equation for force:

$$F = ke = 60 \times 0.019 = 1.14 \text{ N}$$

Tip: There's more on converting between units on pages 18-19.

Example 2

A 12 N force is used to compress a spring with a spring constant of 96 N/kg. Calculate the compression of the spring.

Rearrange the formula to make *e* the subject:

$$F = ke, \text{ so } e = F \div k$$

Then put the right numbers in:

$$e = 12 \div 96 = 0.125 \text{ m}$$

Tip: You might find rearranging equations easier if you use a formula triangle:

The limit of proportionality

There's a limit to the amount of force you can apply to an object for the extension to keep on increasing proportionally. This limit is known as the **limit of proportionality**. Figure 3 shows a graph of force against extension for an object. The limit of proportionality is the point at which the graph starts to curve. It's marked as point P on the graph in Figure 3.

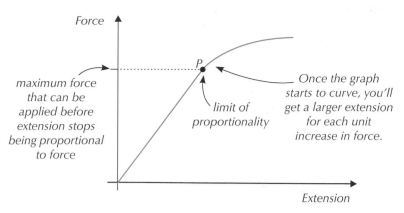

maximum force that can be applied before extension stops being proportional to force

limit of proportionality

Once the graph starts to curve, you'll get a larger extension for each unit increase in force.

Figure 3: *A graph showing the force applied to an object against its extension. The force is proportional to extension up to the limit of proportionality.*

Tip: The graph below point P shows a linear relationship, and the graph above point P shows a non-linear relationship. See pages 339-341 for more on linear and non-linear graphs.

Elastic potential energy store

Work is done when a force stretches or compresses an object and causes energy to be transferred. If the object is elastically deformed, ALL this energy is transferred to the object's elastic potential energy store.

As long as a spring is not stretched past its limit of proportionality, the work done in stretching or compressing the spring can be found using:

Tip: You may also see extension-force graphs (similar, but with the axes swapped around). The graph will curve upwards instead:

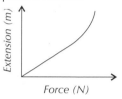

E_e = elastic potential energy (J)

$$E_e = \tfrac{1}{2}ke^2$$

k = spring constant (N/m)

e = extension (m)

Tip: Take a look at page 22 for more on energy stores.

For elastic deformation, this formula can be used to calculate the energy stored in a spring's elastic potential energy store. It's also the energy transferred to the spring as it's deformed (or transferred by the spring as it returns to its original shape).

The work done, or energy stored in the elastic potential energy store, for a particular force (or extension) can also be found by calculating the area under the linear force-extension graph up to that force (or extension).

Tip: If a stretched spring is released, the energy stored in its elastic potential energy store will be transferred to its kinetic energy store as it springs back to its original size and shape.

Tip: See page 341 for more on the area under a graph.

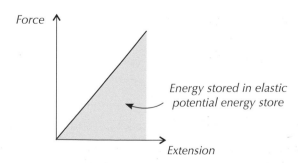

Energy stored in elastic potential energy store

Figure 4: *The area under the force-extension graph of an elastic object is equal to the energy stored in the elastic potential energy store of the object.*

Tip: If an object is deformed inelastically, some of the work done will transfer energy into other energy stores (e.g. the thermal energy store of the object), not just the elastic potential energy store.

Example

A spring has a spring constant of 1.2 N/m. Assuming the spring deforms elastically, calculate the total energy transferred to its elastic potential energy store when it is extended by 0.20 m.

Substitute the numbers into the equation:

$$E_e = \tfrac{1}{2}ke^2 = \tfrac{1}{2} \times 1.2 \times 0.20^2 = 0.024 \text{ J}$$

Tip: Don't forget to square the extension when calculating the energy transferred to the elastic potential energy store.

Practice Questions — Fact Recall

Q1   Explain why more than one force needs to be applied to a spring in order to stretch it.

Q2   What is an elastic object?

Q3   Give the equation that relates the force applied to a spring, the spring constant of the spring and the extension of the spring. Give the units of each term.

Q4   What is the limit of proportionality?

Q5   A spring is elastically deformed. How much of the energy transferred to the spring is released when the spring is released?
Explain your answer.

Practice Questions — Application

Q1   A rubber band has a spring constant of 34 N/m. Calculate the force required to stretch the rubber band from 0.50 m to 0.75 m.

Q2   The graph shows a force-extension graph for a spring. Which point marks the spring's limit of proportionality?

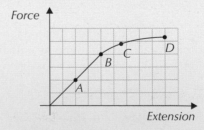

Q3   a) Spring A is compressed from 16 cm to 12 cm when a force of 0.80 N is applied to it. Calculate the spring constant of spring A.

b) Calculate the work done on spring A as it is compressed.

c) The same force is then applied to spring B. It compresses spring B by a smaller amount than it compressed spring A. Is the spring constant of spring B higher than, lower than or equal to the spring constant of spring A?

Exam Tip
Remember to make sure all your values are in the right units before you substitute them into an equation. For example, extension, e, should be in metres for Q3.

2. Investigating Springs

You can investigate the behaviour of a spring using a simple experiment. From the experiment, you can find the limit of proportionality of the spring and its spring constant.

Apparatus and setup

Set up the apparatus as shown in Figure 1.

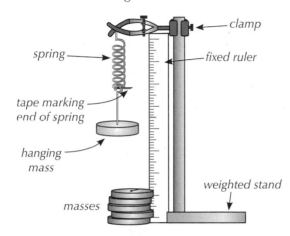

Figure 1: Experimental setup used to investigate a spring.

You could do a quick pilot experiment first to check your masses are a suitable size. Using the spring you'll be testing (or an identical one), load it with masses one at a time, up to a total of five masses. Using the ruler, check that the spring extends by the same amount each time. If adding one of these masses causes the spring to stretch by more than the previous ones, you have gone beyond the spring's limit of proportionality. If this happens, you'll need to use smaller masses. Otherwise, you won't end up with enough values to plot your graph later on.

If the spring does go past its limit of proportionality during the pilot experiment, it could start to deform inelastically and so you'll have to replace it with another (identical) spring in the real thing.

Carrying out the experiment

Make sure you have plenty of masses and calculate their weights (the force that will be applied to the spring) using $W = mg$ (p. 150).

Using the ruler, measure the natural length of the spring (the length when no hanging mass is attached). Make sure you take the reading at eye level and use a marker (e.g. a thin strip of tape, as shown in Figure 1) to make the reading more accurate.

Add a mass to the spring and allow the spring to come to rest. Measure the spring's new length. Record the weight added and work out the extension (the change in length). Repeat this process, recording the total weight attached and calculating the total extension (total length minus natural length) each time, until you have enough measurements (no fewer than 6).

Learning Objectives:

- Be able to investigate the relationship between the force applied to a spring and its extension (Required Practical 6).
- Be able to find the spring constant from a force-extension graph for a linear force-extension relationship.
- Know the difference between linear and non-linear force-extension relationships.

Specification Reference 4.5.3

Tip: You must make sure the ruler and spring are vertical to make sure your measurements are as accurate as possible.

Tip: It's okay to go past the limit of proportionality when you're doing the real thing, but you need to make sure you've recorded enough measurements to plot the linear part of the graph beforehand.

Tip: You should wear safety goggles when carrying out this experiment to protect your eyes in case the spring snaps. Make sure you do a risk assessment to identify any other hazards before you start.

Tip: You first saw force-extension graphs on page 161.

Tip: You could also plot the force on the x-axis and the extension on the y-axis. You would still get an initial straight line graph, but then it would start to curve upwards:

You can still find the spring constant from the linear part of this graph — it's equal to $\frac{1}{\text{gradient}}$.

Tip: Once you've got your graph, you could use it to find the weight of an object. Hang the object from the spring, measure the extension, then use your graph to find the corresponding weight (force). This will only work if you didn't deform the spring inelastically during the experiment and the object's weight is within the range you plotted.

Tip: To check whether the deformation is elastic or inelastic, you can remove each mass temporarily and check whether the spring goes back to the previous extension.

Analysing the results

Plot a force-extension graph of your results and draw a line of best fit. For each measurement, the force you should plot is the total weight of the masses attached to the spring. The extension is the difference between the spring's length with that total weight attached and its natural length. Take a look at pages 16-17 for more information on how to plot graphs.

The graph will only start to curve if you exceed the limit of proportionality when you're adding the masses. Don't worry if yours doesn't start to curve — as long as you've got the straight line bit, you'll be able to find the spring constant. But you will need the start of the curve if you want to find the limit of proportionality. Figure 2 shows how your results might look.

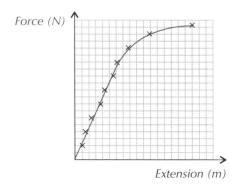

Figure 2: *Graph of force against extension for a spring.*

When the line of best fit is a straight line, it means there is a linear relationship between force and extension (they're directly proportional, see page 159).

You know that $F = ke$, and the gradient of the straight line is:

$$\frac{\text{change in } y}{\text{change in } x} = \frac{F}{e} = k$$

So to find the spring constant, you just need to work out the gradient of the straight line.

When the line begins to bend, the relationship between force and extension is non-linear — the spring stretches more for each unit increase in force. So the point where the line starts to curve is the spring's limit of proportionality.

Practice Questions — Fact Recall

Q1 When carrying out an experiment to find the spring constant of a spring, explain why it is a good idea to carry out a pilot experiment first on an identical spring.

Q2 Give a safety procedure that should be carried out when doing an experiment to find the spring constant of a spring.

Q3 How is the spring constant of a spring found from a force-extension graph?

3. Moments

If you apply a force to an object suspended from a point, like the hands on a clock, it will start turning around the suspension or pivot point. This turning effect is important for understanding how levers and gears work.

What is a moment?

A **moment** is the turning effect of a force. A moment is given by the force multiplied by the perpendicular distance from the **line of action** of the force to the pivot. The line of action of a force, *F*, is a straight line passing through the point at which *F* acts and in the same direction as *F* (see Figure 1).

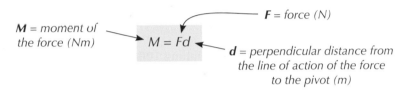

M = moment of the force (Nm)

$M = Fd$

F = force (N)

d = perpendicular distance from the line of action of the force to the pivot (m)

The perpendicular distance means the distance along a line that makes a right angle with the line of action of the force — see Figure 1.

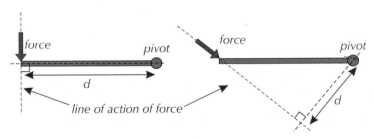

Figure 1: *The same force applied to the same pivoted object at two different angles. Changing the angle at which the force is applied changes the value of d, the perpendicular distance from the force's line of action to the pivot.*

Learning Objectives:
- Know that a force can cause a rotation, and be able to describe examples.
- Know that a moment is the turning effect of a force.
- Know and be able to use the equation $M = Fd$, where d is the perpendicular distance from the line of action of the force to the pivot.
- Know that an object will be balanced if the sum of all clockwise moments acting on it is equal to the sum of all anticlockwise moments acting on it.
- Be able to find missing forces or distances for an object in equilibrium.
- Know and understand how levers and gears are used to transmit the rotational effect of a force.

Specification Reference 4.5.4

Examples

The force on this spanner causes a turning effect or moment on the nut (which acts as pivot). A larger force would mean a larger moment.

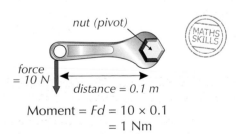

nut (pivot)

force = 10 N

distance = 0.1 m

Moment = Fd = 10 × 0.1
= 1 Nm

Using a longer spanner, the same force can exert a larger moment because the distance from the pivot is greater.

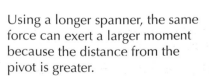

force = 10 N

distance = 0.2 m

Moment = Fd = 10 × 0.2
= 2 Nm

Tip: The distance *d* in these examples is measured to the centre of the nut/pivot.

Figure 2: A spanner being used to tighten a nut on a construction site. The spanner acts as a force multiplier, ensuring the nut can be screwed in as tightly as possible.

Tip: To work out which moments are clockwise and which are anticlockwise, think about which direction the force would cause the object to turn if it was the only force acting.

Tip: The centre of mass is the point at which you can assume the girder's weight acts. See page 149 for more on this.

To get the maximum moment (or turning effect) you need to push at right angles (perpendicular) to the spanner.

Pushing at any other angle means a smaller moment because the perpendicular distance between the line of action and the pivot is smaller.

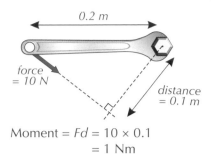

$$\text{Moment} = Fd = 10 \times 0.1$$
$$= 1 \text{ Nm}$$

If the total anticlockwise moment equals the total clockwise moment about a pivot, the object is balanced and won't turn. You can use the equation from the previous page to find a missing force or distance in these situations.

Example

A uniform 6 m long steel girder weighing 1000 N rests horizontally on a pole 1 m from one end. What is the tension in a supporting cable attached vertically to the other end?

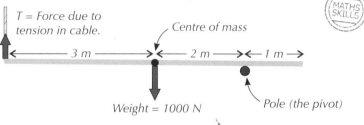

Take moments about the pole:
Anticlockwise: force = weight = 1000 N, distance = 2 m.
So anticlockwise moment = Fd = 1000 × 2 = 2000 Nm.

Clockwise: force = tension = T, distance = 3 + 2 = 5 m.
So clockwise moment = Fd = T × 5.

For the girder to balance, the total anticlockwise moment should equal the total clockwise moment. So: 2000 = T × 5

Rearrange for tension: T = 2000 ÷ 5 = 400 N

Levers

Levers make it easier for us to do work (e.g. lift an object). The moment is equal to force × perpendicular distance from the pivot. So the amount of force needed to produce a particular moment depends on the distance the force is applied from the pivot.

Figure 3: Using a screwdriver to prise open a tin of paint is an example of a lever. The pivot is the lip of the tin.

Levers increase the distance from the pivot at which the force is applied — so this means less force is needed to get the same moment.

That's why levers are known as force multipliers — they reduce the amount of force that's needed to get the same moment by increasing the distance.

In lever systems, the load is placed a small distance from the pivot to reduce the moment/force needed to lift it.

- Using long sticks or bars makes it easier to lift a heavy object off the ground.

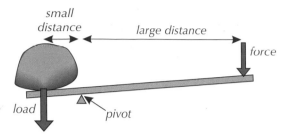

- Wheelbarrows let you move a load around that would normally be too heavy to lift with just your arms.

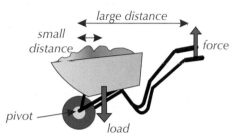

- Scissors let you cut through objects that would be too tough to get through with just a knife.

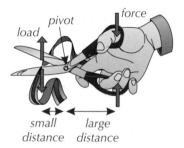

Gears

Gears are circular discs with 'teeth' around their edges. Their teeth interlock so that turning one causes another to turn in the opposite direction. They are used to transmit the rotational effect of a force from one place to another.

Different sized gears can be used to change the moment of a force. A force transmitted to a larger gear will cause a bigger moment, as the distance from the edge to the pivot is greater. A larger gear will turn slower than a smaller gear. Bicycles use gears to transmit the turning effect of the pedals to the back wheel — you can change gear to alter the ratio between how fast you pedal and how fast the wheels turn.

Figure 4: Gears on a bicycle. They're connected by a chain, rather than being in contact with each other.

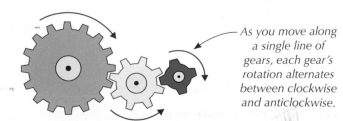

As you move along a single line of gears, each gear's rotation alternates between clockwise and anticlockwise.

Figure 5: Three linked gears all turn together when one gear is turned.

Practice Questions — Fact Recall

Q1 What is the moment of a force?

Q2 Give the equation for calculating the moment of a force, including the units of any variables.

Q3 What condition must be satisfied in order for an object with a pivot to be balanced?

Q4 How does a lever work as a force multiplier?

Q5 Describe what a gear is and what function it serves in machinery.

Practice Questions — Application

Q1 In the setup of gears below, which gears will turn clockwise if gear A is turned anticlockwise?

Tip: It can help to imagine that the gears are turning with this type of question.

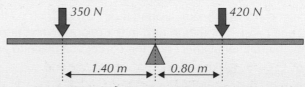

Q2 A plank of wood of length 2.70 m has a pivot at one end. If a force of 60.0 N is applied to its other end (perpendicular to the plank of wood), calculate the moment of the force about the pivot.

Q3 Two children with weights of 350 N and 420 N are sat on a seesaw as shown in the diagram. The pivot is at the seesaw's centre of mass.

350 N   *420 N*

1.40 m   *0.80 m*

Tip: For Q3, think about whether a clockwise or anticlockwise moment is needed to make the seesaw balance.

Where should a third child, with a weight of 308 N, sit in order to balance the seesaw?

Q4 The diagram shows a lever being used to lift a large load. Give two changes that could be made in order to reduce the force needed to lift the load.

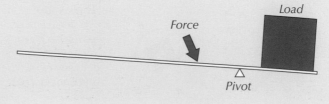

Load

Force

Pivot

4. Fluid Pressure

You might not be able to feel it, but the surrounding air is always exerting a force on you. That's because air is a fluid and has a pressure.

What is pressure?

If a force is applied to a surface, then there is a **pressure** on the surface. Pressure is defined as the force exerted per unit area.

Fluids are substances that can 'flow' because their particles are able to move around. A fluid is either a liquid or a gas. As these particles move around, they collide with surfaces and other particles.

Particles have mass, and so when particles in a fluid collide with a surface, they exert a force on it. As pressure is force per unit area, this means a pressure is exerted on the surface by the fluid. The pressure, and so the force, is exerted normal (at right angles) to the surface.

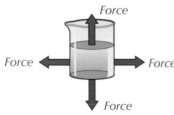

Figure 1: Water in a beaker will exert a force normal to all its surfaces, including the surface in contact with the air.

You can calculate the pressure at the surface of a fluid by using:

$$\boldsymbol{p} = \text{pressure (Pa)} \longrightarrow p = \frac{F}{A} \begin{cases} \longleftarrow \boldsymbol{F} = \text{force normal to a surface (N)} \\ \longleftarrow \boldsymbol{A} = \text{area of that surface (m}^2\text{)} \end{cases}$$

This equation gives the pressure caused by any force which is applied at right angles to an area, not just in a fluid. The unit for pressure is the pascal, Pa.

Example

Two tubes filled with gas, each with a piston at one end, are shown below. In each tube, the gas exerts a force of 1.6 N on the piston. The area of each piston is different. Calculate the pressure exerted by the gas on each piston.

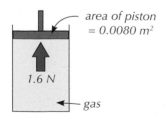

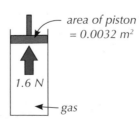

$p = \frac{F}{A} = \frac{1.6}{0.0080} = 200 \text{ Pa}$ $p = \frac{F}{A} = \frac{1.6}{0.0032} = 500 \text{ Pa}$

Learning Objectives:
- Know that a fluid is either a liquid or a gas.
- Know that the force due to the pressure of a fluid acts normal (at right angles) to a surface.
- Know and be able to use the equation that relates pressure, force and area.
- **H** Know how and why the pressure of a liquid is dependent on density and depth.
- **H** Be able to calculate the pressure due to a column of liquid, using the equation $p = h\rho g$.
- **H** Be able to find the difference in pressure between two points in a liquid.
- Know what the Earth's atmosphere is and that its density decreases with altitude.
- Know what atmospheric pressure is and what it is caused by.
- Know how and why atmospheric pressure changes with altitude.

Specification References
4.5.5.1.1, 4.5.5.1.2, 4.5.5.2

Tip: Particles colliding to cause pressure in gases is covered in more detail on pages 114-115.

Tip: So a force will exert a greater pressure if it's concentrated over a smaller area.

Pressure in a liquid Higher

Density is a measure of how close together the particles in a substance are (p. 106). For a given liquid, the density is uniform (the same everywhere) and it doesn't vary with shape or size.

A denser liquid will have more particles in a given volume. This means there are more particles that are able to collide — which means more collisions, a higher total force exerted and so a higher pressure.

Another factor that can affect the pressure exerted by a liquid is depth. As depth increases, the number of particles above that point increases. The weight of these particles adds to the pressure experienced at that point, so liquid pressure increases with depth.

You can calculate the pressure at a certain depth due to the column of liquid above that point using:

h = height of the column of liquid above that point (i.e. the depth) (m)

p = pressure due to column of liquid (Pa) → $p = h\rho g$ ← *g* = gravitational field strength (N/kg)

ρ = density of the liquid (kg/m³) (the symbol is the Greek letter 'rho')

> **Example** **Higher**
>
> **Calculate the change in pressure between a point 20 m below the surface of water and a point 40 m below the surface. The density of water is 1000 kg/m³. Take the gravitational field strength of the Earth to be 9.8 N/kg.**
>
>
>
> Calculate the pressure caused by the water at a depth of 20 m:
> $p = h\rho g = 20 \times 1000 \times 9.8 = 196\,000$ Pa
>
> Do the same for a depth of 40 m:
> $p = h\rho g = 40 \times 1000 \times 9.8 = 392\,000$ Pa
>
> Take away the pressure at 20 m from the pressure at 40 m:
> $392\,000 - 196\,000 = 196\,000$ Pa $= 1.96 \times 10^5$ Pa (or 196 kPa)
>
> See page 333 for more on standard form.

You can show that the pressure of a liquid increases with depth using a tube with equally-sized holes cut into the side of it (see Figure 3). When the tube is filled with water, the pressure is greatest at the bottom (i.e. at the deepest point), and so the water spurting out of the hole at the bottom travels faster than the water spurting out of the hole at the top.

Figure 3: Water spurting out of holes in a tube at different speeds due to differences in pressure at different depths.

Tip: H Unlike a liquid, the density of a gas can vary (see the next page).

Tip: H For example, water has a density of 1000 kg/m³ and mercury has a density of 13 500 kg/m³. So the pressure exerted by mercury is higher as it has a higher density.

Tip: H Be careful — this equation only gives you the pressure due to the column of liquid above a given depth. You'll mainly use it to find the difference in pressure between two different depths.

Figure 2: Divers can only dive to a certain depth before the pressure gets too high and is dangerous.

Tip: H The higher pressure at the bottom of the tube means the force acting on the sides of the tube is higher at the bottom (per unit area).

Atmospheric pressure

The **atmosphere** is a layer of air that surrounds Earth. It is thin compared to the size of the Earth. **Atmospheric pressure** is created on a surface by air molecules colliding with the surface. Figure 4 shows that as altitude (height above Earth) increases, atmospheric pressure decreases. This is due to:

1. Density (i.e. how close together the molecules of the atmosphere are). As altitude increases, the atmosphere gets less dense. This means there are fewer air molecules that are able to collide with a surface, which in turn means a lower atmospheric pressure.

2. How much air there is above a certain point. An increasing altitude means fewer air molecules above a surface. This means that the weight of the air above, which contributes to atmospheric pressure, decreases.

Tip: People need to take oxygen tanks with them when climbing high mountains — the density of air decreases with height, so there is less air at the top of the mountain which makes it harder to breathe.

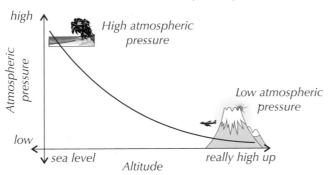

Figure 4: *A graph to show how the atmospheric pressure varies with the height above Earth.*

Practice Questions — Fact Recall

Q1 What is pressure?

Q2 What is a fluid?

Q3 The pressure of a fluid exerts a force at the surfaces of the fluid. In what direction does this force act?

Q4 Give the equation that relates pressure, force and area. Give the units of each variable.

Q5 Explain how and why the pressure exerted by a liquid changes with increasing depth.

Q6 What is the atmosphere?

Q7 What happens to atmospheric pressure as altitude increases and why?

Practice Questions — Application

Tip: Take a look at pages 18-19 for how to convert between different units.

Q1 A balloon that is full of water has a surface area of 320.0 cm². The water inside the balloon has a pressure of 101 000 Pa. Calculate the total force that the water exerts on the balloon.

Q2 A diver swims from the surface of the sea to a depth of 5.0 m. Calculate the change in pressure that the diver experiences. The density of the seawater is 1020 kg/m³ and the gravitational field strength can be assumed to be 9.8 N/kg.

Exam Tip H
The equation needed for Q2 will be given on your equation sheet in the exam (see page 394).

Learning Objectives:

- **H** Understand that upthrust is the resultant force acting on a submerged object due to the greater pressure at the bottom of the object than the top.
- **H** Understand that an object will float if its weight is equal to the upthrust and sink if its weight is greater than the upthrust.
- **H** Understand the factors affecting whether an object floats or sinks.

Specification Reference 4.5.5.1.2

5. Upthrust Higher

The reason that you feel a lot lighter in a swimming pool than on solid ground is all to do with upthrust. It's the force that pushes up on you and keeps you afloat as you're bobbing along.

What is upthrust?

When an object is submerged in a fluid (either partially or completely), the pressure of the fluid exerts a force on it from every direction.

As pressure increases with depth (see page 170), the force exerted on the bottom of the object is larger than the force acting on the top of the object (because the bottom of the object is deeper in the fluid than the top of it). This causes a resultant force (p. 151) upwards, known as **upthrust**. The upthrust is equal to the weight of fluid that has been displaced (pushed out of the way) by the object.

Example Higher

The upthrust on a pineapple in water is equal to the weight of a pineapple-shaped volume of water.

Pressure

Pineapple displaces this much water.

Upthrust is equal to the weight of this amount of water.

Pressure

Tip: Take a look back at page 169 for a reminder on how pressure relates to force.

Tip: Take a look at page 106 for more on density.

Floating and sinking

If the upthrust on an object is equal to the object's weight, then the forces balance and the object floats. If an object's weight is more than the upthrust, the object sinks. Whether or not an object floats depends on its density:

- An object that is less dense than the fluid it is placed in weighs less than the equivalent volume of fluid. This means it displaces a volume of fluid that is equal to its weight before it can become completely submerged. At this point, the upthrust is equal to the object's weight, so the object floats.

- An object that is denser than the fluid it is placed in weighs more than the equivalent volume of fluid. This means it is unable to displace enough fluid to equal its weight (because it will always weigh more than the fluid it displaces). This means its weight is larger than the upthrust, so it sinks.

- An apple will float in water.

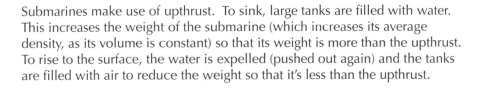

This much water weighs the same as the whole apple (because the apple is less dense than water).

The apple has displaced a volume of water equal to its weight, so it floats.

- A potato will sink in water.

This much water weighs less than a potato (because the potato is denser than water).

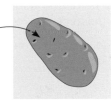

The potato can never displace a volume of water equal to its weight so it sinks.

Figure 1: *Rubber ducks float really well on water as they're mostly made of air, and air has a much lower density than water.*

Tip: **H** Saltwater has a higher density than water, so a potato will float in the water if enough salt is added.

Submarines make use of upthrust. To sink, large tanks are filled with water. This increases the weight of the submarine (which increases its average density, as its volume is constant) so that its weight is more than the upthrust. To rise to the surface, the water is expelled (pushed out again) and the tanks are filled with air to reduce the weight so that it's less than the upthrust.

Figure 2: *A submarine changes its weight in order to move up and down in the water.*

Practice Questions — Fact Recall

Q1 Explain why an object will experience an upthrust in a liquid.

Q2 Will an object float or sink if its weight is equal to the upthrust acting on it?

Q3 Will an object float or sink if its weight is greater than the upthrust acting on it?

Q4 Explain why an object with a higher density than water will sink when it is put in water.

Practice Question — Application

Q1 a) An object with a volume of 0.50 m³ and a uniform density of 1500 kg/m³ is put in water. The density of the water is 1000 kg/m³. Will the object float or sink?

b) The object is then cut in half so that is has a volume of 0.25 m³. Will the object then float or sink?

Topic 5b Checklist — Make sure you know...

Forces and Elasticity

☐ That at least two forces must be applied to an object in order to stretch, compress or bend it.

☐ Know that an object undergoing elastic deformation will return to its original size and shape once all forces have been removed from it.

☐ Know that an object undergoing inelastic deformation will not return to its original size and shape once all forces have been removed from it.

☐ That the force applied to an elastic object is directly proportional to the extension or compression of the object up to the limit of proportionality.

☐ The equation $F = ke$ (force (N) = spring constant (N/m) × extension (m)) and how to use it.

☐ That the limit of proportionality is the point beyond which the extension of an elastic object no longer increases proportionally with the force applied to the object.

☐ That work must be done on a spring to stretch or compress it.

☐ That for elastic deformation, the work done to stretch or compress a spring is all transferred to the spring's elastic potential energy store.

☐ That the equation $E_e = \frac{1}{2}ke^2$ (elastic potential energy = 0.5 × spring constant × extension²) is used to calculate the energy transferred to the elastic potential energy store when an elastic object is stretched, up to the limit of proportionality.

Investigating Springs

☐ The experiment that can be used to investigate the relationship between the force applied to a spring and the extension of the spring, including the setup and how to record and analyse the results.

☐ That the spring constant of a spring can be found by calculating the gradient of the linear part of the spring's force-extension graph.

Moments

☐ That the turning effect of a force is known as the moment of the force and can be calculated using $M = Fd$ (moment (Nm) = force (N) × perpendicular distance from the pivot to the line of action of the force (m)).

☐ Examples of how forces can cause rotation.

☐ That the force applied to an object should be perpendicular to a straight line drawn between the force and the pivot in order to get the largest moment.

☐ That when the total clockwise moments acting on an object equal the total anticlockwise moments acting on the object, then the object will be balanced.

☐ How to find missing forces and distances for an object in equilibrium.

☐ How levers and gears are used to transmit the rotational effect of a force.

cont...

Fluid Pressure

☐ That a fluid can be either a liquid or a gas.

☐ That any surface in contact with a fluid will experience a force normal to it due to the pressure of the fluid.

☐ The equation $p = F \div A$ (pressure (Pa) = force normal to a surface (N) ÷ area of that surface (m^2)).

☐ **H** How and why the pressure of a liquid is dependent on the liquid's density and the depth.

☐ **H** That the height of a column of liquid above a point in a liquid will increase the pressure at that point, and that the pressure due to the column of liquid can be calculated with the equation $p = h\rho g$ (pressure (Pa) = height of column (m) × density of liquid (kg/m^3) × gravitational field strength (N/kg)).

☐ **H** How to calculate the differences in pressure at different depths in a liquid.

☐ That the atmosphere is a relatively thin layer of air around the Earth.

☐ How atmospheric pressure is the pressure due to the air particles around us.

☐ That atmospheric pressure decreases with increasing altitude, because the density of the air and the weight of the air above both decrease with increasing altitude.

Upthrust

☐ **H** That when an object is totally or partially submerged in a liquid, there will be a force acting on all sides of the object due to the pressure of the liquid.

☐ **H** That the pressure on the bottom of a submerged object is greater than the pressure at the top of the object, and the resultant force acting upwards from this is known as upthrust.

☐ **H** That the upthrust on an object is equal to the weight of the fluid displaced by the object.

☐ **H** That an object will float in a liquid if its density is less than that of the liquid.

☐ **H** That the weight of the liquid displaced by a floating object is equal to the weight of the object.

☐ **H** That an object will sink in a liquid if its density is greater than that of the liquid.

☐ **H** That an object sinks because it cannot displace an amount of liquid that is equal to its weight.

1   A student carries out an experiment to investigate how the extension of a spring changes as an increasing force is applied to it. She plots her results on a force-extension graph and draws a line of best fit. Her results are shown below. When drawing the line of best fit, the student assumes the data point recorded when a force of 0.70 N is applied to the spring is anomalous.

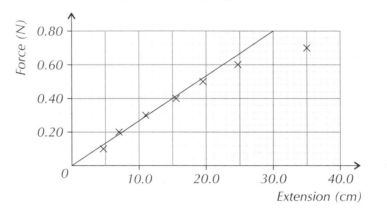

1.1   Explain why the student is wrong to assume the data point is anomalous. Suggest how the student could find out whether it is anomalous or not.

(2 marks)

1.2   The student decides to repeat the experiment. Suggest why a second identical spring should be used rather than the same spring.

(2 marks)

1.3   Using the graph, calculate the spring constant of the spring.

(3 marks)

2   A mass is put on a spring as shown in the diagram. The mass compresses the spring, which remains vertical during the compression. The spring has a spring constant of 160 N/m, and the length of the spring changes from 165 mm to 140 mm when the mass is put on top of it.

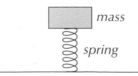

2.1   Calculate the weight of the mass. Give your answer in N.

(3 marks)

2.2   Calculate the energy stored in the elastic potential energy store of the spring. Use the correct equation from those listed on page 394. Give your answer in J.

(2 marks)

3   A plank of wood is resting on a pivot, located at its centre of mass.
Three forces are applied to it, as shown in the diagram.

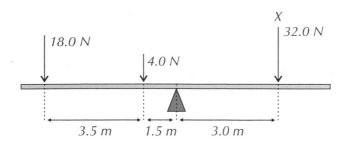

3.1   Calculate the moment of force X about the pivot. Give your answer in Nm.

(2 marks)

3.2   Is the plank of wood balanced? Write down any calculations you use.

(5 marks)

3.3   Explain what would happen if force X was applied at a different angle.

(2 marks)

4   Two identical blocks with a uniform density of 980 kg/m³ and weight 0.58 N are put in
separate beakers. One beaker contains water and one beaker contains oil. The density
of water is 1000 kg/m³ and the density of the oil is 900 kg/m³.

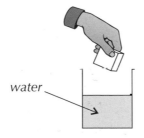
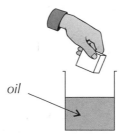

4.1   State whether each block will sink or float.

(2 marks)

One of the blocks is then put in a beaker containing
saltwater, which has a density of 1100 kg/m³.

4.2   What is the upthrust acting on the block when it is in the saltwater?
Explain your answer.

(2 marks)

4.3   The pressure of the saltwater on the base of the beaker is 101 000 Pa.
The total surface area of the base of the beaker is 0.00800 m².
Calculate the total force acting on the base of the beaker due to the saltwater pressure.
Give your answer in N.

(3 marks)

1. Distance, Displacement, Speed and Velocity

Some of your bread and butter physics here — it's all about things that are moving. Make sure you really understand it, otherwise things could get tricksy.

Distance and displacement

You met vectors and scalars on page 147. A scalar quantity has magnitude but no direction, whereas a vector quantity has magnitude and direction.

Distance is a scalar quantity — it's just how far an object has moved. **Displacement** is a vector quantity. It measures the distance and direction in a straight line from an object's starting point to its finishing point. The direction could be relative to a point, e.g. towards the school, or a bearing (which is a three-digit angle from north, e.g. 035°).

> **Example**
>
> **A person walks 5 m north and then 5 m south.**
> **Calculate the distance they travel and their displacement.**
>
> Distance travelled = 5 + 5 = 10 m
> Displacement = 0 m as they have ended up back at their starting position.

Speed and velocity

Speed and **velocity** both measure how fast you're going, but speed is a scalar and velocity is a vector:

1. Speed is just how fast you're going with no regard to the direction, for example 30 mph or 20 m/s.

2. Velocity is how fast you're going and in which direction, for example 30 mph north or 20 m/s on a bearing of 060°.

> **Example**
>
> The cars below are all travelling at the same speed of 0.5 m/s, but they're all moving in different directions, so all have a different velocity.
>
>

Moving in a circle `Higher`

You can have objects travelling at a constant speed with a changing velocity. This happens when the object is changing direction whilst staying at the same speed.

An object moving in a circle at a constant speed has a constantly changing velocity, as the direction is always changing, for example a car going around a roundabout.

Figure 1: *People on a Ferris wheel will have a constant speed but a changing velocity as the wheel turns.*

Everyday speeds

Objects rarely travel at a constant speed. E.g. when you walk, run or travel in a car, your speed is always changing. Whilst every person, train, bus etc. is different, there is usually a typical speed that each object travels at. Remember these typical speeds for everyday objects:

> A person walking — 1.5 m/s   A car — 25 m/s
> A person running — 3 m/s   A train — 55 m/s
> A person cycling — 6 m/s   A plane — 250 m/s

Tip: It might help to remember that running is about half the speed of cycling, and walking is about half the speed of running.

Lots of different things can affect the speed something travels at. For example, the speed at which a person can walk, run or cycle depends on their fitness, their age, the distance travelled and the terrain (what kind of land they're moving over, e.g. roads, fields), as well as many other factors.

It's not only the speed of objects that varies. The speed of sound is 330 m/s in air, and changes depending on what the sound waves are travelling through. Similarly, wind speed can be affected by things like changes in temperature and atmospheric pressure as well as whether there are any large buildings or structures nearby (for example, forests can reduce the speed of the air travelling through them).

Calculating speed

If an object is travelling at a constant speed, then its distance, speed and time are related by the formula:

s = distance travelled (m)　　v = speed (m/s)

$$s = vt$$

t = time (s)

Tip: This equation can be written in a formula triangle:

$$\frac{s}{v \times t}$$

Example

A cat is walking at a speed of 0.4 m/s. Calculate how far the cat walks in 50 s and how long it takes to walk 32 m.

To find how far the cat walks in 50 s:
$$s = vt = 0.4 \times 50 = 20 \text{ m/s}$$

To find the time it takes the cat to walk 32 m, rearrange the equation:
$$t = s \div v = 32 \div 0.4 = 80 \text{ s}$$

If you want to measure the speed of an object that's moving with a constant speed, you should time how long it takes the object to travel a certain distance, e.g. using a ruler and a stopwatch, and use the equation to find the speed.

For an object that isn't travelling at a constant speed, you can use the equation to calculate the average speed. Just use the total distance travelled by the object and the total time taken to travel that distance.

Example

A lorry moves at a steady speed and travels 24 m in 30 s. The lorry then slows down and travels a further 45 m in 70 s before stopping. Calculate the average speed of the lorry for the whole time that it's moving.

Total distance travelled = 24 + 45 = 69 m
Total time taken to travel 69 m = 30 + 70 = 100 s

Rearrange the equation and substitute in the values
for distance and time to find the average speed:
$$s = vt, \text{ so } v = s \div t = 69 \div 100 = 0.69 \text{ m/s}$$

MATHS SKILLS

Practice Questions — Fact Recall

Q1   State whether the following quantities are scalars or vectors.

a) displacement   b) speed   c) velocity   d) distance

Q2   How can an object have a constant speed and a changing velocity?

Q3   Give a typical speed for when someone is:

a) cycling,   b) running,   c) walking.

Q4   Give the equation that relates speed, distance and time, for an object travelling at a constant speed. Give the units of each term.

Practice Questions — Application

Q1   A car drives 16 m east, then 25 m south, then 16 m west.

a) Calculate the distance travelled by the car.

b) Calculate the displacement of the car.

Tip: For Q2, think about the typical speed that a person will walk at.

Q2   A person walks for 18 s. Estimate the distance that they will have walked during this time.

Q3   A train travels from station A to station B at a constant speed of 45 m/s. It takes the train 120 s to reach station B. The train then travels at a constant speed of 60 m/s to station C. Stations B and C are 16.8 km apart.

a) Calculate how far apart stations A and B are.

b) Calculate how long it takes for the train to travel from station B to station C.

c) Find the average speed of the train between stations A and C.

2. Acceleration

How quickly an object changes its speed is all to do with its acceleration.

What is acceleration?

Acceleration is definitely not the same as velocity or speed:

- Acceleration is how quickly the velocity is changing.

- This change in velocity can be a change in speed, or a change in direction, or both.

You can calculate the acceleration of an object using the formula below. If acceleration isn't constant, this will give you the average acceleration over that period.

$$a = \frac{\Delta v}{t}$$

$\boldsymbol{a}$ = acceleration (m/s^2) $\boldsymbol{\Delta v}$ = change in velocity (m/s) $\boldsymbol{t}$ = time taken (s)

Δv, the change in velocity , is just 'final velocity – initial velocity' (see page 337 for more on the Δ symbol). So if an object is slowing down, the change in velocity will be negative, giving a negative acceleration. A negative acceleration is just a deceleration.

Any calculations you do will involve objects travelling in a straight line, so you won't need to worry about a change in direction. The units of acceleration are m/s^2 — don't get them confused with the units for speed and velocity, m/s.

Example

Find the average acceleration of a dog whose velocity goes from 2 m/s to 6 m/s in 5 s.

Calculate Δv first: Δv = final velocity – initial velocity = 6 – 2 = 4 m/s

Then substitute the numbers into the acceleration formula:
$a = \frac{\Delta v}{t} = 4 \div 5 = 0.8$ m/s^2

Everyday accelerations

You might have to estimate the acceleration (or deceleration) of an object. To do this, you'll need to use the typical speeds from page 179.

Example

A car is travelling along a road, when it collides with a tree and comes to a stop. Estimate the deceleration of the car.

- First, give a sensible speed for the car to be travelling at and estimate how long it would take the car to stop. The ~ symbol just means it's an approximate value (or answer).

A typical speed for a car is ~25 m/s. The car comes to a stop in ~1 s.

- Put these numbers into the acceleration equation.

$a = \Delta v \div t = (-25) \div 1 = -25$ m/s^2 — so the deceleration is ~25 m/s^2

Learning Objectives:

- Know what acceleration is.

- Know and be able to use the equation for calculating the average acceleration of an object.

- Know that an object that is slowing down is decelerating.

- Be able to estimate everyday accelerations.

- Be able to use the equation $v^2 - u^2 = 2as$.

- Know that the acceleration of a falling object near the Earth's surface is about 9.8 m/s^2.

Specification Reference 4.5.6.1.5

Figure 1: *When cheetahs start running, they can get to a high velocity very quickly, which means they have a very high acceleration.*

Tip: Δv is negative as the car is slowing down. You can ignore the minus sign in the answer, because you've been asked for a deceleration.

Uniform acceleration

Constant acceleration is sometimes called uniform acceleration. You can use this equation for uniform acceleration:

Tip: If you struggle to remember which velocity is which, think of it this way — u comes before v in the alphabet, so u is initial velocity and v is final velocity.

v = final velocity (m/s)
a = acceleration (m/s²)

$$v^2 - u^2 = 2as$$

u = initial velocity (m/s)
s = distance (m)

Acceleration due to gravity (g) is uniform for objects falling freely under gravity. It's roughly equal to 9.8 m/s² near the Earth's surface and has the same value as gravitational field strength (p. 150).

Tip: You'll come across air resistance, and how it affects motion, a bit later (see page 190).

Example

A ball has been dropped from the top of a building. The velocity of the ball when it is 2.25 m from the ground is 6.0 m/s. Calculate the velocity of the ball when it reaches the ground. You can assume there is no air resistance.

- First, rearrange the equation so v^2 is on one side, and then put the numbers in.

$$v^2 = 2as + u^2 = (2 \times 9.8 \times 2.25) + 6.0^2 = 80.1$$

Tip: See page 15 for more on significant figures.

- Finally, square root the whole thing.

$$v = \sqrt{80.1} = 8.94... = 8.9 \text{ m/s (to 2 s.f.)}$$

Practice Questions — Fact Recall

Q1   What is acceleration?

Q2   What are the units of acceleration?

Practice Questions — Application

Q1   A car is travelling forwards at 25 m/s and the driver applies the brakes. The car's velocity drops steadily for 5 seconds until it becomes 10 m/s. Find its acceleration during this time.

Tip: If an object starts from rest, its initial velocity is 0 m/s.

Q2   A cheetah begins from rest and accelerates at an average acceleration of 4 m/s² over a period of 5 s.
Calculate the speed of the cheetah after this time.

Q3   A runner is running at a steady velocity of 5.6 m/s. When she is 38.1 m away from the finish line, she accelerates at a constant rate. If her velocity at the finish line is 7.1 m/s, what was her acceleration as she approached the finish line?

Tip: Remember acceleration due to gravity is about 9.8 m/s².

Q4   An apple falls from a tree and hits the ground at a speed of 6.0 m/s. Calculate the height the apple fell from. You can assume there is no air resistance.

3. Distance-Time Graphs

Distance-time graphs show how far an object is from a point at a given time. They're useful in physics because they help keep track of an object's motion.

What are distance-time graphs?

Distance-time graphs are a good way of describing the motion of something travelling in a straight line. They have time on the horizontal axis and distance on the vertical axis.

Speed = distance ÷ time (see page 179), so the gradient (slope) of a distance-time graph tells you how fast your object is travelling. This is because the gradient is the change in the distance (vertical axis) divided by the change in time (horizontal axis).

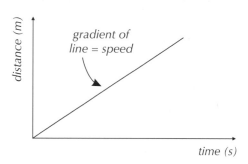

gradient of line = speed

Figure 1: *A basic distance-time graph.*

Drawing and interpreting distance-time graphs

You need to be able to draw and interpret distance-time graphs in the exam. Here are some important points to remember for an object's distance-time graph:

1. Gradient = speed.

2. Straight uphill sections mean it is travelling at a steady speed.

3. The steeper the graph, the faster it's going.

4. Flat sections are where it's stationary — it's stopped.

5. Curves represent acceleration (speeding up) or deceleration (slowing down) (page 181).

6. A steepening curve means it's accelerating/speeding up — the gradient is increasing.

7. A levelling off curve means it's decelerating/slowing down — the gradient is decreasing.

Learning Objectives:

- Know that the distance travelled by an object moving in a straight line can be shown on a distance-time graph.
- Be able to draw and interpret distance-time graphs.
- Know that the gradient of a distance-time graph for a moving object is equal to the speed of the object.
- **H** Know that the speed of an accelerating object at a certain time can be found from the object's distance-time graph, by calculating the gradient of the tangent to the graph at that time.

Specification Reference 4.5.6.1.4

Tip: Take a look at page 339 for how to calculate the gradient of a straight-line graph.

This distance-time graph shows an object that moves off from its starting point at a steady speed for 20 s, then stops for 20 s. It then accelerates for 25 s and decelerates for 25 s before resuming a steady speed for 30 s.

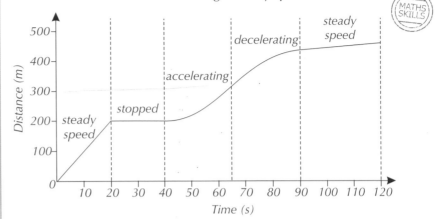

The object's speed during the first 20 s can be found from the gradient:

$$\text{gradient} = \frac{\text{change in the vertical}}{\text{change in the horizontal}} = \frac{200 - 0}{20 - 0} = 10 \text{ m/s}$$

Tip: The speed of the object is faster in the first 20 s compared to the speed of the object in the last 30 s. You can tell because the gradient is steeper in the first 20 s.

Tip: Be careful with the units on the axes of these graphs. Here you have distance in m and time in s, so you get speed in m/s. But you may have to deal with graphs with axes in other units of distance (e.g. kilometres) and time (e.g. hours).

Calculating speed for accelerating objects `Higher`

If an object is changing speed (accelerating), you can find its speed at a point by finding the gradient of the tangent to the curve at that point.

Tip: Take a look at page 341 for more on drawing tangents.

Example — Higher

The graph below is the distance-time graph for a bike accelerating for 30 s and then travelling at a steady speed for 5 s.

The speed of the bike at 25 s can be found by drawing a tangent to the curve (shown by the red line) at 25 s and then finding the gradient of the tangent:

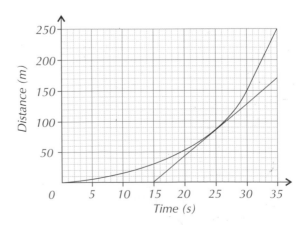

Tip: When drawing a tangent, always make it as long as possible, so it's easier to work out the gradient.

$$\text{gradient} = \frac{\text{change in the vertical}}{\text{change in the horizontal}} = \frac{170 - 0}{35 - 15} = 8.5 \text{ m/s}$$

Practice Questions — Fact Recall

Q1   What does the gradient of a distance-time graph represent?

Q2   What does a flat section on a distance-time graph tell you?

Q3   What does a curved section on a distance-time graph represent?

Q4   How would you calculate the speed of an accelerating object at a given time from its distance-time graph?

Practice Questions — Application

Q1   For this distance-time graph, say what's happening to the speed of the object in each of the sections labelled A, B, C and D.

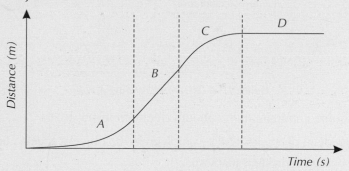

Q2   An object moving in a straight line accelerates for 10 seconds and then moves at a steady speed for 5 seconds. Describe the distance-time graph for the object during this time.

Q3   Look at this table showing the distance travelled (in a straight line) by a toy car over a period of time. The car initially moves with a constant speed, then stops, accelerates for 6 seconds, then stops.

Time (s)	0	2	4	6	8	10	12	14	16
Distance (m)	0.0	2.0	4.0	4.0	4.0	7.0	7.8	8.0	8.0

a) Plot these values and join the points on a distance-time graph to represent the car's motion.

b) Calculate the speed of the car between 0 and 4 seconds.

c) Estimate the speed of the car at 10 seconds.

- Be able to draw and interpret velocity-time graphs for moving objects.
- Be able to find an object's acceleration by calculating the gradient of its velocity-time graph.
- **H** Know that the area under a velocity-time graph is equal to distance travelled (or displacement), and be able to calculate this.
- **H** Be able to find the area under a velocity-time graph by counting the squares under the graph.

Specification Reference 4.5.6.1.5

4. Velocity-Time Graphs

Velocity can be plotted against time to help find out about an object's motion.

Velocity-time graphs

You can plot a **velocity-time graph** to show an object's motion. Time goes on the horizontal axis and velocity goes on the vertical axis.

Acceleration is the change in an object's velocity over time (see page 181). So the gradient of a velocity-time graph tells you the acceleration of your object. This is because the gradient is the change in the velocity (vertical axis) divided by the change in time (horizontal axis).

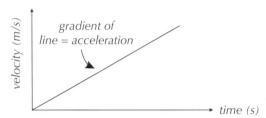

Figure 1: A basic velocity-time graph.

Here are some important things about velocity-time graphs:

1. The gradient of a velocity-time graph gives the object's acceleration.
2. Flat sections represent steady speed.
3. The steeper the graph, the greater the acceleration or deceleration.
4. Uphill sections are acceleration.
5. Downhill sections are deceleration.
6. A curve means changing acceleration.

Tip: Make sure you don't get confused between velocity-time graphs and distance-time graphs (which you met on page 183).

Tip: If an object wasn't moving, its velocity-time graph would just be a straight line along the x-axis, i.e. velocity = 0 m/s.

> **Example**
>
> This velocity-time graph shows an object that accelerates from rest to 20 m/s in 20 seconds then travels at a steady velocity for 20 seconds. It then accelerates at an increasing rate for 30 seconds, travels at a steady 40 m/s for a further 30 seconds and finally decelerates back to rest in 20 seconds.
>
>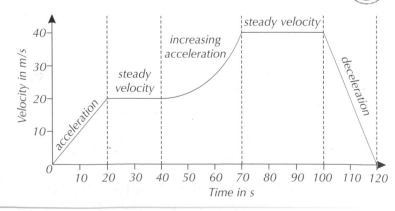

Acceleration on a velocity-time graph

The acceleration of an object can be found by calculating the gradient of its velocity-time graph.

Tip: This is the same method as finding the speed from a distance-time graph (see page 184).

Example

This is a velocity-time graph for a race car accelerating from 0 to 50 m/s. Calculate the acceleration of the car between 3 and 6 seconds.

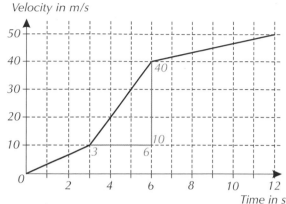

$$\text{acceleration} = \text{gradient}$$
$$= \frac{\text{change in vertical}}{\text{change in horizontal}} = \frac{(40 - 10)}{(6 - 3)} = 10 \, \text{m/s}^2$$

Tip: If the graph is curved, the acceleration is changing and you can find the acceleration at a point using a tangent, in the same way as on page 184.

Distance travelled on a velocity-time graph   Higher

The area beneath a velocity-time graph gives the distance travelled (or displacement). The distance travelled in any time interval is equal to the area under the velocity-time graph in that interval.

Tip: **H** If you know the direction of the velocity, you can find the displacement using the area.

Example 1   Higher

For the same car as in the previous example, calculate the distance travelled between 3 and 6 seconds.

- The distance travelled is equal to the area under the graph, so look at the graph between 3 and 6 seconds. It might seem difficult to work out the area of this part of the graph, but you can make it easier by splitting the area into a triangle (**A**) and a rectangle (**B**).

Tip: **H** You could also use the 'counting the squares method' (see the next page) to find the area under the graph.

Velocity in m/s

- You can then calculate the area of each shape individually:

$$\text{Area}_A = \frac{1}{2} \times \text{base} \times \text{height} \qquad \text{Area}_B = \text{base} \times \text{height}$$
$$= \frac{1}{2} \times 3 \times 30 = 45 \qquad\qquad = 3 \times 10 = 30$$

- Then just find the total area by adding Area_A and Area_B together:

   distance travelled = total area under graph = 45 + 30 = 75 m

You can also find the area under a graph using the 'counting the squares method'. First you need to find out the distance each square of the grid represents. To do this, multiply the width of one square (in seconds) by the height of one square (in metres per second).

Then you just multiply this by the number of squares under the graph. If there are multiple squares that are partly under the graph, you can add them together to make whole squares (see the example below).

This is the best method to use when working with an irregularly-shaped area under a graph.

Example 2 — Higher

The graph below is a velocity-time graph. You can estimate the distance travelled in the first 10 s by counting the number of squares under the graph (shown by the shaded area).

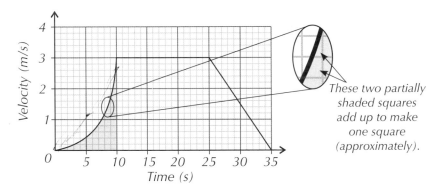

These two partially shaded squares add up to make one square (approximately).

Total number of shaded squares ≈ 32

Distance represented by one square = width of square × height of square
= 1 s × 0.2 m/s = 0.2 m

So total distance travelled in 10 s = 32 × 0.2 = 6.4 m

Practice Questions — Fact Recall

Q1   What does the gradient of a velocity-time graph represent?

Q2   What does the area under a velocity-time graph represent?

Practice Questions — Application

Q1 For this velocity-time graph, say what's happening to the acceleration of the object in each of the sections labelled A, B, C and D.

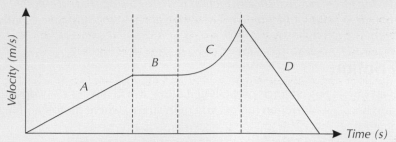

Tip: Make sure you talk about what's happening to the <u>acceleration</u> in Q1, not the velocity.

Q2 Below is a velocity-time graph for a cyclist during a race.

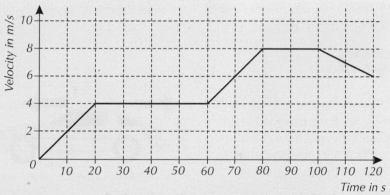

a) What's the cyclist's velocity at 40 seconds?

b) During which part(s) of the race is the cyclist decelerating?

c) What's the cyclist's acceleration between 60 and 80 seconds?

d) How far does the cyclist travel in the first 60 seconds of the race?

Tip: Remember, acceleration can have a negative value (in which case it can be called deceleration).

Q3 A car travels forwards at 25 m/s for 3 s before the driver applies the brakes. The car decelerates steadily for 5 seconds until its velocity becomes 10 m/s. Plot a velocity-time graph to show this and, using the graph, find its acceleration whilst it's slowing down.

Q4 The velocity-time graph for a helicopter travelling in a straight line is shown below. Estimate the distance travelled by the helicopter in the first 20 s of its motion.

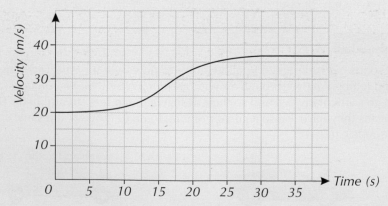

In an ideal world, things would just move freely without needing any external driving force, but realistically friction is there to stop that. It's not all bad though — without friction we'd never be able to stop and take a break.

Friction

If an object has no force propelling it along it will always slow down and stop because of **friction** (unless you're in space, where there's nothing to rub against).

Friction always acts in the opposite direction to movement. To travel at a steady speed, the driving force needs to balance the frictional forces (this will be covered in more detail on page 194).

Example

If a car is travelling at a steady speed, the force provided by the engine is exactly the same as the resistive forces acting on the car.

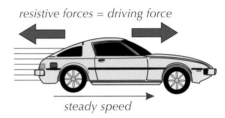

resistive forces = driving force

steady speed

You get friction between two surfaces in contact or when an object passes through a fluid — in which case it's usually called drag. A fluid is just a gas or a liquid — e.g. air, water, oil, etc.

Drag

Most of the resistive forces are caused by **air resistance** or "**drag**". The most important factor by far in reducing drag in fluids is keeping the shape of the object streamlined. A streamlined object is one that allows fluids to flow over it easily, so they don't slow down the object much as it passes through them.

The opposite extreme is a parachute which is about as high a drag as you can get — which is, of course, the whole idea.

Figure 1: *Fish have streamlined bodies that reduce drag from the water (a fluid).*

Example

A sports car is designed to allow fluids to flow over it easily, reducing drag and letting it move through air without much effort. Vans aren't designed to go particularly fast and so their design is much less streamlined. This means they have to use a greater driving force to move at the same speed as a sports car.

Frictional forces from fluids always increase with speed. A car has much more friction to work against when travelling at 70 mph compared to 30 mph. So at 70 mph the engine has to work much harder just to maintain a steady speed.

Figure 2: *A car moving at 30 mph and 70 mph. As the speed of a car increases, the frictional forces acting on it increase, so the engine works harder to maintain a steady speed.*

Ways of increasing the top speed of a vehicle

There are two main ways of changing a vehicle to increase its top speed:

1. Reducing drag.
    This can be done by altering the shape of the vehicle to make it more streamlined.

2. Increasing the power of the vehicle's engine.
    This way, the driving force becomes larger and so the drag force on the vehicle will equal the driving force at a higher speed.

Figure 3: *Racing cars have incredibly powerful engines as well as a low, streamlined body. This is what helps them shoot round the track at breakneck speeds.*

Terminal velocity

Generally when an object falls through a very large distance, it won't just keep accelerating at the same rate until it hits the ground. Its acceleration will decrease until it reaches a steady velocity, called its **terminal velocity**.

When a falling object first sets off, the force of gravity is much greater than the frictional force slowing it down, so it accelerates. As the object moves faster, the frictional forces that act on it become greater.

This gradually reduces the acceleration until eventually the frictional force is equal to the accelerating force — the resultant force will be zero. If there is zero resultant force, then the object will no longer accelerate (see page 194). It will have reached its maximum speed — or terminal velocity — and will fall at a steady speed.

The velocity-time graph for a falling object starts with a steep gradient (i.e. a large acceleration), which gradually decreases until the object reaches its terminal velocity. From this point, the graph becomes flat — the forces are balanced and the object is no longer accelerating, so the gradient is zero.

Tip: If you'd like a refresher on velocity-time graphs, head to page 186.

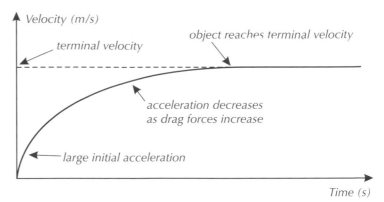

Figure 4: *The velocity-time graph for a falling object.*

Within the graph:
- Velocity (m/s)
- terminal velocity
- object reaches terminal velocity
- acceleration decreases as drag forces increase
- large initial acceleration
- Time (s)

Factors affecting terminal velocity

The terminal velocity of a falling object depends on its shape and area. The accelerating force acting on all falling objects is gravity and it would make them all fall at the same rate, if it wasn't for resistance.

This means that on the Moon, where there's no air, rocks and feathers dropped simultaneously from the same height will hit the ground together. However, on Earth, air resistance causes things to fall at different speeds, and the terminal velocity of any object is determined by its drag in comparison to its weight. The frictional force depends on its shape and area.

Figure 5: *Apollo 15 astronaut David Scott demonstrating that a hammer and a feather fall to the ground at the same speed on the Moon.*

Example

The most important example is the human skydiver. Without their parachute open they have quite a small area and a force of "$W = mg$" pulling them down. They reach a terminal velocity of about 120 mph.

But with the parachute open, there's much more air resistance (at any given speed) and still only the same force "$W = mg$" pulling them down. This means their terminal velocity comes right down to about 15 mph, which is a safe speed to hit the ground at.

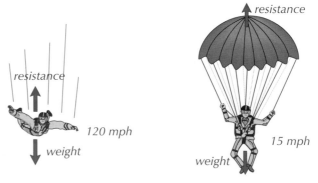

Figure 7: *A skydiver at terminal velocity with and without a parachute. Resistance has become equal to weight in both cases — the difference is the speed at which this has happened.*

Figure 6: *A BASE jumper using a parachute to slow down his free fall so that he can land safely.*

Practice Questions — Fact Recall

Q1   What must the resistive forces acting on a car be equal to if the car is travelling at a steady speed over flat ground?

Q2   What's drag?

Q3   What's the relationship between an object's speed and the drag it experiences?

Q4   Describe how a free-falling object reaches terminal velocity.

Q5   Sketch a velocity-time graph for an object that reaches terminal velocity falling through a fluid.

Practice Questions — Application

Q1   Why does a feather fall slower than a rock on Earth but not on the Moon?

Q2   Explain, in terms of forces, why using a parachute reduces a skydiver's terminal velocity.

Tip: When a skydiver is at terminal velocity, the resultant force acting on them is zero (see p. 194 for more on resultant forces and motion).

Learning Objectives:

- Know that Newton's First Law says that an object will remain stationary or moving at a constant velocity if there is no resultant force acting on it.
- Know that a vehicle that has a driving force equal to the resistive forces acting on it will be moving at a steady speed.
- Understand that when a non-zero resultant force is acting on an object, it will change the object's speed or direction of motion or both (the velocity).
- Be able to explain an object's motion using Newton's First Law.

Specification Reference 4.5.6.2.1

6. Newton's First Law

Resultant forces are often described as being 'non-zero' (i.e. there is one) or 'zero' (i.e. there isn't one) — and you need to know their effects.

Resultant forces

The size and direction of a resultant force (see page 151) acting on an object will dictate what happens to its motion. **Newton's First Law** says that a resultant force is needed to make something start moving, speed up, slow down or change direction. So an object will remain stationary or moving at a constant velocity unless a resultant force acts upon it.

Zero resultant force

Objects don't just start moving on their own — if there's no resultant force acting on a stationary object, there's no acceleration and it will just stay put.

> If the resultant force on a stationary object is zero, the object will remain stationary.

Example 1

A ball is being held stationary by two taut strings. The force to the left (F_1) is the same size as the force to the right (F_2). So, even though forces are acting on it in both directions, the resultant force on it is zero. Consequently, it remains stationary.

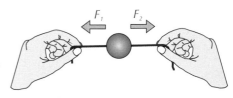

If there's no resultant force acting on an object, it won't change velocity. That means if it's already moving, it will just keep moving at the same velocity.

> If the resultant force on a moving object is zero, it'll just carry on moving at the same velocity.

If any object (be it a train, a car, a horse, or anything else really) is moving at a constant velocity then the forces on it must all be balanced. Never let yourself stray down the path of thinking that things need a constant resultant force to keep them moving — this is a common misconception.

To keep going at a steady velocity, there must be zero resultant force. This doesn't mean there must be no driving force — it means the driving force is balanced by other forces, like friction and air resistance.

Example 2

A van travelling at a steady velocity has zero resultant force acting on it. This is because the driving force from the engine is balanced by friction forces.

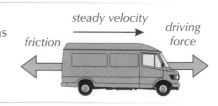

Tip: Don't forget, velocity is a vector (see page 147). So in order for an object to have a constant velocity, the object must keep moving at the same speed <u>and</u> in the same direction.

Non-zero resultant force

In most situations, there will be a resultant force acting on an object.

> If there is a non-zero resultant force on an object, its velocity will change (it will accelerate in the direction of the force).

This applies to both stationary objects and ones that are already moving. When a resultant force acts on an object, the change in velocity it experiences can take five different forms:

- starting
- speeding up
- stopping
- slowing down
- changing direction

Tip: Remember — when an object accelerates, its velocity changes, and a change in velocity means a change in speed or direction of motion (or both) (see page 181).

Tip: **H** On a free body diagram (see page 151), if there's a resultant force, then the total size of the arrows in one direction will be different to the total size of the arrows in the opposite direction.

Example 1 — continued

One of the hands lets go of the string that's keeping the ball stationary. Now the ball experiences a resultant force of F_2 to the right, and accelerates in that direction.

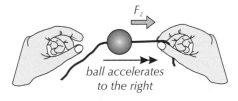

ball accelerates to the right

Practice Question — Fact Recall

Q1 Say what will happen to the object in each of the following cases:

a) No resultant force acting on a stationary object.

b) A resultant force acting on a stationary object.

c) No resultant force acting on a moving object.

d) A resultant force acting on a moving object in the same direction as its motion.

e) A resultant force acting on a moving object in the opposite direction to its motion.

Practice Question — Application

Q1 If a rocket is moving through space at a steady velocity, what can you say about the resultant force acting on the rocket?

Figure 1: *When a rocket is launched, there needs to be a force acting upwards that is larger than the weight of the rocket, so that it accelerates away from Earth.*

- Know that Newton's Second Law says that an object's acceleration is directly proportional to the resultant force acting on it and inversely proportional to its mass.

- Know and be able to use the equation $F = ma$, where F is the resultant force acting on an object.

- Be able to estimate the resultant forces acting in everyday accelerations and be able to use the '~' symbol.

- **H** Know that the inertia of an object is the tendency for its motion to remain unchanged.

- **H** Know that an object's inertial mass is given as the ratio of force over acceleration, and that it measures how difficult it is to change the object's velocity.

**Specification References
4.5.6.2.1, 4.5.6.2.2**

Tip: If you need to, you can use this formula triangle for $F = ma$:

Tip: Once the car has started moving, it will experience a drag force, so the resultant force will be the driving force minus the drag force.

7. Newton's Second Law and Inertia

So you've seen from Newton's First Law that an object will accelerate if there's a non-zero resultant force acting on it. You can find the size of the acceleration using Newton's Second Law.

What is Newton's Second Law?

Newton's Second Law has two points that you need to know:

- The larger the resultant force acting on an object, the more the object accelerates — the force and the acceleration are directly proportional. You can write this as $F \propto a$.

- Acceleration is also inversely proportional to the mass of the object — so an object with a larger mass will accelerate less than one with a smaller mass (for a fixed resultant force).

There's an incredibly useful formula that describes Newton's Second Law:

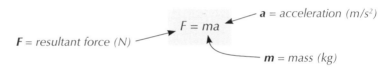

a = acceleration (m/s^2)

F = resultant force (N)

m = mass (kg)

This formula can be used to calculate the acceleration produced by a resultant force, or the size of the resultant force producing an acceleration.

Example 1

A car with a mass of 1250 kg has an engine that provides a driving force of 5200 N. At 70 mph the drag force acting on the car is 5100 N. Find its acceleration at 70 mph.

- First calculate the resultant force acting on the car:

    Resultant force = 5200 − 5100 = 100 N

- Then work out the acceleration of the car:

    $a = F \div m = 100 \div 1250 = 0.080$ m/s^2

Example 2

A car with a mass of 900 kg accelerates from rest with an initial acceleration of 2.5 m/s^2. Calculate the resultant force required to produce this acceleration.

At the point that the car's just about to start moving, the driving force of the engine provides an acceleration of 2.5 m/s^2. As the car is still stationary (just), there's no drag, and so the resultant force is just the driving force. So:

    $F = ma = 900 \times 2.5 = 2250$ N

Estimating forces

You can use Newton's Second Law to get an idea of the forces involved in everyday transport. Large forces are needed to produce large accelerations:

Tip: For more on estimating typical speeds, check out page 179. For more on estimating typical accelerations, head to page 181.

Example

Estimate the resultant force on a car as it accelerates from rest to a typical speed in 10 s.

- First you need to estimate the acceleration of the car. To do this, you should use typical speeds from page 179.

  A typical speed of a car is ~25 m/s.
  It takes 10 s to reach this.

  So $a = \frac{\Delta v}{t} = 25 \div 10 = 2.5$ m/s^2

- Then estimate the mass of the car.

  Mass of a car is ~1000 kg.

- Finally, put these numbers into Newton's Second Law.

  $F = ma = 1000 \times 2.5 = 2500$ N
  So the resultant force is ~2500 N.

The '~' sign used in the example above means 'approximately'. Make sure you use it when asked to estimate the value of something.

Exam Tip
Knowing some typical vehicle masses could come in very handy in the exam:
A car ~ 1000 kg
A single-decker bus ~ 10 000 kg
A loaded lorry ~ 30 000 kg.

Inertia Higher

You saw on page 194 that until acted upon by a resultant force, objects at rest stay at rest and objects moving at a steady speed will stay moving at that speed. This is Newton's First Law. This tendency to continue in the same state of motion is called **inertia** (or in other words, it's the tendency to continue moving at the same velocity).

An object's **inertial mass** measures how difficult it is to change the velocity of the object. Imagine that a bowling ball and a golf ball roll towards you with the same velocity. It would require a larger force to stop the bowling ball than the golf ball in the same time. This is because the bowling ball has a larger inertial mass.

Inertial mass can be found using Newton's Second Law, $F = ma$.
Rearranging this gives:

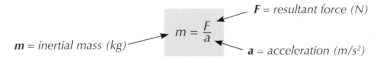

m = inertial mass (kg)
$$m = \frac{F}{a}$$
F = resultant force (N)
a = acceleration (m/s^2)

So inertial mass is just the ratio of force over acceleration.

Figure 1: A full trolley will accelerate less for a given pushing force then an empty one would, as it has a larger mass.

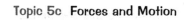

Practice Questions — Fact Recall

Q1 What is Newton's Second Law?

Q2 Write down the formula used for calculating the resultant force acting on an object from its mass and its acceleration. Say what each term represents and the units it's measured in.

Q3 What is inertia?

Q4 What is inertial mass?

Practice Questions — Application

Q1 Two identical remote-control cars take part in a straight race across flat ground. One of the cars is loaded with a large rock. Which of the two will win the race? Explain your answer.

Q2 When a catapult is released, it applies 303 N of force to a rock with a mass of 1.5 kg. Find the acceleration of the rock at this point.

Q3 Two cars are starting from rest at full power, as shown below. Which engine is providing a larger driving force? Assume there are no resistive forces acting on the cars.

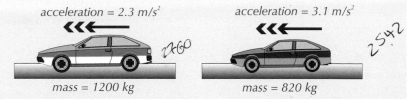

acceleration = 2.3 m/s²   acceleration = 3.1 m/s²

mass = 1200 kg   mass = 820 kg

Q4 A car accelerates from a speed of 5 m/s to its typical speed in 20 s. Estimate the resultant force on the car.

8. Investigating Motion

Learning Objectives:
- Be able to investigate how the force on, and the mass of, an object affect its acceleration. (Required Practical 7)

Specification Reference 4.5.6.2.2

Here's an experiment you can do to investigate Newton's Second Law. You just need to get your hands on a trolley, a light gate and some masses.

Setting up the experiment

REQUIRED PRACTICAL **7**

This experiment makes use of a light gate. A light gate is an arch-shaped piece of equipment which sends a beam of light from one side of the arch (or 'gate') to the other. When something passes through the gate, it interrupts this beam of light. When used with a computer or data logger, the light gate can detect an interruption and measure how long the interruption lasted.

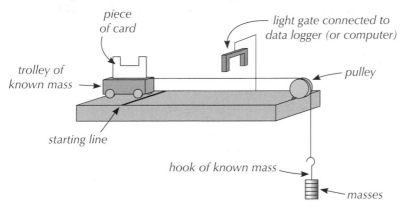

Figure 1: *The experimental setup for investigating Newton's Second Law.*

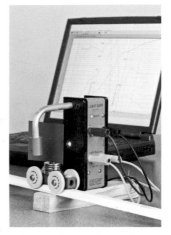

Figure 2: *The beam in a light gate being interrupted by a piece of card attached to a trolley.*

Set up the apparatus as shown in Figure 1. The trolley should hold a piece of card with a gap in the middle that will interrupt the signal on the light gate twice. If you measure the length of each bit of card that will pass through the light gate and input this into the software, the light gate can measure the velocity for each bit of card. It does this using $v = s \div t$ (page 179), where s is the length of each bit of card and t is the duration of the interruption.

The software can also work out the acceleration of the trolley, using $a = \Delta v \div t$ (page 181). Here, Δv is the difference between the two velocities it has measured, and t is the amount of time that has passed between the first and second interruptions of the light gate signal.

Tip: You could also do this experiment with two light gates and a card without a gap.

Connect the trolley to a piece of string that goes over a pulley and is connected on the other side to a hook (that you know the mass of and can add more masses to). Mark a starting line on the table the trolley is on, and place the trolley so that its front end is lined up with it. This way, when the trolley moves it will always have travelled the same distance when it reaches the light gate. You should make sure the string and table are the right length so that your trolley passes through the light gate before the masses hit the floor or the trolley hits the pulley, otherwise the accelerating force will be removed before the acceleration has been recorded.

Tip: The friction between the trolley and the bench might affect your acceleration measurements. You could use an air track to reduce this friction (a track which hovers a trolley on jets of air), or use a slightly sloped surface to use gravity to compensate for friction.

The weight of the hook and any masses attached to it will provide the accelerating force that causes the trolley to move when released. This force is equal to the total mass of the hook and masses (m) × acceleration due to gravity (g) — which is just $W = mg$ (see page 150).

Tip: Remember — always carry out a risk assessment before doing a practical. In this experiment, watch out for the string snapping or the masses landing on people's toes.

Carrying out the experiment

Hold the trolley so the string between the trolley and pulley is horizontal and taut (not loose and touching the table). Next, release the trolley and record the acceleration measured by the light gate as the trolley passes through it. This is the acceleration of the whole system (trolley, hook and masses). Repeat this twice more and calculate the average acceleration (see page 14).

You can use this setup to investigate the effect that varying the force and mass has on the trolley's acceleration.

Varying the mass

To investigate the effect of varying the mass, add masses to the trolley one at a time. This will increase the mass of the system. Don't add masses to the hook, or you'll change the force. Record the average acceleration each time you add a mass.

Varying the force

Tip: Remember, the force acting on the system (the trolley, hook and masses) is equal to the weight of the hook and the masses attached to it. So adding masses to the hook changes the weight, which changes the force.

To investigate the effect of varying the force, you need to keep the total mass of the system the same, but change the weight on the hook. To do this, start with all the masses loaded onto the trolley, and transfer the masses to the hook one at a time, recording the new acceleration each time you transfer one of the masses. This will increase the accelerating force while keeping the mass of the system the same, as you're only transferring the masses from one part of the system (the trolley) to another (the hook).

The results

Tip: $F = ma$ was introduced on page 196.

You can use Newton's Second Law, $F = ma$, to explain the results. In this case, F = weight of the hanging masses and hook, m = mass of the whole system (trolley, hook and any added masses) and a = acceleration of the system.

By adding masses to the trolley, you increase the mass of the whole system, but keep the force applied to the system the same. This should lead to a decrease in the acceleration of the trolley, as acceleration is inversely proportional to mass ($a = F \div m$).

By transferring masses from the trolley to the hook, you are increasing the accelerating force without changing the mass of the system. Increasing the force should lead to an increase in the acceleration of the trolley, because a is directly proportional to F. If you were to plot a graph of acceleration against force, the graph should be a straight line through the origin.

Practice Question — Fact Recall

Q1 A trolley has a constant force acting on it which causes it to travel in a straight line.

a) What would happen to the acceleration of the trolley if a mass was added to it?

b) What would happen to the acceleration of the trolley if the force acting on it decreased?

9. Newton's Third Law

Newton's Third Law is just as important as the other two. If it wasn't true, then we'd never be able to sit anywhere without falling to the centre of the Earth.

What is Newton's Third Law?

When two objects interact, they apply forces to each other. These forces always act in the opposite direction to each other. You've probably heard the law in physics that "every action has an equal and opposite reaction" — this is **Newton's Third Law**, which is defined as:

> When two objects interact, the forces they exert on each other are equal and opposite.

If you push something, say a shopping trolley, the trolley will push back against you, just as hard. And as soon as you stop pushing, so does the trolley.

So far so good. The slightly tricky thing to get your head round is this — if the forces are always equal, how does anything ever go anywhere? The important thing to remember is that the two forces are acting on different objects.

Example

Think about a pair of ice skaters.

Skater A   Skater B

mass = 55 kg

mass = 65 kg

When skater A pushes on skater B with a force, she feels an equal and opposite force from skater B's hand (the normal contact force). Both skaters feel the same sized force, in opposite directions, and so accelerate away from each other.

Skater A will be accelerated more than skater B, though, because she has a smaller mass — $a = F \div m$ (see page 196). It's the same sort of thing when you go swimming. You push back against the water with your arms and legs, and the water pushes you forwards with an equal-sized force in the opposite direction.

Learning Objectives:

- Know that Newton's Third Law states that two interacting objects will exert equal and opposite forces on each other.
- Be able to use Newton's Third Law to explain different situations in which objects are in equilibrium.

Specification Reference 4.5.6.2.3

Figure 1: *British physicist Sir Isaac Newton, who is famous for his three laws.*

Tip: In this example, the two objects interacting are the two skaters.

Equilibrium

It's easy to get confused with Newton's Third Law and an object in equilibrium. You need to be careful about which forces you say are equal and opposite.

Example 1

An example of Newton's Third Law in an equilibrium situation is a man pushing against a wall. As the man pushes the wall, there is a normal contact force acting back on him. These two forces are the same size. As the man applies a force and pushes the wall, the wall 'pushes back' on him with an equal force.

Example 2

A book resting on a table is in equilibrium. The weight of the book is equal to the normal contact force. But this is NOT Newton's Third Law because the two forces are different types, and both acting on the book.

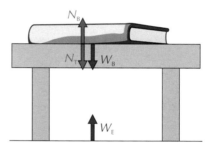

In this situation, Newton's Third Law is:

- The normal contact force acting on the book (from the table), N_B, is equal to the normal contact force acting on the table (from the book), N_T.
- The weight of the book being pulled down by Earth, W_B, is equal to the weight of the Earth being pulled up by the book, W_E.

Practice Question — Fact Recall

Q1   What is Newton's Third Law?

Practice Question — Application

Q1   A ball hanging on the end of a piece of string is in equilibrium. It only has two forces acting on it — its weight acting downwards and tension in the string acting upwards. State whether this is an example of Newton's Third Law or not. Explain your answer.

Distance, Displacement, Speed and Velocity

- [ ] That distance and speed are scalar quantities.
- [ ] That displacement is a vector quantity and is the distance and direction in a straight line from an object's starting point to its finishing point.
- [ ] How to give the displacement of an object in terms of its magnitude and direction.
- [ ] That velocity is a vector quantity and is the speed of an object and its direction.
- [ ] **H** Examples of when an object has a constant speed and a changing velocity, e.g. an object moving in a circle.
- [ ] That things don't often travel at a constant speed.
- [ ] That a typical speed of someone walking is 1.5 m/s, of someone running is 3 m/s and of someone cycling is 6 m/s, and the factors that these values depend on.
- [ ] The typical speeds of everyday vehicles.
- [ ] That the speed of sound and the speed of wind vary.
- [ ] The equation distance (m) = velocity (m/s) × time (s) ($s = vt$) and how to use it for uniform and non-uniform motion.

Acceleration

- [ ] That acceleration is a measure of how quickly velocity is changing.
- [ ] The equation acceleration (m/s^2) = change in velocity (m/s) ÷ time (s) ($a = \Delta v \div t$) and how to use it.
- [ ] That an object that is slowing down is decelerating (and will have negative acceleration).
- [ ] How to make calculations to estimate everyday accelerations.
- [ ] How to use the equation $v^2 - u^2 = 2as$, including what all the letters represent.
- [ ] That when an object is free falling on Earth, it has an acceleration of about 9.8 m/s^2.

Distance-Time Graphs

- [ ] How to sketch a distance-time graph for an object from a description of its motion.
- [ ] How to interpret a distance-time graph.
- [ ] That the speed of an object can be found by calculating the gradient of its distance-time graph.
- [ ] **H** That finding the gradient of a tangent to a curve on a distance-time graph will give the speed of an accelerating object at that point.

Velocity-Time Graphs

- [ ] How to sketch a velocity-time graph for an object from a description of its motion.
- [ ] How to interpret a velocity-time graph.

cont...

- [ ] That the acceleration of an object can be found by calculating the gradient of its velocity-time graph.
- [ ] **H** That the distance travelled by an object can be found by calculating the area under its velocity-time graph, and how to find the area under a graph by counting squares.

Terminal Velocity

- [ ] That the faster an object moves through a fluid, the larger the frictional force acting on the object.
- [ ] That terminal velocity is the velocity at which the resultant force acting on a falling object is equal to zero.
- [ ] How the forces acting on a falling object affect the object's motion.
- [ ] How to sketch a velocity-time graph for a falling object that is reaching its terminal velocity.
- [ ] How certain factors affect the terminal velocity of an object.

Newton's First Law

- [ ] That an object will remain stationary or at a constant velocity if the resultant force on it is zero.
- [ ] That a vehicle will move at a steady speed if the driving forces and resistive forces are equal.
- [ ] That an object will change its velocity (speed and/or direction) if there is a non-zero resultant force acting on it.
- [ ] How to explain the motion of an object using Newton's First Law.

Newton's Second Law and Inertia

- [ ] That Newton's Second Law says that acceleration is directly proportional to the resultant force and inversely proportional to mass.
- [ ] The equation force (N) = mass (kg) × acceleration (m/s^2) ($F = ma$) and how to use it.
- [ ] How to estimate the resultant forces acting in everyday accelerations.
- [ ] **H** That the tendency of an object to stay at a constant velocity is known as inertia.
- [ ] **H** That the inertial mass of an object is a measure of how difficult it is to change its velocity, and that it is given as the ratio of force over acceleration.

Investigating Motion

- [ ] How to carry out an experiment that investigates the effect of changing the resultant force on, and the mass of, an object on its acceleration.

Newton's Third Law

- [ ] That Newton's Third Law says that when two objects interact, they will exert equal and opposite forces on each other.
- [ ] How to explain objects in equilibrium in terms of Newton's Third Law.

Exam-style Questions

1   This is a velocity-time graph for a car during a journey to the local shops.

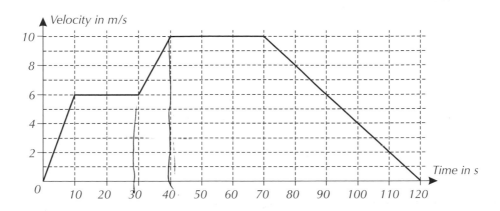

1.1   During which time period(s) does the graph show a negative acceleration?

(1 mark)

1.2   Calculate the acceleration of the car between 30 and 40 seconds.
Show clearly how you work out your answer. Give your answer in m/s^2.

(3 marks)

1.3   If the car has a mass of 980 kg, calculate the resultant force acting on the car between 30 and 40 seconds. Give your answer in N.

(2 marks)

1.4   Use the graph to calculate how far the car travels between 30 and 40 seconds.
Show clearly how you work out your answer. Give your answer in m.

(3 marks)

1.5   The graph below shows how the distance the car has travelled changes for the same journey. Explain three things about the shape of the graph that are incorrect.

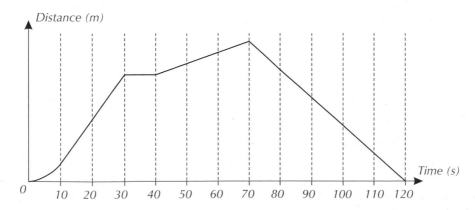

(3 marks)

2   A student sets up the experiment shown below in order to investigate the effect of varying the force acting on a trolley.

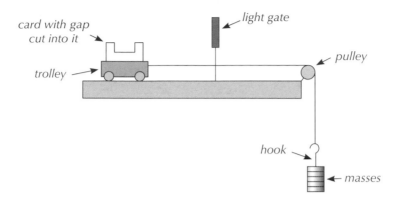

2.1   State what provides the force that causes the trolley to accelerate.

(1 mark)

2.2   Explain how the force can be decreased whilst keeping the mass of the system constant.

(1 mark)

2.3   Explain why there is a gap cut into the piece of card.

(2 marks)

2.4   The force acting on the trolley is doubled. Explain what will happen to the acceleration of the trolley.

(2 marks)

3   A 25 000 kg truck is travelling along a straight, flat road at its top speed of 28 m/s.

3.1   What is the resultant force acting on the truck?

(1 mark)

3.2   Calculate how long it takes the truck to travel 448 m.
Write down any equations you use. Give your answer in s.

(3 marks)

3.3   Explain why changing the body of the truck to make it more streamlined will increase its top speed.

(2 marks)

        The truck stops applying a driving force when it is travelling at 28 m/s.
The total frictional force acting on the truck is 35 000 N.

3.4   Calculate how far the truck travels before coming to a stop, assuming the frictional force is constant. Use the correct equation from the equations listed on page 394.

(5 marks)

4   A skydiver jumps from a plane and reaches terminal velocity after 15 seconds.

4.1   Choose the correct answer to complete the sentence.

Before reaching terminal velocity, the force due to gravity is
(greater than / smaller than / the same as) the resistive force due to air resistance.

(1 mark)

4.2   Choose the correct answer to complete the sentence.

After reaching terminal velocity, the force due to gravity is
(greater than / smaller than / the same as) the resistive force due to air resistance.

(1 mark)

4.3   Sketch a velocity-time graph showing the skydiver's vertical motion during the
first 40 seconds of his dive.

(3 marks)

After 40 seconds the skydiver opens a parachute.

4.4   State the effect opening the parachute will have on the skydiver's terminal velocity.

(1 mark)

4.5   State the effect opening the parachute will have on the resistive
forces acting on the skydiver at a given speed.

(1 mark)

5   An astronaut is on a spacewalk where gravitational forces and air resistance acting on
the astronaut are assumed to be 0. He pushes against a rock with a force F, as shown.

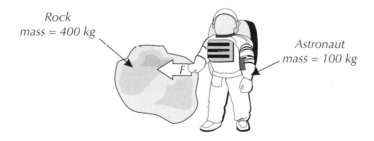

Rock
mass = 400 kg

Astronaut
mass = 100 kg

F

5.1   The rock applies a force on the astronaut.
State the size and direction, relative to F, of this force.

(1 mark)

The astronaut pushes against the rock for 1.2 seconds and he accelerates during this
time. After this point the astronaut moves away from the rock at a velocity of 3 m/s.

5.2   Calculate the acceleration of the astronaut while he's in contact with the rock.
Give your answer in m/s^2.

(2 marks)

5.3   Calculate the size of force F. Give your answer in N.

(2 marks)

Learning Objectives:

- Know the definitions of thinking distance and braking distance.

- Know that a vehicle's stopping distance is the sum of the thinking distance and the braking distance.

- Know and be able to evaluate the factors that can affect reaction time and your ability to react.

- Know that the greater the speed of a vehicle, the greater the stopping distance for a given braking force.

- Know the typical range of values for a person's reaction time.

- Know the factors that affect braking distance, be able to explain how they can affect the distance covered in an emergency stop and the implications this could have on safety.

- Be able to estimate how the stopping distance in an emergency stop varies with vehicle speed.

Specification References
4.5.6.3.1, 4.5.6.3.2,
4.5.6.3.3

Tip: Bad visibility and distractions can delay when you spot a hazard. It doesn't directly affect thinking distance, but it means you won't start to think about stopping until later, making you more likely to crash.

1. Stopping Distances

Stopping distances are important — awareness of them can make the difference between crashing and not. It's easy to get stopping, braking and thinking distances confused, so make sure you learn what each one means.

Stopping distance

In an emergency (e.g. a hazard ahead in the road), a driver may perform an emergency stop. This is where maximum force is applied by the brakes in order to stop the car in the shortest possible distance. The longer it takes to perform an emergency stop, the higher the risk of crashing into whatever's in front.

The total **stopping distance** of a vehicle is the distance covered in the time between the driver first spotting a hazard and the vehicle coming to a complete stop. The total stopping distance is the sum of the thinking distance and the braking distance.

> **stopping distance = thinking distance + braking distance**

- The **thinking distance** is the distance the vehicle travels during the driver's reaction time (the time between seeing a hazard and applying the brakes).

- The **braking distance** is the distance the vehicle travels after the brakes are applied until it comes to a complete stop, as a result of the braking force.

Example

A driver sees a hazard on the road and brakes. His thinking distance is 11 m and his braking distance is 32 m. Find his stopping distance.

Stopping distance = thinking distance + braking distance
= 11 + 32 = 43 m

Many factors affect your total stopping distance — and you can break it down into thinking distance and braking distance to look at the factors that affect each of these.

Thinking distance

Thinking distance is affected by two main factors:

1. How fast you're going — whatever your reaction time, the faster you're going, the further you'll go in that time.

2. How quick to respond you are, i.e. your **reaction time** — this can be affected by tiredness, drugs, alcohol and a lack of concentration. A typical reaction time is between 0.2 and 0.9 s.

Braking distance

Braking distance is affected by four main factors:

1. How fast you're going — the faster you're going, the further it takes to stop. (See the next page for more on this.)

2. How good your brakes are — all brakes must be checked and maintained regularly. Worn or faulty brakes won't be able to apply as much force as well-maintained brakes and could let you down catastrophically just when you need them the most, i.e. in an emergency.

3. How good the tyres are — tyres should have a minimum tread depth of 1.6 mm. In wet conditions, the tread pattern helps to stop water getting trapped between the tyres and the road — they provide a channel through which the water can 'escape'. With too little tread, the tyres may lose contact with the ground, causing the vehicle to slide.

4. How good the grip is — as well as the condition of the tyres, this depends on the weather conditions and the road surface. Water, ice, leaves, diesel spills, muck on the road etc. can greatly increase the braking distance. They can result in reduced friction between the tyres and the road — so you travel further before stopping and may skid. Often you only discover this when you try to brake hard.

You need to be able to describe the factors affecting stopping distance and how this affects safety — especially in an emergency stop. For example, icy conditions increase the chance of skidding (and so increase the stopping distance) so the driver needs to leave more space between their vehicle and the vehicle in front of them. The longer your stopping distance, the more space you need to leave in front in order to stop safely.

Figure 1: A good tyre tread depth helps reduce braking distance.

Figure 2: Petrol spills on roads can reduce grip and cause tyres to skid, increasing the braking distance.

Typical stopping distances

Looking at things simply — for any given braking force, the faster you're going, the greater your stopping distance. Figure 3 shows typical stopping distances at different speeds taken from the Highway Code.

The actual stopping distance will depend on the vehicle and the driver, but if the hazard is closer than the distances shown then it's likely there will be a collision.

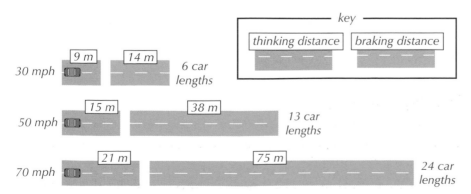

Figure 3: Typical stopping distances taken from the Highway Code.

Tip: Don't forget — things like bad weather and road conditions will make stopping distances even longer.

Tip: The speed limits for lorries in the UK are lower than for cars — their increased weight means they take longer to stop.

To avoid an accident, drivers need to leave enough space between their car and the one in front so that if they had to stop suddenly they would have time to do so safely. 'Enough space' means the stopping distance for whatever speed they're going at.

Speed limits are really important because speed affects the stopping distance so much.

Speed and stopping distance

Speed affects braking distance more than thinking distance. As a car speeds up, the thinking distance increases at the same rate as the speed — so they're directly proportional. For example, if speed doubles (increases by a scale factor of 2), thinking distance also doubles (increases by a factor of 2).

This is because the thinking time (how long it takes the driver to apply the brakes) stays pretty constant — but the higher the speed, the more distance you cover in that same time.

Braking distance, however, increases faster the more you speed up. If speed doubles, braking distance increases 4-fold (2^2). And if speed trebles, braking distance increases 9-fold (3^2). So the braking distance increases with the square of the scale factor of the speed increase.

Braking distance increases in this way because when a vehicle brakes, work must be done to transfer energy away from the vehicle's kinetic energy store. The energy in a vehicle's kinetic energy store is $\frac{1}{2}mv^2$ (see page 26). So if speed doubles, kinetic energy increases 4-fold. Therefore, the work that must be done to stop the vehicle increases 4-fold too.

Tip: For more on how work is done to stop a car, see page 215.

Work done is equal to force × distance ($W = Fs$, page 30) and the braking force is constant (at its maximum). So this means that the braking distance also increases 4-fold.

> ### Example
>
> **When travelling at 20 mph, a driver's thinking distance is 6.0 m and their braking distance is 6.0 m. Estimate their total stopping distance at 80 mph.**
>
>
>
> First work out the scale factor of the speed increase: $80 \div 20 = 4$
>
> Look at each of the thinking and braking distances separately. Start with the thinking distance — it's directly proportional to speed, so simply multiply the thinking distance at 20 mph by the scale factor:
>
> $$6.0 \times 4 = 24 \text{ m}$$
>
> Now look at the braking distance — it increases by the square of the scale factor of the speed increase, so multiply the original braking distance by the scale factor squared:
>
> $$6.0 \times 4^2 = 96 \text{ m}$$
>
> Finally, stopping distance is the sum of thinking and braking distance, so:
>
> $$\text{Stopping distance at 80 mph} = 24 + 96 = 120 \text{ m}$$

Tip: So, when travelling four times faster, your thinking distance will be four times longer and your braking distance will be sixteen (4^2) times longer.

Practice Questions — Fact Recall

Q1 Define stopping distance, thinking distance and braking distance.

Q2 Other than speed, name three factors that affect braking distance.

Q3 Say whether each of the following would affect the thinking distance or the braking distance of a vehicle:

a) Ice on the road
b) Alcohol intake of the driver
c) Tiredness of the driver
d) Petrol spills

Practice Questions — Application

Q1 A driver in a car has a thinking distance of 15 m and a braking distance of 38 m. What's the stopping distance of the car?

Q2 Sasha is driving her colleagues home from work on a winter evening. Explain three factors which could affect the distance she travels before stopping after a hazard appears in the road.

Q3 A driver, travelling at 30 mph, makes an emergency stop. Their thinking distance is 9.0 m and their braking distance is 14.0 m. The driver then accelerates to a new speed of 60 mph. Estimate:

a) the thinking distance at 60 mph.

b) the braking distance at 60 mph.

c) the stopping distance at 60 mph.

Tip: With this type of question, the first thing to do is find the scale factor of the speed increase.

Learning Objective:
- Be able to interpret graphs of stopping distance against speed for various vehicles.

Specification Reference
4.5.6.3.1

2. Graphs of Stopping Distance

It's important to consider speed when keeping a safe distance from other vehicles. You can use graphs to see how speed affects stopping distance.

Stopping distance-speed graphs

Because thinking distance increases at the same rate as the speed, a graph of thinking distance against speed would be linear (a straight line). Because braking distance increases with the square of the scale factor (see the example on page 210), a graph of braking distance against speed would be a curve.

Stopping distance is a combination of both of these distances, so a graph of speed against stopping distance for a car looks like this:

> **Tip:** You need to be able to interpret and use graphs like this for a range of vehicles — they're all a similar shape.

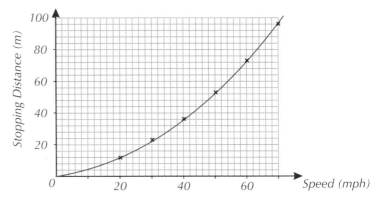

Figure 1: A typical graph of stopping distance against speed for a given vehicle.

Example

The graph below shows stopping distance against speed for two vehicles — A and B. Compare the stopping distance for both vehicles at a speed of 30 mph.

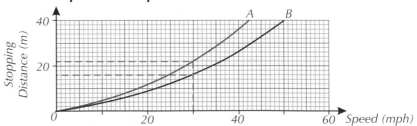

Read off the graph to find the stopping distance for each vehicle at 30 mph:

Vehicle A stopping distance = 22 m   Vehicle B stopping distance = 16 m

Find the difference between these two values: 22 – 16 = 6 m

So the stopping distance for vehicle A is 6 m longer than for vehicle B.

Practice Question — Application

Q1   Using the graph of vehicles A and B in the example above, compare the stopping distance of the two vehicles at 40 mph.

3. Reaction Times

Your reaction time is an important factor in determining your thinking distance. You can measure reaction times in the classroom really easily.

Testing reaction times

Everyone's reaction times are different, and many different factors affect it (see page 208). You can do a simple experiment to investigate your reaction time.

As reaction times are so short, you haven't got a chance of measuring one with a stopwatch. One way of measuring reaction times is to use a computer-based test (e.g. clicking a mouse when the screen changes colour). Another is the ruler-drop test.

The ruler-drop test

Sit with your arm resting on the edge of a table (this should stop you moving your arm up or down during the test). Get someone else to hold a ruler so it hangs between your thumb and forefinger, lined up with zero. You may need a third person to be at eye level with the ruler to check it's lined up.

Without giving any warning, the person holding the ruler should drop it. At this point, close your thumb and finger to try and catch the ruler as quickly as possible.

The measurement on the ruler at the point where it's caught is how far the ruler dropped in the time it took you to react. The longer the distance, the longer the reaction time.

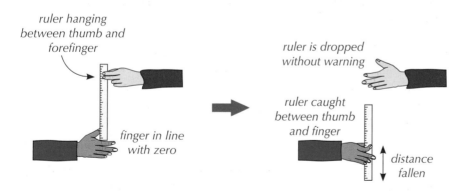

Figure 2: *A diagram of the two main stages of the ruler-drop test.*

You can calculate how long the ruler falls for (the reaction time) because acceleration due to gravity is constant (roughly 9.8 m/s²). See the next page for an example calculation.

Learning Objectives:

- Know that different people have different reaction times.
- Be able to describe an experiment which can be used to test reaction times.
- Be able to use the results of this experiment to calculate a person's reaction time, and recall typical results.

Specification Reference 4.5.6.3.2

Figure 1: *The minimum reaction time for a human is thought to be 0.1 seconds. In high-level races, any athlete starting within 0.1 s of the starting gun is considered to have false-started. This is measured by computerised sensors built into the starting blocks.*

Tip: A 30 cm ruler should work for measuring most people's reaction times. However, you could use a metre ruler if necessary.

Tip: For a refresher on calculations with acceleration due to gravity, see page 182.

Example

Say you catch the ruler at 20 cm.

From page 182, you know: $v^2 - u^2 = 2as$.

$u = 0$, $a = 9.8$ m/s^2 and $s = 0.2$ m.

Rearrange for v: $v = \sqrt{2as + u^2} = \sqrt{2 \times 9.8 \times 0.2 + 0} = 1.97...$ m/s

v is equal to the change in velocity of the ruler.
From page 181 you know that $a = \Delta v \div t$. Rearrange for t.

$t = \Delta v \div a = 1.97... \div 9.8 = 0.202.. = 0.2$ s (to 1 s.f)

This gives you your reaction time of 0.2 s.

> **Tip:** Remember — the distance must be in metres to use this equation. To convert from cm to m, you divide by 100.

Remember, a typical human reaction time is between 0.2 s and 0.9 s, so you should get results in this range.

It's pretty hard to do this experiment accurately, so you should do a lot of repeats. The results will be better if the ruler falls straight down — you might want to add a blob of modelling clay to the bottom to stop it from waving about. Make sure it's a fair test — use the same ruler for each repeat and have the same person dropping it.

> **Tip:** See page 208 for other factors that can affect reaction times and increase the thinking distance of someone driving a vehicle.

You could try to investigate some factors affecting reaction time, e.g. you could introduce distractions by having some music playing or by having someone talk to you while the test takes place. Remember to still do lots of repeats and calculate the mean reaction time with distractions, which you can compare to the mean reaction time without distractions.

Practice Question — Fact Recall

Q1   Describe the main steps involved in measuring your reaction time using the ruler-drop test.

Practice Question — Application

Q1   A student's reaction time is tested using the ruler drop test. She repeats the test three times, grabbing the ruler at 3.0 cm, 5.0 cm and 7.0 cm.

   a) Calculate the average distance, in centimetres, at which the student grabbed the ruler.

   b) Using your answer to part a), calculate their reaction time. Use $g = 9.8$ m/s^2.

   c) The student later repeated the test at the end of the school day. The average reaction time was longer than that from earlier in the day. Suggest a reason for this.

4. Braking and Energy Transfer

Braking works by transferring energy away from the kinetic energy stores of the wheels. This is because work is done against friction.

Braking and friction

Braking relies on friction between the brakes and the wheels. When the brake pedal is pushed, this causes the brake pads to be pressed onto the wheels. This contact causes friction, which causes work to be done.

The work done between the brakes and the wheels transfers energy from the kinetic energy stores of the wheels to the thermal energy stores of the brakes. This means that the temperature of the brakes increases.

The faster a vehicle is going, the more energy it has in its kinetic energy stores, so the more work needs to be done to stop it. This means that a greater braking force is needed to make it stop within a certain distance. A larger braking force means a larger deceleration.

Very large decelerations mean lots of work is done, so lots of energy is transferred to thermal energy stores and the brakes become really hot. If they overheat, they can stop working and can cause the driver to lose control of the vehicle.

Estimating the braking force `Higher`

You can estimate the force involved in accelerations of vehicles using the typical values from page 179.

Learning Objectives:

- Understand the energy transfers which occur when a vehicle brakes.
- Understand why braking increases the temperature of a vehicle's brakes.
- Know that a greater speed means a larger braking force is needed to stop a vehicle in a given distance.
- Know that a greater braking force means a larger deceleration.
- Understand why large decelerations can be dangerous.
- **H** Be able to estimate the force required to produce a deceleration of a vehicle in a typical road situation.

Specification Reference 4.5.6.3.4

`Example` `Higher`

A car travelling at a typical speed makes an emergency stop to avoid hitting a hazard 50 m ahead. Estimate the braking force needed to produce this deceleration.

The typical speed of a car is v = ~25 m/s
and its typical mass is m = ~1000 kg

Assume the deceleration is uniform and rearrange $v^2 - u^2 = 2as$ to find the deceleration.

$$a = (v^2 - u^2) \div 2s = (0^2 - 25^2) \div (2 \times 50) = -6.25 \text{ m/s}^2$$

Then use $F = ma$ to estimate the force using the typical mass.

$$F = 1000 \times 6.25 = 6250 \text{ N}$$

So F is ~6250 N   (The ~ symbol just means it's an approximate value.)

Tip: For more on constant acceleration calculations, see page 182.

Tip: You don't need to keep the minus sign on the acceleration during the force calculation — you just want to know the size of the force, not the direction.

Practice Questions — Application

Q1 A driver makes an emergency stop from a high speed, causing the brakes to overheat. Explain why the brakes overheat.

Q2 A car is travelling at 30 m/s and brakes to avoid hitting a hazard 45 m away. Estimate the braking force needed to produce this deceleration.

Learning Objectives:

- **H** Understand what momentum is.
- **H** Remember and be able to use the formula $p = mv$.
- **H** Know that conservation of momentum says that, in a closed system, the total momentum before and after an event are equal.
- **H** Be able to describe and explain an event (e.g. a collision) in terms of momentum.
- **H** Be able to use conservation of momentum to calculate values in an event, e.g. two objects colliding.

**Specification References
4.5.7.1, 4.5.7.2**

If something is moving along, it'll have some momentum. How much depends on its mass and velocity. If it collides with something, it'll 'share' its momentum with it...

Momentum **Higher**

Momentum is a property of moving objects. The greater the mass of an object and the greater its velocity (see p. 178) the more momentum the object has.

Momentum is a **vector** quantity (see page 147) — it has size and direction (like velocity, but not speed). You can work out the momentum of an object using:

$$p = momentum\ (kg\ m/s) \longrightarrow p = mv \longleftarrow v = velocity\ (m/s)$$
$$m = mass\ (kg)$$

Tip: H Use this formula triangle to help you rearrange the formula:

$$\frac{p}{m \times v}$$

Example 1 **Higher**

A 1800 kg rhino is running north at 9.50 m/s. How much momentum does it have?

$p = mv = 1800 \times 9.50 = 17\ 100$ kg m/s to the north

Example 2 **Higher**

A 40.0 kg rock that is falling off a cliff has 484 kg m/s momentum. What is the rock's velocity?

Rearranging $p = mv$, $v = p \div m = 484 \div 40.0 = 12.1$ m/s downwards

---

Conservation of momentum **Higher**

Tip: H A closed system is just a fancy way of saying that no external forces act.

In a closed system, the total momentum before an event (e.g. a collision or an explosion) is the same as after the event. This is called **conservation of momentum**.

In some collisions, the objects bump into one another and stay stuck (for example, if you throw a lump of clay at a wall). In other collisions, the objects bounce off each other (e.g. when snooker balls hit each other). In both types of collision, the momentum is always conserved (if it's a closed system).

If the momentum before an event is zero, then the momentum after will also be zero. Before an explosion, the momentum is zero. After the explosion, the pieces fly off in different directions, so that the individual momentums of each piece cancel each other out and the total momentum is zero.

In the exam, you might be asked to describe or calculate how the momentums of objects change in simple collisions and explosions.

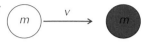 

In snooker, balls of the same size and mass collide with each other. Each collision is an event where the momentum of each ball changes, but the overall momentum stays the same (momentum is conserved).

Before: (m) —v→ (m)   *After:* (m) →(m) ——→

In the diagram, the red ball is initially stationary, so it has zero momentum. The white ball is moving with a velocity v, so has a momentum of $p = mv$.

The white ball then hits the red ball, causing it to move. The red ball now has momentum. The white ball continues moving, but at a much smaller velocity (and so a much smaller momentum).

The combined momentum of the red and white ball is equal to the original momentum of the white ball, mv.

Tip: **H** You can think of the momentum of the white ball being 'shared out' between the two balls after the collision. That's why the white ball's velocity decreases.

Example 2 — Higher

Two skaters, Ed and Sue, approach each other, collide, and move off together as shown. What is their combined momentum before the collision?

2 m/s   1.5 m/s

Ed   *Sue*

80 kg   60 kg

Before collision

After collision

(80 + 60) kg

Figure 1: Ice skaters rely on momentum to keep themselves moving across the ice.

- Choose which direction is positive.

  Let's say "positive" means "to the right".

- Next, work out the total momentum before the collision:

  Total momentum = momentum of Ed + momentum of Sue
  $$= (m_{Ed}\, v_{Ed}) + (m_{Sue}\, v_{Sue})$$
  $$= (80 \times 2) + (60 \times (-1.5)) = 70 \text{ kg m/s to the right}$$

Tip: **H** Sue's velocity has a minus sign in front of it because she's moving to the left.

At what velocity do Ed and Sue move after the collision?

velocity (v)

- Work out what the total momentum after the collision is:

  Total momentum = momentum of Ed and Sue together
  $$= m_{Ed+Sue}\, v_{Ed+Sue}$$
  $$= (80 + 60) \times v$$
  $$= 140v$$

(80 + 60) kg

After collision

- Finally, find out what v is:

  Momentum before collision = momentum after collision
  $$70 = 140v$$
  $$v = 70 \div 140 = 0.5 \text{ m/s to the right}$$

Tip: **H** The final velocity is positive, so it's to the right.

┌─ **Example 3** — **Higher** ─────────────────

Tip: **H** Conservation of momentum is why a gun recoils as it is fired.

A gun fires a bullet as shown. At what speed does the gun move backwards?

velocity of gun (v)   150 m/s

1.0 kg   *0.010 kg*

After bullet is fired

- Say "positive" means "to the right".
- Total momentum before firing = 0 kg m/s
- Total momentum after = momentum of bullet + momentum of gun
  $$= (0.010 \times 150) + (1.0 \times v) = 1.5 + v$$
- $1.5 + v = 0$, i.e. $v = -1.5$ m/s — so the gun moves backwards at 1.5 m/s.

Practice Questions — Fact Recall

Q1 What does the momentum of an object depend on?

Q2 What is the formula for calculating the momentum of an object? Say what each term represents and what its units are.

Q3 How is the momentum before and after a collision linked?

Practice Questions — Application

Q1 Work out the momentum of the following:

a) A 100 g magnet moving north at 0.6 m/s.

b) A 0.80 g bug travelling to the left at 12 m/s.

c) A 5.2 kg rock falling vertically downwards 8.0 m/s.

Q2 Two skaters (A and B) collide and move off together. Skater A, with a mass of 98.0 kg, was moving right at 1.75 m/s. Skater B, with a mass of 53.0 kg, was moving left at 2.31 m/s. What is their combined momentum after the collision?

Q3 A stationary gas canister explodes. What is the momentum of the system before and after the explosion? Explain how you know.

Q4 What is the velocity of the following?

a) A 0.95 kg turtle swimming south with 3.04 kg m/s momentum.

b) A 2000 kg car travelling east with 45 000 kg m/s of momentum.

Q5 What is the mass of the following?

a) A child skiing at 0.75 m/s with 31.5 kg m/s momentum.

b) A dog running at 7.5 m/s with 210 kg m/s of momentum.

Q6 A bullet is fired from a stationary gun. Afterwards, the 1 kg gun moves backwards at 2 m/s and the bullet moves forwards at 200 m/s. What is the mass of the bullet?

Q7 Two skiers (C and D) crash into each other and move off together to the left at 1.50 m/s after the collision. Skier C has a mass of 56.0 kg and was moving right at 1.30 m/s before the collision. Skier D has a mass of 70.0 kg. What was the velocity of skier D before the collision?

6. Changes in Momentum Higher

The faster your momentum changes, the bigger force you'll feel. Lots of safety mechanisms work by slowing down your change in momentum.

Force and change in momentum Higher

You know that when a non-zero resultant force acts on a moving object (or an object that can move), it causes its velocity to change (page 195). This means there is a change in momentum.

You also know $F = ma$, and $a = \frac{\Delta v}{t}$ (change in velocity ÷ change in time) (see pages 196 and 181).

So $F = m \times \frac{\Delta v}{t}$. This can be written as:

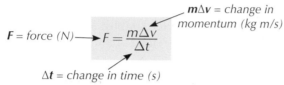

$m\Delta v$ = change in momentum (kg m/s)

F = force (N) ⟶ $F = \dfrac{m\Delta v}{\Delta t}$

Δt = change in time (s)

Remember that Δ just means "change in", so $m\Delta v$ is mass, m, multiplied by change in velocity, Δv. Since the mass of the object is unchanged, this gives the change in momentum.

So the force causing the change is equal to the rate of change of momentum. A larger force means a faster change in momentum.

You need to be able to use these formulas to relate force, mass, velocity and acceleration, so make sure you understand how they link together above.

Example — **Higher**

A 400 g football is travelling horizontally towards a player at 10 m/s. The player kicks the ball back along the same line at a speed of 15 m/s, as shown below. The kick lasts 0.010 s. Calculate the force exerted on the ball by the player.

Before: $u = 10$ m/s   *After:* $v = 15$ m/s

Take right as the positive direction. Then the 15 m/s velocity is negative as the ball is travelling to the left.

$m = 0.40$ kg, $\Delta t = 0.010$ s, $\Delta v = v - u = -15 - 10 = -25$ m/s

Now just substitute the values into $F = \dfrac{m\Delta v}{\Delta t}$:

$F = (0.40 \times -25) \div 0.010 = -1000$ N (i.e. $F = 1000$ N to the left)

Learning Objectives:

- **H** Know that there's a change in momentum when a force acts on a moving object or causes an object to move.
- **H** Be able to use the equation $F = (m\Delta v) \div \Delta t$ and know how it comes from $F = ma$ and $a = \Delta v \div t$.
- **H** Be able to apply these equations to relate force, mass, velocity and acceleration.
- **H** Know that the force experienced by an object is equal to its rate of change of momentum.
- **H** Know and understand some examples of safety features which work by slowing down the change in momentum.

Specification Reference 4.5.7.3

Exam Tip **H**
Don't worry about memorising this equation — it'll be given to you in the exam on the Physics Equation Sheet. You just need to remember the units.

Tip: You might see Δv written as '$v - u$'. This just means final velocity minus initial velocity (see page 57).

Tip: **H** Be careful with negatives here. v is -15 m/s (as it's to the left), and u is 10 m/s (as it's to the right). So $v - u$ is $-15 - 10$.

Safety and rate of change of momentum   Higher

If someone's momentum changes very quickly (like in a car crash or a hard landing), the forces on the body will be very large, and likely to cause injury.

Various safety devices are designed to slow people down over a longer time when there is an impact. The longer it takes for a change in momentum, the smaller the rate of change of momentum, and so the smaller the force. Smaller forces mean any injuries are likely to be less severe.

Figure 1: An inflated air bag in a car. In a crash, the air bag inflates to act as a 'cushion'.

Figure 2: The blue crash mat prevents athletes from getting injured when they land.

Examples   Higher

- Cars have many safety features, such as:

  1. Crumple zones — these are parts of the car's bodywork that are designed to crumple on impact, increasing the time taken for the car to stop.

  2. Seat belts — these stretch slightly, increasing the time taken for the wearer to stop.

  3. Air bags — these inflate during a collision, so that you hit the air bag instead of the dashboard. The compressing air inside it slows you down more gradually than if you'd just hit the hard dashboard.

- Bike helmets contain a crushable layer of foam which helps to lengthen the time taken for your head to come to a stop if you crash. This reduces the impact on your brain.

- Crash mats and cushioned playground flooring increase the time taken for you to stop if you fall or land on them. This is because they are made from soft, compressible (squishable) materials.

Practice Question — Fact Recall

Q1 Explain how air bags in cars help to prevent injury in collisions.

Practice Questions — Application

Q1 A ball undergoes a change in momentum of 27 kg m/s. This change in momentum takes 3 seconds to occur. Calculate the size of the force on the ball which causes this change in momentum.

Q2 A force of 20 N is applied to an object, which experiences a change in momentum of 32 kg m/s. Calculate the time for which the force was applied.

Q3 A paintball fired from a gun experiences a force of 100 N, applied over 0.05 seconds. The mass of the paintball is 50 g. Calculate the speed at which the paintball leaves the gun.

Topic 5d Checklist — Make sure you know...

Stopping Distances

- [ ] That the stopping distance is a measure of the distance it takes a vehicle to stop in an emergency.
- [ ] That the stopping distance of a vehicle is the sum of the thinking distance and braking distance.
- [ ] That the thinking distance is the distance travelled between the driver noticing a hazard and applying the brakes.
- [ ] That the braking distance is the distance a vehicle travels between the brakes being applied and the vehicle coming to a stop.
- [ ] That both the thinking and braking distance depend on the speed at which the vehicle is travelling.
- [ ] That a typical human reaction time is between 0.2 s and 0.9 s.
- [ ] How thinking distance depends upon a person's reaction time and be able to evaluate the factors that affect it, such as drug use and tiredness.
- [ ] How braking distance can be affected by factors such as the weather, road conditions and the conditions of the brakes and tyres, and the implications on safety in an emergency.
- [ ] Some typical stopping distances at typical speeds, and be able to estimate how speed affects the distance required to stop in an emergency.

Graphs of Stopping Distance

- [ ] How to interpret graphs of stopping distance against speed for different vehicles.

Reaction Times

- [ ] How to carry out a simple experiment to measure reaction times.
- [ ] How to calculate reaction times from the results of this experiment.
- [ ] Typical results expected when measuring human reaction times.

Braking and Energy Transfer

- [ ] That when brakes are applied, friction acts against the movement of the wheels, transferring energy from the kinetic energy stores of the wheels to the thermal energy stores of the brakes.
- [ ] How braking causing the temperature of the brakes to increase.
- [ ] That the faster a vehicle is travelling, the greater the braking force required to stop in a given distance.
- [ ] That a greater braking force causes a greater deceleration.
- [ ] That large decelerations can be dangerous, and may lead to brakes overheating and loss of control.
- [ ] H How to estimate the force required to produce a deceleration in a typical road situation.

cont...

Momentum

- [ ] [H] That momentum is a property of every moving object.
- [ ] [H] That momentum (p) has both magnitude and direction, and is a product of the mass (m) and velocity (v) of an object, $p = mv$.
- [ ] [H] That momentum is conserved — in a closed system, the total momentum before a collision is equal to the total momentum after.
- [ ] [H] How to describe and explain events (e.g. collisions) in terms of momentum.
- [ ] [H] How to calculate the momentum, mass or velocity of objects in a collision using conservation of momentum.

Changes in Momentum

- [ ] [H] That when a force acts on an object, causing it to move or change its motion, the object undergoes a change in momentum.
- [ ] [H] How to use the equation $F = \frac{m\Delta v}{\Delta t}$ and how to show that it comes from $F = ma$ and $a = \frac{\Delta v}{t}$.
- [ ] [H] How to relate force, mass, velocity and acceleration with these equations.
- [ ] [H] That the force acting on a moving object is equal to its rate of change of momentum.
- [ ] [H] How various safety mechanisms (e.g. seat belts, air bags, bike helmets and crash mats) decrease the force experienced by a person by increasing the time over which their momentum changes.

Exam-style Questions

1 A truck is travelling along a straight, flat road at its top speed of 30 m/s.
On a clear, dry day, the truck's stopping distance with this driver is 84 m.
The stopping distance is the sum of the thinking distance and the braking distance.

1.1 Explain why the truck's stopping distance when travelling at the same speed in the same conditions may be different for a different driver.

(1 mark)

1.2 The braking distance of the truck is 60 m. The mass of the truck is 2000 kg.
Calculate the force required for it to decelerate to rest in the braking distance given.
Use the correct equation from those listed on page 394. Give your answer in newtons.

(4 marks)

1.3 The driver halves his speed. Estimate the new stopping distance of the truck.

(4 marks)

1.4 To stop the truck, the driver applies the brakes. Explain, in terms of energy transfer, how the brakes cause the car to slow down and come to a stop.

(2 marks)

1.5 Earlier that week, the truck was driving along the same route in heavy rain.
State and explain **two** ways in which this could make the truck more likely to hit an obstacle in the road.

(2 marks)

2 A white snooker ball collides with a stationary blue snooker ball. Both balls have a mass of 0.16 kg. Before the collision, the white ball moves to the right at 0.5 m/s.

Before collision

white ball   stationary blue ball

0.5 m/s

0.16 kg   0.16 kg

2.1 What is meant by the conservation of momentum?

(1 mark)

2.2 Calculate the total momentum of the system before the collision.
Give your answer in kg m/s.

(3 marks)

2.3 After the collision, the white ball moves right at 0.1 m/s.
What is the velocity of the blue ball after the collision?
Give your answer in m/s.

(3 marks)

Exam Tip
In the exam, it's really important that you say the direction <u>of energy transfer</u> of the wave. You might not get full marks for just saying the direction of the wave.

Tip: Another type of wave is a seismic wave. Seismic S-waves are transverse and seismic P-waves are longitudinal. See p. 285.

1. Wave Basics

Waves move through substances carrying energy from one place to another — and once they've gone, it's as if they were never there.

What is a wave?

A **wave** is an oscillation (vibration) that transfers energy without transferring any matter, by making the particles of the substance (or fields) that it is travelling through oscillate.

Waves can be either transverse or longitudinal. These words sound complicated but they describe something simple — the direction of the wave oscillations.

Transverse waves

Waves transfer energy in the same direction that they travel. **Transverse waves** oscillate at right angles to the direction that they travel in.

> In transverse waves the oscillations are perpendicular (at 90°) to the direction of energy transfer of the wave.

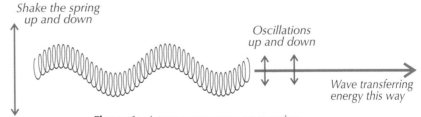

Figure 1: A transverse wave on a spring.

Examples of transverse waves include: light and all other electromagnetic waves (p. 242), ripples on water and waves on strings or springs when wiggled up and down (see Figure 1).

Longitudinal waves

Longitudinal waves have oscillations along the same line as they travel. They have areas of compression, in which the particles are bunched together, and areas of rarefaction, in which the particles are spread out — see Figure 2.

> In longitudinal waves the oscillations are parallel to the direction of energy transfer of the wave.

Examples of longitudinal waves include sound waves (page 279) and a spring when you push the end (see Figure 2).

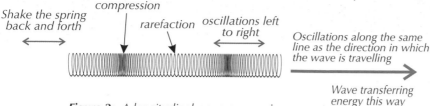

Shake the spring back and forth

compression

rarefaction

oscillations left to right

Oscillations along the same line as the direction in which the wave is travelling

Wave transferring energy this way

Figure 2: *A longitudinal wave on a spring.*

Waves and matter

All waves transfer energy in the direction in which they are travelling, but they don't transfer matter. When waves travel through a medium (a material), such as air or water, the particles of the medium oscillate and transfer energy between each other, but overall the particles stay in the same place — only energy is transferred. You need to be able to describe some observations that provide evidence for this idea, so make sure you know these examples.

Figure 3: *The energy transferred by earthquake waves can be seen in the damage they cause.*

> **Examples**
>
> - Ripples on a water surface cause floating objects, e.g. twigs or birds, to just bob up and down. They don't move the object across the water to the edge. This is evidence that the wave travels but not the water.
>
> - If you strum a guitar string and create sound waves, the sound waves don't carry the air away from the guitar to create a vacuum (completely empty space).

Tip: Waves that need a medium to travel in are classed as mechanical waves. They include water waves, waves in springs and strings, seismic waves and sound waves. Electromagnetic waves are an example of non-mechanical waves (see page 242).

Practice Questions — Fact Recall

Q1 Waves can be transverse or longitudinal.

a) What is meant by a transverse wave?

b) Give an example of a transverse wave.

c) What is meant by a longitudinal wave?

Q2 What is meant by areas of compression and rarefaction in a longitudinal wave?

Q3 Which of the following waves is longitudinal?

    A light   **B** sound   **C** water ripples

Q4 Do water waves cause water molecules to travel across the water's surface? Describe an observation which supports this.

2. Features of Waves

Describing a wave as 'really big and fast' might work for surfers, but physicists need a more accurate way — and that's where amplitude, wavelength, frequency and period come in.

Representing waves

A wave can be represented on a set of axes — a line is drawn to show the displacement of the particles from their undisturbed positions at a moment in time. It is as if a transverse wave had been set up in a piece of string and a 'snapshot' taken of it. Crests and troughs are just points of maximum positive and maximum negative displacement from the particle's rest position — see Figure 1.

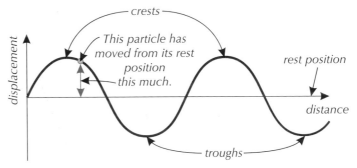

Figure 1: A diagram showing the displacements of particles along a wave.

Wave measurements

There are a few measurements that you can use to describe waves...

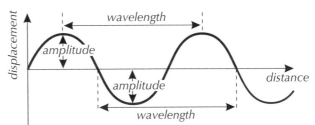

Figure 2: A diagram showing the amplitude and wavelength of a wave.

- The **amplitude** of a wave is the maximum displacement of a point on the wave from its undisturbed (or rest) position. In other words it's the distance from the undisturbed position to a crest or a trough.

- The **wavelength** is the distance between the same point on two adjacent waves. So on a transverse wave it may be the distance between the crest of one wave and the crest of the next wave.

- **Frequency** is the number of complete waves passing a certain point per second. Frequency is measured in hertz (Hz). 1 Hz is 1 wave per second.

Learning Objectives:

- Know what the amplitude, wavelength, frequency and period of a wave are and be able to describe a wave's motion in terms of these properties.

- Be able to show the amplitude and wavelength on a diagram of a wave.

- Be able to calculate the period of a wave given its frequency.

Specification Reference 4.6.1.2

Tip: You might see a wave represented with the horizontal axis showing time instead of distance.

This shows the displacement of a single particle as time passes. Oscilloscope traces are often like this (page 332).

Tip: Amplitude is <u>not</u> the distance from a trough to a crest — it's an easy mistake to make, so watch out...

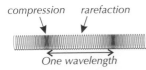

Figure 3: On a longitudinal wave, the distance between the centres of two adjacent compressions is the wavelength.

The period of a wave

The **period** of a wave is the amount of time it takes for a full cycle of the wave to be completed. In other words, it's the length of time between one crest passing a point and the next crest passing the same point.

You can find the period of a wave from the frequency using this equation:

$$\text{Period} = \frac{1}{\text{frequency}}$$

or:

$T = \text{period (s)} \rightarrow T = \frac{1}{f} \leftarrow f = \text{frequency (Hz)}$

Exam Tip
You'll find this equation on the equation sheet in your exam, but the units won't be given.

Example

A buoy measures the frequency of an ocean wave as 0.2 Hz. Calculate the period of this wave.

The frequency is in the correct unit (Hz), so 0.2 Hz can be substituted directly into the formula $T = \frac{1}{f}$:

$$T = \frac{1}{f} = \frac{1}{0.2} = 5 \text{ seconds}$$

Tip: You can use a formula triangle to rearrange the equation.

Practice Question — Fact Recall

Q1 What is:

a) the amplitude of a wave?   b) the wavelength of a wave?

c) the frequency of a wave?   d) the period of a wave?

Practice Questions — Application

Q1 The diagram below shows a man shaking a spring up and down to produce a wave. What is the wavelength of the wave?

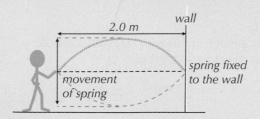

2.0 m

wall

spring fixed to the wall

movement of spring

Q2 An oscilloscope is used to display the wave below.

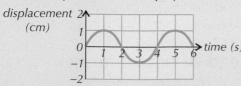

displacement (cm)

time (s)

a) What is the amplitude and the period of the wave shown?

b) Calculate the frequency of the wave.

Figure 4: *An oscilloscope screen showing a waveform. The amplitude, wavelength and frequency can be found from it.*

- Know what the term 'wave speed' means.
- Be able to remember and use the wave equation, $v = f\lambda$, and know that it applies to all waves.
- Be able to describe how the speed of sound waves in air can be measured.
- Be able to measure the speed of ripples in water, and of waves in a solid using suitable apparatus to take appropriate measurements. (Required Practical 8)

Specification Reference 4.6.1.2

3. Wave Speed

A wave's speed is how fast it moves. Finding it is a bit trickier than finding the speed of, say, a car, so that's what the next few pages are all about. Luckily, there's a formula for it though.

The wave equation

The wave speed is the speed at which energy is being transferred (or the speed the wave is moving at). You use this equation, called the wave equation, to work it out:

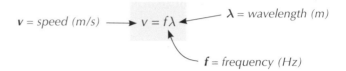

$v = speed\ (m/s) \longrightarrow v = f\lambda \longleftarrow \lambda = wavelength\ (m)$

$f = frequency\ (Hz)$

The wave equation applies to all waves, so it's really useful.

Tip: The symbol for wavelength is the Greek letter λ, which is called 'lambda'.

Example 1

A paddle vibrating up and down in a pool is used to produce waves on the water. The wavelength of each wave is 1.2 m and exactly 2 complete waves are produced per second. Calculate the speed of the wave.

The number of waves produced per second is the frequency.

So $f = 2$ Hz and $\lambda = 1.2$ m.

Substitute these into the wave equation:

$v = f\lambda = 2 \times 1.2 = 2.4$ m/s

Tip: Make sure you can rearrange the formula to calculate either frequency or wavelength. Here's the formula triangle.

Example 2

A wave has a frequency of 4.0×10^7 Hz and a speed of 3.0×10^8 m/s. Find its wavelength.

You're trying to find λ using f and v, so you've got to rearrange the equation.

So $\lambda = v \div f = (3.0 \times 10^8) \div (4.0 \times 10^7) = 7.5$ m

---

Tip: When you state the speed of sound, you should say what it's travelling through. Sound moves at different speeds through different materials — see page 279.

Measuring the speed of sound in air

To measure the speed of sound, you need to find the frequency and the wavelength of a sound wave — you can then use the wave equation $v = f\lambda$.

You can generate a sound wave with a specific frequency by attaching a signal generator to a speaker. This sound wave can then be detected by microphones, which convert it to a trace on an oscilloscope.

Method

Set up your equipment as shown below, with both microphones next to the speaker. The detected wave at each microphone can be seen as a separate wave on the oscilloscope.

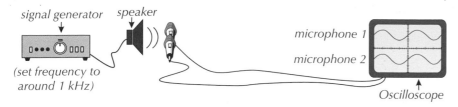

Figure 1: *The initial set-up of apparatus when measuring the speed of sound in air.*

Tip: The waves line up with each other as both microphones are the same distance from the speaker.

Slowly move one microphone away from the speaker. Its wave will shift sideways on the oscilloscope. Keep moving it until the two waves on the oscilloscope are aligned once more. At this point the microphones will be exactly one wavelength apart, so measure the distance between them.

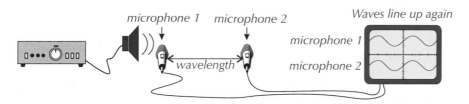

Figure 2: *Arrangement of apparatus when the microphones are one wavelength apart.*

Tip: The traces are both of the same wave, just one has travelled further.

You can then use the formula $v = f\lambda$ to find the speed (v) of the sound wave passing through the air — the frequency (f) of the wave will be equal to the frequency set by the signal generator.

The speed of sound in air is around 330 m/s, so check that your results roughly agree with this.

Tip: Don't forget you need the wavelength in metres and the frequency in Hz to use the wave equation.

Measuring the speed of water ripples

REQUIRED PRACTICAL **8**

You can see water ripples, so they're really useful for investigating wave properties. Again, you use a signal generator to produce waves of a known frequency, but this time you attach it to the dipper of a ripple tank.

A ripple tank is a shallow glass tank used to show the properties of waves. The glass bottom of a ripple tank means a light can be shone on the tank from above to project the wave pattern onto a screen below. This makes it much easier to measure the waves without disturbing them (see Figure 3).

Method

Set up your equipment as shown in Figure 3. Fill your ripple tank with water to a depth of around 5 mm. Connect your dipper to the signal generator and set it off at a known frequency. Dim the lights and turn on the lamp — you'll see a wave pattern made by the shadows of the wave crests on the screen below the tank.

Tip: You should carry out a risk assessment before you do a practical. Take care when using electronics like the lamp and signal generator near the water in the ripple tank.

Tip: Measure the wavelength by looking at the screen from underneath the ripple tank — don't look at it through the water. Your view will be distorted by refraction (see page 232).

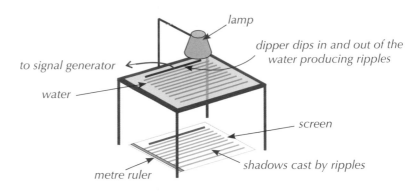

Figure 3: Apparatus for measuring the speed of ripples using a ripple tank.

Tip: If you're struggling to measure the moving waves, you can take a digital photo of the shadows and the ruler and use that to measure the length of 10 wavelengths instead. It should improve your accuracy.

The distance between each shadow line is equal to one wavelength. As the wavelength is small, and the waves are constantly moving, it's tricky to measure just one of these accurately. Instead, measure the distance between shadow lines that are 10 wavelengths apart using a metre ruler. Then divide this distance by 10 to find the average wavelength. This is a suitable method for accurately measuring the wavelength of moving waves, or measuring small wavelengths (page 327).

You can then use the formula $v = f\lambda$ to find the speed (v) of the ripples, where f is the frequency of the signal generator.

Observing a wave on a string

You can also measure waves on a string. To do this, you need a taut string and a piece of kit called a vibration transducer. A vibration transducer will convert an electrical signal from a signal generator into vibrations. When it's attached to a taut string, waves are formed on the string.

Tip: Remember to carry out a risk assessment for each investigation. Be careful of the string snapping and the masses falling.

You can ensure your string is taut by connecting one end to the vibration transducer and passing it over a pulley with masses hanging from the other end. This will keep it pulled nice and tight.

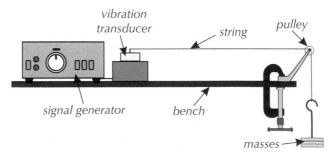

Figure 4: The experimental set-up for measuring the speed of waves along a string.

Set up the equipment as shown, then turn on the signal generator and vibration transducer. The string will start to vibrate.

Adjust the frequency of the signal generator until there's a clear wave on the string (see Figure 5). The frequency you need will depend on the length of string between the pulley and the transducer, and the masses you've used.

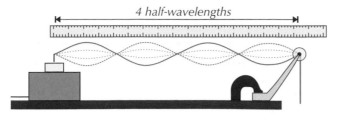

Figure 5: A diagram showing how to measure the wavelength of a clear wave shape on your apparatus.

Tip: This set-up is suitable for investigating waves on a string because it's easy to see and measure the wavelength (and frequency).

You need to measure the wavelength of these waves. The best way to do this accurately is to measure the length of four or five half-wavelengths (or as many as you can) in one go, then divide to get the average half-wavelength. You can then double this average to get a full wavelength.

The frequency of the wave is whatever the signal generator is set to. You can then find the speed of the wave using $v = f\lambda$.

Practice Questions — Fact Recall

Q1 Write down the wave equation. Say what each symbol stands for and the units that it is measured in.

Q2 Describe a way of measuring the speed of sound in air.

Q3 What is the approximate speed of sound in air?

Q4 A wave is generated on a taut string using a vibration transducer. Describe how the speed of the waves on the string can be measured.

Practice Questions — Application

Q1 Calculate the speed of a wave with a wavelength of 0.45 m and a frequency of 15 Hz.

Q2 The speed of sound in a seawater sample is known to be 1500 m/s. A sound wave with a frequency of 3 kHz is produced under water. Calculate the wavelength of this sound wave.

Q3 A light wave has a wavelength of 7.5×10^{-7} m and travels at 3.0×10^8 m/s. What is its frequency?

Q4 A signal generator connected to the paddle in a ripple tank is set to 100 Hz. A lamp is used to cast shadows of the ripples onto a screen underneath the ripple tank. The diagram below shows part of the shadow pattern.

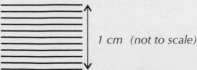

1 cm (not to scale)

Calculate the speed of the ripples.

- Know that waves can be reflected, absorbed or transmitted at a boundary and what these terms mean.

- Know that waves are transmitted and refracted differently by different materials.

- Be able to draw ray diagrams to show the refraction of waves as they move from one material to another.

- **H** Know that refraction is caused by the differences in wave speeds in different substances.

- **H** Be able to draw wavefront diagrams to explain refraction.

Specification References
4.6.1.3, 4.6.2.2

Figure 1: *Refraction of light waves causes objects to look distorted when they are underwater. The light is bent when it passes through boundaries between media.*

Tip: Ray diagrams are mainly used to show the refraction of light, but all types of waves can be refracted. Different types of waves will just be refracted differently.

4. Refraction of Waves

When a wave hits the boundary between the substance it's moving through and a new substance, it sometimes carries on moving through the new substance — but its direction may change.

Waves at a boundary

Waves travel through different materials, e.g. air, water, glass. When they arrive at a boundary between two different materials, three things can happen:

Absorption — The waves may be absorbed by the material the wave is trying to cross into. This transfers energy to the material's energy stores (see p. 22).

Reflection — The waves may bounce back (see p. 235).

Transmission — The waves may be transmitted. This means that they carry on travelling through the new material. However, they often undergo **refraction** — there's more on this to follow.

In reality, a combination of all three of these will usually occur at a boundary.

Refraction basics and ray diagrams

When a wave reaches the boundary between two different materials, it can be transmitted through the new material. If it hits the boundary at an angle, it changes direction — it's refracted.

A **ray** is a straight line showing the path a wave, such as light, travels along. You can show the path taken by a light ray on a ray diagram.

1. Draw the boundary between the two materials and then add in the **normal** (a straight line that is at 90° to the boundary). The normal is usually shown as a dotted line.

2. Draw an incoming (incident) ray that meets the normal at the boundary. The angle between the ray and the normal is the **angle of incidence**. (If you're given this angle, make sure to draw it carefully with a protractor.)

3. Now draw the refracted ray on the other side of the boundary. The angle that the refracted ray makes with the normal is called the **angle of refraction**.

4. The angle of refraction could be smaller than the angle of incidence (e.g. when light moves from air into a glass block — see Figure 2) or bigger than the angle of incidence (e.g. when light moves from glass into air — see Figure 3).

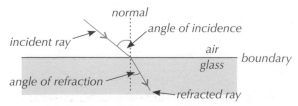

Figure 2: *Refraction of a light ray moving from air into glass. When the light is refracted, it bends towards the normal.*

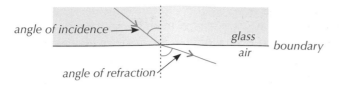

*Figure 3: Refraction of a light ray moving from glass into air.
When the light refracts, it bends away from the normal.*

If a ray travels along the normal and hits the boundary at 90°, it'll pass through without changing direction. The angles of incidence and refraction will both be 0°.

Tip: A ray that hits the boundary at 90° won't change direction.

incident ray

boundary

More on refraction   [Higher]

Refraction occurs because waves travel faster in some materials than others, so the speed of a wave can change as it crosses a boundary. When a wave refracts, its speed changes, but its frequency remains the same.

Since the speed, wavelength and frequency of a wave are all related by the wave equation ($v = f\lambda$ — see page 228), if the speed changes but the frequency is constant, the wavelength must change. If the speed of the wave increases, its wavelength increases. If the speed decreases, so does the wavelength.

The optical density of a material is a measure of how quickly light travels through it — the higher the optical density, the slower light travels.

Tip: **H** You might also hear the term 'refractive index'. This is just a measure of a material's optical density. The bigger the refractive index, the more optically dense it is.

Wavefronts

Wavefronts are imaginary lines drawn through certain points on waves, e.g. through each crest. They're perpendicular (at right angles) to the direction in which the wave is moving (and so perpendicular to the line you'd draw in a ray diagram).

Figure 4 is a wavefront diagram showing light waves travelling along a normal to a boundary. The wavefronts are closer together in the more optically dense material, showing the decrease in wavelength.

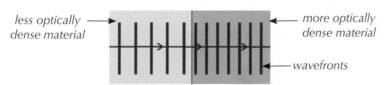

less optically dense material → ... ← more optically dense material

← wavefronts

*Figure 4: A wavefront diagram showing light
travelling along a normal to a boundary.*

Tip: **H** The wave travelling along the normal isn't refracted, but its speed and wavelength still change.

Wavefront diagrams are useful for explaining why refraction happens. When a wave crosses a boundary into a new substance at an angle to the normal, one end of it enters the new material before the rest of the wave. This means that end changes speed before the rest of the wave, causing the wave to change direction.

Which way the wave bends depends on whether it moves faster or slower in the new medium.

- If a wave slows down at a boundary, it bends towards the normal.

- If a wave speeds up at a boundary, it bends away from the normal.

Figure 5 shows a light wave entering a more optically dense material. The wave moves more slowly in the new material, so it bends towards the normal. The angle of refraction is smaller than the angle of incidence.

Tip: [H] Imagine driving a go-kart in the direction of the wave, and that the denser medium is a pool of mud. One side of the go-kart will reach the mud first and slow down. The other side of the go-kart isn't yet in the mud so it carries on moving quickly, causing the go-kart to swing round, changing direction.

more optically dense material

normal

less optically dense material

Figure 5: *A light wave entering a more optically dense material at an angle to the normal.*

Figure 6 shows a light wave entering a less optically dense material. It moves faster in the new substance, so it bends away from the normal. The angle of refraction is greater than the angle of incidence.

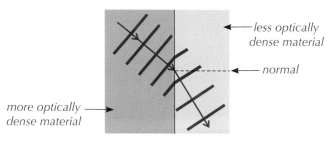

less optically dense material

normal

more optically dense material

Figure 6: *A light wave entering a less optically dense material at an angle to the normal.*

Practice Questions — Fact Recall

Q1   What is refraction?

Q2   Draw a ray diagram to show the refraction of a light wave crossing from air into a glass block at an angle to the normal.

Q3   What happens to the path of a wave if it crosses the boundary between two media travelling along the normal?

Q4   What happens to the speed, wavelength and frequency of a light wave as it crosses into a medium with a lower optical density?

Q5   How is decreased wavelength shown on a wavefront diagram?

Q6   Describe the refraction of a wave that crosses a boundary at an angle to the normal and speeds up.

Practice Questions — Application

Q1   Draw a ray diagram to show a light wave hitting a boundary between two substances with an angle of incidence of 40° and an angle of refraction of 60°.

Q2   Choose the correct word to make this statement true.
The wave in Q1 **slows down / speeds up** as it enters the new substance.

Tip: You don't need to know how to work out the amount that waves change direction, just that they do.

5. Reflection of Waves

When light waves hit a mirror they bounce back, or 'reflect', in such a way that you can see yourself looking back. Other types of waves can reflect too.

Ray diagrams of reflections

As well as (or instead of) being refracted, a wave could be reflected when it meets a boundary between two different materials. Just like with refraction, you can show reflection on a ray diagram.

To draw a ray diagram showing reflection, you need to draw the incoming ray and the reflected ray. You also need to draw in the normal.

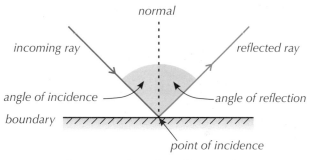

Figure 1: *A ray diagram to show the reflection of a wave at a surface.*

The **angle of incidence** is the angle between the incoming wave and the normal. The **angle of reflection** is the angle between the reflected wave and the normal. The **law of reflection**, which applies to every reflected wave, is:

> angle of incidence = angle of reflection

So, a reflected wave always bounces off a boundary at the same angle to the normal as it hit the boundary (but on the other side of the normal).

Types of reflection

Waves are reflected at different boundaries in different ways.

Specular reflection happens when parallel waves are reflected in a single direction by a smooth surface. E.g. when light is reflected by a mirror you get a nice clear reflection.

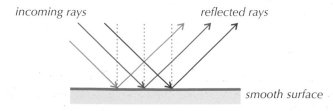

Figure 2: *A ray diagram showing specular reflection.*

Tip: The dull side of a mirror is always shown with diagonal lines like in Figure 1.

Tip: It's not just light that follows this rule, it's true for all waves.

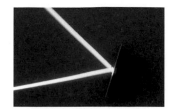

Figure 3: *A beam of light being specularly reflected by a mirror.*

Tip: There's an investigation into the different types of reflection on page 238.

Figure 5: *A beam of light being diffusely reflected by a piece of paper.*

Diffuse reflection is when parallel waves are reflected by a rough surface (e.g. a piece of paper) and the reflected rays are scattered in lots of different directions.

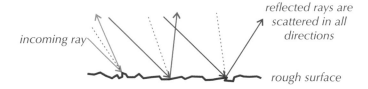

Figure 4: *A ray diagram showing diffuse reflection.*

The angle of incidence is still equal to the angle of reflection for each ray. However, the tiny bumps on the surface mean that the normal for each ray is different. Therefore, the angle of incidence is different for each ray.

When light is reflected by a rough surface, the surface appears matt (not shiny) and you don't get a clear reflection of objects.

Practice Questions — Fact Recall

Q1 Sketch a diagram showing a reflected ray. Label the normal, the angle of incidence and the angle of reflection.

Q2 Explain the difference between specular and diffuse reflection.

Practice Questions — Application

Q1 Explain why you may be able to see your reflection in the glass part of a door, but not in the wooden part of the door.

Q2 Construct a ray diagram to show the reflection of a ray with an angle of incidence of 40°.

6. Investigating Light

Light is a good thing to use to investigate the behaviour of waves. You can produce it easily, you can see it, and it's much safer than, say, gamma rays.

Experimenting with light

Both of these experiments use rays of light, so it's best to do them in a dim room so you can clearly see the light rays. You need to use either a ray box or a laser to produce thin rays of light. This is so you can easily see the middle of the ray when tracing it and measuring angles from it.

Investigating refraction

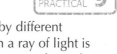

The boundaries between different substances refract light by different amounts. You can investigate this by looking at how much a ray of light is refracted when it passes from air into different transparent materials, such as glass or clear plastic.

The first thing to do is place a transparent rectangular block on a piece of paper and trace around it. Then use a ray box or a laser to shine a ray of light at the middle of one side of the block.

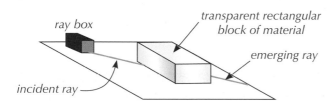

Figure 2: *Set-up of the equipment needed to investigate refraction.*

Trace the incident ray on the paper, and do the same for the light ray that emerges on the other side of the block. Remove the block and, with a straight line, join up the lines for the incident ray and the emerging ray. This shows the path of the refracted ray through the block.

Draw the normal at the point where the light ray entered the block. Use a protractor to measure the angle between the incident ray and the normal (the angle of incidence, *I*) and the angle between the refracted ray and the normal (the angle of refraction, *R*).

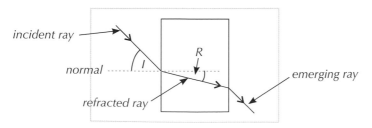

Figure 3: *Rays and angles marked on the piece of paper.*

Repeat this experiment using rectangular blocks made from different materials, keeping the incident angle the same throughout. Record your results in a table. You should find that the angle of refraction changes for different materials.

Learning Objectives:

- Be able to investigate how much different substances refract light. (Required Practical 9)
- Be able to investigate how different substances reflect light. (Required Practical 9)

Specification Reference 4.6.1.3

Tip: You'll need to do a risk assessment before you start.

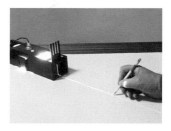

Figure 1: *A ray box is basically a lamp with a metal box around it. In one of the sides you can insert a metal plate with one or more narrow slits to produce narrow beams of light.*

Tip: Make sure you trace the light rays (and join them up) using a ruler and sharp pencil.

Tip: **H** This difference in angle of refraction between materials is due to their different optical densities — the different materials slow down the light ray by different amounts (see p. 233).

Investigating reflection

Different surfaces reflect light in different ways, depending on how smooth they are (see pages 235-236). To investigate this, take a piece of paper and draw a straight line across it. Place a straight-edged object on the paper so that it lines up with this line.

Shine a ray of light at the object's surface and trace the incoming and reflected light beams. Add in a normal line that meets the surface at the point of incidence.

Draw and label the incident and reflected rays. Use a protractor to measure the angle of incidence and the angle of reflection and record these values in a table. Also measure and record the width of the reflected ray and make a note of how bright it is.

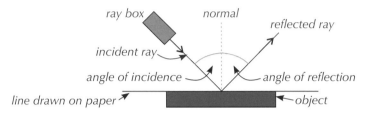

Figure 4: *Set-up of reflection investigation with angles marked.*

Repeat this experiment for a range of objects. You should see that smooth surfaces like mirrors give clear reflections (the reflected ray is as thin and bright as the incident ray). Rough surfaces like paper will cause diffuse reflection (p. 236), where the reflected beam is wider and dimmer (or not observable at all). You should also find that the angle of incidence always equals the angle of reflection.

Practice Questions — Fact Recall

Q1   Describe an experiment you could do to measure how much different materials refract light.

Q2   Describe an experiment you could do to measure how light is reflected by different surfaces.

Practice Question — Application

Q1   A beam of light is shone onto a piece of polystyrene. A second identical beam of light is shone onto a sheet of aluminium foil. In both cases, the angle of incidence is 30°. Describe the difference you would expect to see between the two reflected rays.

Topic 6a Checklist — Make sure you know...

Wave Basics

☐ That waves are vibrations that transfer energy by causing particles (or fields) to vibrate.

☐ That there are two types of wave — transverse (e.g. water ripples) and longitudinal (e.g. sound waves).

☐ That the vibrations in transverse waves are perpendicular to the direction of energy transfer.

☐ That longitudinal waves have regions of compression and rarefaction, and that their vibrations are parallel to the direction of energy transfer.

☐ Examples showing that waves transfer energy as they travel through matter, but do not transfer the matter itself.

Features of Waves

☐ What the amplitude, wavelength, frequency and period of a wave are, and how to describe a wave using them.

☐ How to represent amplitude and wavelength on a diagram of a wave.

☐ How to calculate the period of a wave from the frequency, and vice versa.

Wave Speed

☐ That the wave equation, $v = f\lambda$, applies to all waves and how to use it.

☐ How to measure the speed of sound in air and the speed of water ripples and waves on strings, using suitable apparatus to make measurements.

Refraction of Waves

☐ That waves can be reflected, absorbed or transmitted when they meet the boundary between two different materials.

☐ How to draw ray diagrams to show the refraction of waves as they move between types of matter.

☐ **H** That the difference in the speed of waves in different substances causes refraction.

☐ **H** How to use wavefront diagrams to explain refraction.

Reflection of Waves

☐ How to draw ray diagrams to show the reflection of a wave.

☐ That reflection can be specular or diffuse, depending on the smoothness of the surface.

Investigating Light

☐ How to investigate the refraction of light by different substances.

☐ How to investigate how different surfaces reflect light.

Exam-style Questions

1 The diagram below shows a transverse wave on a rope.

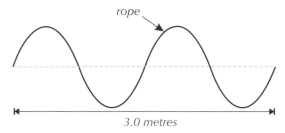

rope

3.0 metres

1.1 What is the wavelength of the wave on the rope?

(1 mark)

1.2 Complete the sentence below by selecting the correct answer from the options given.

Increasing the (frequency / wavelength / amplitude) of the wave on the rope will increase the number of complete waves passing a point on the rope each second.

(1 mark)

The frequency of the wave is 2 Hz.

1.3 Calculate the period of the wave. Give your answer in seconds.
Use the correct equation from the equations listed on page 394.

(2 marks)

1.4 Write down the equation linking wave speed, wavelength and frequency.

(1 mark)

1.5 Calculate the speed of the waves. Give your answer in m/s.

(2 marks)

1.6 A wave on a rope is a transverse wave. Waves can be transverse or longitudinal.
Give **one** similarity between transverse and longitudinal waves.

(1 mark)

1.7 List A, shown below, gives the names of three different waves. List B gives two possible wave types.

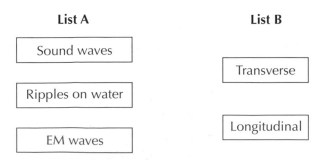

List A	List B
Sound waves	
	Transverse
Ripples on water	
	Longitudinal
EM waves	

Match each wave in List A to the correct wave type(s) in List B.

(2 marks)

2   A light ray hits a mirror as shown in the diagram below.

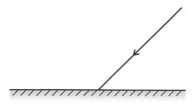

2.1   Copy and complete the ray diagram to show the reflection.
Label the angle of incidence and the angle of reflection.

(2 marks)

2.2   A student draws the ray diagram below to show reflection from a rough surface.

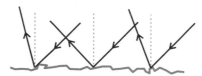

Explain why the student's diagram is incorrect.

(1 mark)

2.3   Name the type of reflection that occurs at a rough surface.

(1 mark)

3   A student is investigating refraction and shines a light ray onto the side of a transparent glass block. She traces the light ray and marks the point where the light leaves the glass block. She removes the glass block and draws the path of the refracted ray through the block.

3.1   Suggest what the student might be using to produce the light ray. Explain your choice.

(2 marks)

3.2   The student measures the angle of incidence as 47° and the angle of refraction as 32°. Construct a ray diagram to show the refraction of the light ray.

(2 marks)

The wavefront diagram below shows the wave during this investigation.

3.3*   Explain what this diagram shows happening to the wave in terms of wavelength, speed and direction.

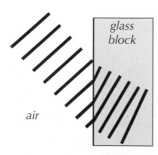

(6 marks)

1. What are Electromagnetic Waves?

Electromagnetic waves are a group of waves that all travel at the same speed in a vacuum. All the different wavelengths of EM waves form a spectrum.

The electromagnetic (EM) spectrum

Electromagnetic waves are a group of transverse waves (see page 224). They are sometimes called electromagnetic radiation. They consist of vibrating electric and magnetic fields, which is why they can travel through a vacuum — they don't rely on vibrating matter.

EM waves form a spectrum of waves with different wavelengths. The spectrum is continuous — meaning there's no gaps in it.

> The **electromagnetic (EM) spectrum** is a continuous spectrum of all the possible wavelengths of electromagnetic waves.

As electromagnetic waves with longer wavelengths have lower frequencies, you can also say that it's a spectrum of all the possible frequencies of electromagnetic waves.

EM waves are grouped into seven basic types according to their wavelength (or frequency), but remember the spectrum is continuous — so the different regions actually merge into each other, see Figure 1.

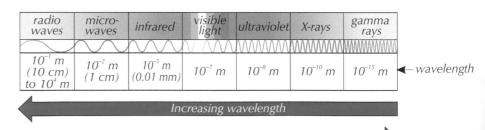

radio waves	micro- waves	infrared	visible light	ultraviolet	X-rays	gamma rays
10^{-1} m (10 cm) to 10^4 m	10^{-2} m (1 cm)	10^{-5} m (0.01 mm)	10^{-7} m	10^{-8} m	10^{-10} m	10^{-15} m

← wavelength

Increasing wavelength

Increasing frequency

Figure 1: The electromagnetic spectrum, with decreasing wavelength and increasing frequency from left to right.

All the different types of EM wave travel at the same speed (about 3×10^8 m/s) in a vacuum (e.g. space). They travel slightly slower in air, but still all at the same speed as each other.

Electromagnetic waves with different wavelengths (or frequencies) have different properties. The properties of the different types of electromagnetic waves are covered later on in the topic. Our eyes can only detect electromagnetic waves that fall into the visible light part of the spectrum — there's more on visible light on pages 253-254.

Electromagnetic waves and energy

All electromagnetic waves transfer energy from a source to an absorber.

Example 1

A hot object (the source) transfers energy by emitting infrared radiation, which is absorbed by the surroundings.

Example 2 **Higher**

Oscillating electrons (the source) in a radio transmitter produce radio waves. These transfer energy to a receiver where it causes the electrons in the receiver to oscillate — see page 245.

Exam Tip
Make sure you can give some examples of EM energy transfer. Another example is microwaves transferring energy to cook food (see p. 247 for more on this).

Atomic changes and electromagnetic waves

Electromagnetic waves can be produced by changes inside atoms. These could be changes to the arrangement of electrons or changes within the nucleus.

Electron changes

If an atom absorbs energy, some of its electrons move to higher energy levels within the atom. When each electron falls back down to its original level, an electromagnetic wave is produced.

Tip: There are several ways of transferring energy into the energy stores of atoms and their electrons. E.g. energy can be transferred electrically or by heating.

The electron absorbs energy and is excited to a higher energy level. *The electron falls back down and the excess energy is released as an electromagnetic wave.*

Tip: See page 122 for more on atomic structure and energy levels.

Figure 2: Electromagnetic waves are produced when excited electrons fall back down.

Different atoms have different energy levels, so there are lots of possible changes that can happen within atoms and molecules. This is why electromagnetic waves produced in this way have such a large range of frequencies. The higher the frequency of an electromagnetic wave, the more energy it transfers. So an atomic change that releases a lot of energy will produce a high-frequency electromagnetic wave.

The huge number of possible changes within atoms also allows a wide range of frequencies of electromagnetic radiation to be absorbed by electrons.

Nuclear changes

Gamma rays are high-energy electromagnetic waves which come from changes in the nuclei of atoms. Basically, unstable nuclei of radioactive atoms decay, giving out particles — there's more on the types of particles on pages 126-127. After spitting out a particle, the nucleus might need to get rid of some extra energy. It does this by giving out a gamma ray.

Figure 3: Neon signs work by passing electricity through neon gas. This excites electrons in the neon atoms. When the electrons fall back down, waves from the visible light part of the EM spectrum are produced.

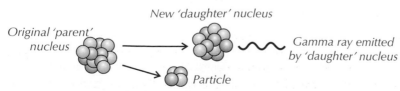

New 'daughter' nucleus

Original 'parent' nucleus

Gamma ray emitted by 'daughter' nucleus

Particle

Figure 4: Radioactive decay resulting in gamma ray emission.

Electromagnetic waves and matter

Tip: See pages 253-254 for how differences in behaviour when waves meet a boundary determine the colour or transparency of an object.

When any wave meets a boundary between two different materials, it can be reflected, absorbed or transmitted (see page 232). Electromagnetic waves of different wavelengths are reflected, absorbed or transmitted differently when they meet a boundary. This is one of the reasons why different types of electromagnetic waves have different uses (see pages 245-250).

Also, EM waves with different wavelengths are refracted (see page 232) by different amounts — see below for an example of this.

Example

White light is made up of all the wavelengths of electromagnetic radiation in the visible part of the spectrum. Each narrow band of wavelengths corresponds to a different colour, from red to violet. The different wavelengths all bend (refract) by different amounts when they enter and leave a prism, so the colours are separated out.

Practice Questions — Fact Recall

Q1 Are electromagnetic waves transverse or longitudinal?

Q2 What is the electromagnetic spectrum?

Q3 Write down the seven types of electromagnetic radiation in order of increasing wavelength.

Q4 Which type of electromagnetic wave has the highest frequency?

Q5 What can you say about the speed of different types of electromagnetic wave travelling in a vacuum?

Q6 Give an example of a situation in which electromagnetic waves transfer energy from a source to an absorber.

Q7 Describe two types of atomic change that produce electromagnetic waves.

Q8 What can happen to an electromagnetic wave when it meets the boundary between two substances? Will the exact same thing always happen to any type of electromagnetic wave when it meets the same boundary?

Tip: Remember, you might hear electromagnetic waves called electromagnetic radiation — they're the same thing.

Practice Question — Application

Q1 An electromagnetic wave source produces microwaves.

a) The wave source is adjusted so that it produces EM waves with a lower frequency. What effect does this have on the wavelength of the waves produced?

b) The wave source is now giving out a different type of electromagnetic radiation. What type of electromagnetic radiation is it now giving out?

2. Radio Waves

Radio waves are the electromagnetic waves with the longest wavelengths (between about 10 cm and 10 000 m — see page 242). This huge range is split up further into short-, medium- and long-wave radio signals.

How are radio waves produced? `Higher`

You can produce **radio waves** using an alternating current (ac) in an electrical circuit. Alternating currents are made up of oscillating charges (electrons). As the charges oscillate, they produce oscillating electric and magnetic fields, i.e. electromagnetic waves, which is what radio waves are (see page 242). The frequency of the radio waves produced will be equal to the frequency of the alternating current. The object in which charges oscillate to create the radio waves is called a transmitter.

How are radio waves received? `Higher`

When transmitted radio waves reach a receiver, the radio waves are absorbed. The energy carried by the waves is transferred to the kinetic energy stores of the electrons in the material of the receiver. This causes the electrons to oscillate and, if the receiver is part of a complete electrical circuit, it generates an alternating current. This current has the same frequency as the radio wave that generated it.

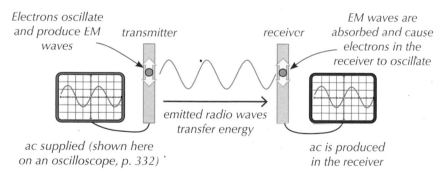

Electrons oscillate and produce EM waves   transmitter   receiver   *EM waves are absorbed and cause electrons in the receiver to oscillate*

emitted radio waves transfer energy

ac supplied (shown here on an oscilloscope, p. 332)

ac is produced in the receiver

Figure 1: *A diagram showing how radio waves are produced, transmitted and received.*

Using radio waves for communication

Radio waves are used mainly for radio and TV signals. They are sent out by transmitters and received by TV or radio aerials (receivers). Different wavelengths of radio wave are used in different ways.

If you're doing Higher tier exams, you'll need to be able to recall the properties that make radio waves suited to each of their uses. But even if you're not doing Higher tier, you could still be asked to apply anything you already know about EM waves to their use in a practical context. So make sure you understand the stuff on the next page.

Learning Objectives:

- **H** Know that oscillations in electric currents can produce radio waves.

- **H** Know that a radio wave causes oscillations in a receiver, which can produce an alternating current of the same frequency as the radio wave.

- Know some practical applications of radio waves.

- **H** Know why radio waves are suited to these uses.

Specification References 4.6.2.3, 4.6.2.4

Tip: Oscillating means moving back and forth, i.e. vibrating.

Tip: See page 89 for more on alternating current.

Long-wave radio

Long-wave radio (wavelengths of 1 – 10 km) can be transmitted and received halfway round the world because long wavelengths diffract (bend) around the curved surface of the Earth (see Figure 3). Long-wave radio wavelengths can also diffract around hills, into tunnels and all sorts. This diffraction effect makes it possible for radio signals to be received even if the receiver isn't in the line of sight of the transmitter.

Short-wave radio

Short-wave radio signals (wavelengths of about 10 m – 100 m) don't diffract around the Earth's curve, but they can still be received at large distances from the transmitter. They are reflected between the Earth and the ionosphere — an electrically charged layer in the Earth's upper atmosphere (see Figure 3).

Bluetooth® uses short-wave radio waves to send data over short distances between devices without wires (e.g. wireless headsets so you can use your phone while driving a car).

Medium-wave radio

Medium-wave signals (well, the shorter ones) can also reflect from the ionosphere, depending on atmospheric conditions and the time of day.

TV signals and FM radio

The radio waves used for TV and FM radio transmissions have very short wavelengths (10 cm – 10 m). To get reception, you must be in direct sight of the transmitter — the signal doesn't bend around hills or travel far through buildings.

Figure 2: A television aerial raised high above a house to increase the signal quality. The aerial needs to be in direct sight of the transmitter as the radio waves used for TV signals do not diffract around large obstacles.

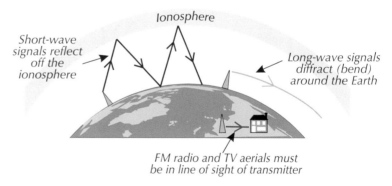

Figure 3: A diagram showing how different wavelength radio waves travel.

Practice Questions — Fact Recall

Q1 What type of current is used to produce radio waves?

Q2 Give two uses of radio waves.

Q3 Short-wave and medium-wave radio signals can reflect off layers in the Earth's atmosphere. Suggest why this lets them be used to broadcast radio over long distances.

3. More EM Waves and their Uses

Because of their different properties, different electromagnetic waves are used for a range of different purposes, from communications to treating cancer.

Useful properties of EM waves

Just as with radio waves, if you're doing Higher tier exams, you'll need to be able to recall the properties that make each of the following EM waves suited to each of their uses. But even if you're not doing Higher tier, you could still be asked to apply anything you already know about EM waves to their use in a practical context. So make sure you understand everything over the next few pages.

Microwaves

The **microwave** region of the EM spectrum covers a range of different wavelengths. The short wavelength microwaves have different properties from the long wavelength microwaves, which means they have different uses.

Satellite communications

Communication to and from satellites usually uses microwaves. This is because certain wavelengths of microwave can pass easily through the Earth's atmosphere without really being reflected, refracted, diffracted or absorbed, which means they can reach satellites. Satellite communications have a range of applications, e.g. satellite TV, satellite phones and internet, and military communications.

For all types of satellite communications, the signal from a transmitter is transmitted into space, where it's picked up by the satellite's receiver orbiting high above the Earth. The satellite transmits the signal back to Earth in a different direction, where it's received by a satellite receiver on the ground. There is a slight time delay between the signal being sent and received because of the long distance the signal has to travel.

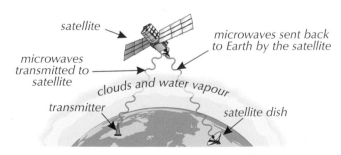

Figure 1: A diagram showing how microwaves are used for satellite communication.

Cooking

The wavelengths of microwaves used in satellite communications pass through the Earth's watery atmosphere. However, the microwaves used in microwave ovens have wavelengths which allow them to be absorbed by water molecules in food.

Learning Objectives:

- Know that microwaves are used for satellite communication and for cooking.
- Know that infrared radiation is used in cooking and that it allows infrared cameras and electric heaters to work.
- Know that visible light is used for communications via fibre optic cables.
- Know that ultraviolet light allows energy efficient lamps and sun tanning lamps to work.
- Know some of the medical applications of X-rays and gamma rays.
- **H** Know why each type of EM wave is suited to its uses.

Specification Reference 4.6.2.4

Tip: Satellite TV uses microwaves but terrestrial TV (normal TV) uses radio waves (see previous page).

Figure 2: The TV satellite dishes on these flats point in specific directions in order to receive the microwave signal from a space satellite.

These microwaves penetrate up to a few centimetres into the food before being absorbed and transferring the energy they are carrying to the thermal energy stores of the water molecules in the food, causing the water to heat up. The water molecules then transfer this energy to the rest of the molecules in the food, which quickly cooks it.

Infrared radiation

Going along the EM spectrum in order of decreasing wavelength, **infrared (IR) radiation** comes next. Its uses are related to temperature.

Infrared cameras

Infrared (IR) radiation is given out by all objects — and the hotter the object, the more IR radiation it gives out. Infrared cameras can be used to detect infrared radiation and monitor temperature. The camera detects the IR radiation and turns it into an electrical signal, which is displayed on a screen as a picture.

Figure 3: An infrared camera being used to monitor the skin temperature of people arriving at an airport. Raised skin temperature can be a sign of illnesses such as bird flu.

Examples

- The heat loss through different parts of a house can be detected using an infrared camera. Colour coding is used to show different amounts of infrared.

hot

cold

- Infrared cameras can also be used as night-vision equipment by the military and by the police to spot criminals. Often the image isn't colour-coded — the hotter an object is compared to its surroundings, the brighter it appears. It works best at night when the surroundings are colder.

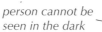

person cannot be seen in the dark

night-vision camera senses heat difference between person and surroundings

Cooking

Absorbing IR radiation causes objects to get hotter. This means that food can be cooked using IR radiation — the temperature of the food increases when it absorbs IR radiation, e.g. from a toaster's heating element, see Figure 4.

Electric heaters

Electric heaters heat a room in the same way. They contain a long piece of wire that heats up when a current flows through it. This wire then emits lots of infrared radiation (and a little visible light — the wire glows). The emitted IR radiation is absorbed by objects and the air in the room — energy is transferred by the IR waves to the thermal energy stores of the objects, causing their temperature to increase.

Figure 4: The element of a toaster heats up and emits infrared radiation. This is absorbed by the bread, causing it to 'toast'.

Visible light

You could argue that the most important use of **visible light** is allowing us to see. However, make sure you know about the following use too.

Fibre optic cables

Optical fibres are thin glass or plastic fibres that can carry data (e.g. from telephones or computers) over long distances. Pulses of visible light are used, because when the light hits the walls of the fibres, it's reflected back into the fibre. This means the light is bounced back and forth until it reaches the end of the fibre.

Visible light is also suited to this use as it will travel down the fibre without being absorbed or scattered much, so the signal hardly weakens at all.

Figure 5: A fibre optic cable which contains a bundle of optical fibres.

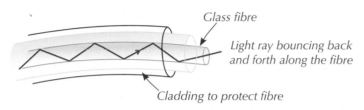

Glass fibre

Light ray bouncing back and forth along the fibre

Cladding to protect fibre

Figure 6: An optical fibre transmitting a pulse of visible light.

Ultraviolet radiation

Ultraviolet (UV) radiation is produced by the Sun, and exposure to it is what gives people a suntan.

Sun tanning lamps

When it's not sunny, some people go to tanning salons where UV lamps are used to give them an artificial suntan. However, overexposure to UV radiation can be dangerous (see page 251 for more details).

Security ink

Fluorescence is a property of certain chemicals. When UV radiation hits them, it's absorbed and visible light is emitted. That's why fluorescent colours look so bright — they actually emit light. Security pens can be used to mark valuable items, such as laptops, with fluorescent ink. Under UV light the ink will glow (fluoresce), but it's invisible otherwise. This can help the police identify your property if it's stolen. Fluorescent ink is also used as a security feature in banknotes — see Figure 7.

Figure 7: Banknotes contain a fluorescent ink pattern, making them harder to forge. It only shows up under UV light of the correct frequency.

Energy efficient lamps

Fluorescent lamps and bulbs use fluorescence too. They have glass tubes which are coated on the inside with a fluorescent material and then filled with a mixture of mercury and noble gases such as neon and argon. When the electrical current is switched on, electrons in the mercury atoms are excited to higher energy levels. When they fall back down, the energy is released as UV radiation (see page 243). The UV radiation hits the fluorescent coating and is converted into visible light. A lamp of this type is far more energy efficient than an older-style incandescent lamp containing a wire filament.

Figure 8: This compact fluorescent lamp has a glass tube which is coiled so that it fits in the same space as an incandescent bulb.

X-rays and gamma rays

X-rays and gamma rays are really useful in medicine.

Medical imaging

Radiographers in hospitals take X-ray 'photographs' of people to see if they have any broken bones. X-rays pass easily through flesh but not so easily through denser material like bone or metal. So it's the amount of radiation that's absorbed (or not absorbed) that gives you an X-ray image.

X-rays are directed through a patient towards a detector plate. The plate starts off white, and the bits that are exposed to the fewest X-rays (i.e. the areas where X-rays have been absorbed by bone on the way through the patient) remain white. The bits where the X-rays have passed through soft tissue without being absorbed turn black. So an X-ray image is really a negative image.

Gamma rays can also be used as a medical tracer (see page 138). A gamma-emitting isotope can be traced as it travels through the body because gamma rays can easily pass through the body to be detected.

Medical treatment

Radiographers use X-rays (and also gamma rays) to treat people with cancer — this is **radiotherapy**. High doses of these rays kill living cells, so they are directed towards cancer cells. However, great care has to be taken to avoid killing too many normal, healthy cells.

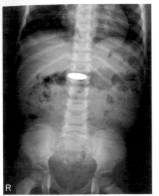

Figure 9: *An X-ray showing a swallowed coin. No X-rays have passed through the coin, so that part of the plate has remained bright white.*

Tip: X-rays can also be used to sterilise medical equipment. X-rays don't damage the equipment or its packaging, they leave no residue, and the source of the X-rays can be switched on and off easily.

Practice Questions — Fact Recall

Q1 What two types of EM radiation are commonly used in cooking?

Q2 How does an electric heater heat a room?

Q3 How is visible light sent through optical fibres?

Q4 Give two ways in which ultraviolet radiation is used.

Q5 Why are X-rays useful in medical imaging?

Practice Question — Application

Q1 A wildlife documentary team used night-vision equipment to detect a rhino hiding in the bushes. They took two images, A and B, as shown. The temperature scale of the images is also shown.

image A   *image B*

cold

hot

One image was taken during the night and one was taken during the day, but they forgot to label them. Which image was taken during the night, image A or image B? Explain how you know.

4. Dangers of EM Radiation

As you've seen on pages 245-250, electromagnetic radiation is really useful. However, it can be pretty harmful too.

Which types of EM radiation are most damaging?

When EM radiation enters living tissue — like you — it's often harmless, but sometimes it creates havoc. The effects of each type of radiation are based on how much energy the wave transfers. Low frequency waves, like radio waves, don't transfer much energy and mostly pass through soft tissue without being absorbed. High frequency waves like UV, X-rays and gamma rays all transfer lots of energy and so can cause lots of damage.

What harm does radiation cause?

UV radiation damages surface cells, which can lead to sunburn and cause skin to age prematurely. More serious effects include damage to the eyes, which can lead to blindness, and an increased risk of skin cancer.

X-rays and gamma rays are types of **ionising radiation**. This means that they carry enough energy to knock electrons off of atoms. This can kill cells or cause gene mutations which can lead to cancer.

Measuring the risk of harm from radiation

Exposure to ultraviolet radiation, X-rays and gamma rays can be harmful to the human body, but these types of radiation are very useful in medical imaging and treatment. Before any of these types of EM radiation are used, people look at whether the benefits outweigh the health risks.

For example, the risk of a person involved in a car accident developing cancer from having an X-ray photograph taken is much smaller than the potential health risk of not finding and treating their injuries.

Radiation dose

The **radiation dose**, measured in sieverts, is a measure of the risk of harm from the body being exposed to radiation (it's not just a measure of the quantity of radiation that hits someone). Radiation dose takes into account:

- The total amount of radiation absorbed.
- How harmful the type of radiation is.
- The type of body tissue absorbing the radiation. Some types of body tissue are more easily damaged by radiation than others.

Figure 1: *UV radiation has been shown to be a cause of cataracts, a leading cause of blindness across the world. A cataract is a clouding of the lens in the eye.*

Figure 2: *UV radiation is used to treat skin conditions such as psoriasis. Special goggles must be worn to prevent damage to the eyes.*

Example

A CT scan uses X-rays and a computer to build up a picture of the inside of a patient's body. The table below shows the radiation dose received by two different parts of a patient's body when those body parts are scanned.

	Radiation dose (mSv)
Head	2.0
Chest	8.0

If a patient has a CT scan on their chest, it is four times more likely to cause harm than if they had a head scan.

Tip: A sievert is pretty big, so you'll often see doses in millisieverts (mSv), where 1000 mSv = 1 Sv.

Practice Questions — Fact Recall

Q1 Describe two ways in which UV radiation can damage the body.

Q2 What is meant by ionising radiation?

Q3 Name two types of ionising radiation.

Q4 What effect can gamma rays have on the body?

Q5 What does the radiation dose tell you?

Q6 What does the radiation dose depend on?

Practice Question — Application

Q1 A patient's pelvis is being examined. It can either be examined with a single X-ray photograph or with a CT scan.
An X-ray of the pelvis gives a radiation dose of 0.7 mSv.
A CT scan of the pelvis gives a radiation dose of 7 mSv.

a) How much larger is the added risk of harm if the patient has a CT scan?

b) A single dental X-ray photograph gives a radiation dose of 0.004 mSv. Suggest why this dose is different from the dose from an X-ray of the pelvis.

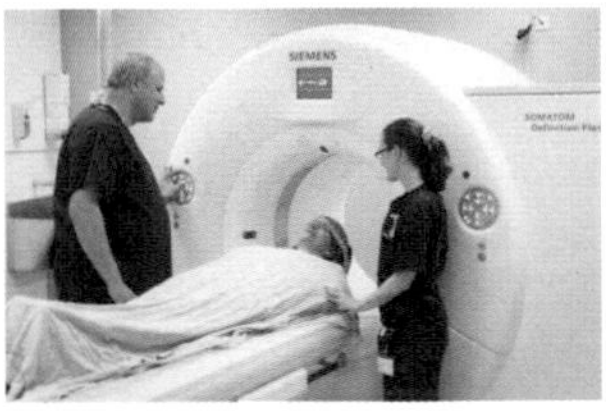

Figure 3: *In a CT scan, lots of X-rays are taken from different angles to build up a detailed image of inside the body.*

5. Visible Light and Colour

Visible light is the only type of EM radiation that our eyes can detect. Like the other types of EM radiation, visible light is actually a band of wavelengths. It's these differing wavelengths that makes the world a colourful place.

The visible light spectrum

As you saw on page 242, EM waves cover a very large spectrum. We can only see a tiny part of this — the visible light spectrum. Within this, narrow bands of wavelengths (and frequencies) correspond to certain colours. For example, violet is at one end, with wavelengths in the range of around 380-450 nm. Red is at the other end, with wavelengths in the range of around 620-750 nm.

Colours can also mix together to make other colours. The only colours you can't make by mixing are the primary colours: pure red, green and blue.

When all of the different colours in the visible light spectrum are put together, it creates white light.

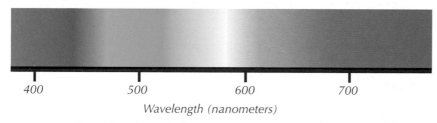

Wavelength (nanometers)

Figure 1: *Diagram showing the colours corresponding to the different wavelengths across the visible light spectrum.*

Colour of opaque objects

When any wave (including visible light) meets an object, it can either be transmitted, absorbed or reflected by it (p. 232). Opaque objects are objects that do not transmit light. When visible light waves hit them, they absorb some wavelengths of light and reflect others. The colour of an object depends on which wavelengths of light are most strongly reflected.

> **Example**
>
> A red apple appears to be red because the wavelengths corresponding to the red part of the visible spectrum are most strongly reflected. The other wavelengths of light are absorbed.
>
>
>
> *White light (a combination of all the colours) hits the apple*
>
> *Red light is reflected — all other colours are absorbed. So the apple looks red.*

Learning Objectives:

- Know that each colour of visible light corresponds to a certain wavelength/frequency range.
- Understand how an opaque object's colour depends on which wavelengths of light are reflected most strongly.
- Know that some objects appear white because all wavelengths are reflected equally and some objects appear black because all wavelengths are absorbed.
- Know that transparent and translucent objects transmit light.
- Know that colour filters absorb some wavelengths and transmit others and understand how this affects the appearance of objects viewed through them.

Specification Reference 4.6.2.6

Tip: 1 nanometre (nm) is equal to 1×10^{-9} m. So 400 nm = 4×10^{-7} m.

Tip: Waves with longer wavelengths have lower frequencies. So the violet end of the visible light spectrum is the high frequency end.

Tip: The reflection of light from a surface might be specular (in one direction) or diffuse (scattered), depending on how smooth the surface is — see pages 235-236.

For opaque objects that aren't a primary colour, they may be reflecting either the wavelengths of light corresponding to that colour or the wavelengths of the primary colours that can mix together to make that colour. So a banana may look yellow because it's reflecting yellow light or because it's reflecting both red and green light.

Objects that appear white do so because they reflect all of the wavelengths of visible light equally. Objects that appear black are those that absorb all the wavelengths of visible light. Your eyes see black as the lack of any visible light (i.e. the lack of any colour).

Figure 2: *Frosted glass is translucent to provide privacy while still letting light through.*

Transparent and translucent objects

Objects which transmit light are either transparent or translucent. Transparent objects transmit light in straight lines so you can see an image clearly through them, e.g. still water or clear glass. Translucent objects transmit light, but also scatter it, so you can't see clearly through them, e.g. frosted glass.

Transparent and translucent objects do not necessarily transmit all wavelengths of light — they may still absorb or reflect some. A transparent or translucent object's colour is related to the wavelengths of light transmitted and reflected by it.

Colour filters

A colour filter only transmits certain wavelengths (colours) and absorbs all other wavelengths. This has the effect of only letting some colours through.

If you look at a blue object through a blue colour filter, it will still look blue. This is because the blue light reflected from the object's surface is transmitted by the filter. However, if the object was e.g. red (or any colour not made from blue light), the object would appear black when viewed through a blue filter. All of the light reflected by the object will be absorbed by the filter.

Example

A red hat appears black when viewed through a blue filter.

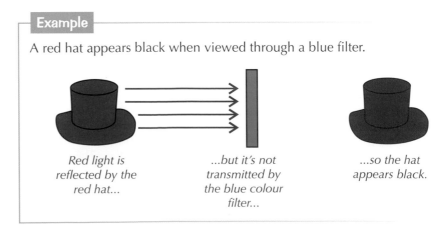

Red light is reflected by the red hat...

...but it's not transmitted by the blue colour filter...

...so the hat appears black.

Filters that aren't for primary colours let through both the wavelengths of light for that colour and the wavelengths of the primary colours that can be added together to make that colour.

Practice Questions — Fact Recall

Q1 What determines the colour of a wave of visible light?

Q2 What happens to visible light when it hits an opaque object?

Q3 Which wavelengths are reflected by an object that appears:

a) black?

b) white?

Q4 What is the difference between an opaque material and a translucent material, in terms of what happens to visible light incident on it?

Q5 What effect does a coloured filter have on visible light?

> **Tip:** If a question talks about radiation incident on something, it just means radiation that hits it.

Practice Questions — Application

Q1 Explain why an opaque green cup appears green.

Q2 A boy is wearing glasses with red filters for lenses. He looks at a set of lights at a pedestrian crossing, as shown below. What colour will each of the lights (red and green) appear to him?

Q3 A student looks at a red bag with a blue buckle through a blue colour filter. Describe and explain the appearance of the bag and the buckle when the student looks through the filter.

- Understand that all objects are continually absorbing and emitting infrared radiation.
- Know that the hotter an object is, the more infrared radiation it will emit in a given time.
- **H** Understand how an object's temperature depends on the amount of radiation it is absorbing compared to the amount it is emitting.
- **H** Know that an object's temperature will rise if it absorbs more radiation than it emits, and fall if it emits more than it absorbs.
- **H** Know that for an object to maintain a constant temperature, it must absorb and emit radiation at the same rate.
- Be able to investigate how surface colours and textures affect the amount of radiation a material emits or absorbs. (Required Practical 10)
- **H** Understand how the Earth's temperature depends on the amount of radiation it absorbs, reflects and emits, and be able to use diagrams to show this.

Specification References
4.6.2.2, 4.6.3.1,
4.6.3.2

Tip: So a matt black surface would be the best absorber and emitter.

6. Infrared Radiation

When you feel the warmth of the sun, it's actually the infrared (IR) radiation emitted by it that you're feeling. The warmth you feel from other hot objects is infrared radiation too — it just hasn't had such a long journey to reach you.

Emission and absorption of infrared radiation

All objects are continually emitting and absorbing infrared (IR) radiation. The hotter an object is, the more infrared radiation it radiates in a given time. So, say if you have two cups of tea, one really hot and the other lukewarm, the really hot one will emit more infrared radiation in one minute than the lukewarm one will. They'll both be continually absorbing some infrared radiation too.

IR radiation and temperature Higher

The temperature of an object depends on the balance between the amount of infrared radiation it absorbs and the amount of infrared radiation it emits.

An object that's hotter than its surroundings, such as a cup of tea, emits more IR radiation than it absorbs — so its temperature falls. However, an object that's cooler than its surroundings absorbs more IR radiation than it emits (e.g. a cold glass of water on a sunny day). This means its temperature rises.

> **Example**
>
> The hot chocolate (and the mug) is warmer than the air around it, so it gives out more IR radiation than it absorbs, which cools it down.
>
>

When an object is at the same temperature as its surroundings, it will be emitting infrared radiation at the same rate as it is absorbing it, and its temperature will be constant.

IR radiation and type of surface

It's the surface of an object that absorbs and emits infrared radiation. Some colours and surfaces absorb and emit radiation better than others. For example, a black surface is better at absorbing and emitting radiation than a white one, and a matt (non-shiny) surface is better at absorbing and emitting radiation than a shiny one.

Investigating IR emissions

REQUIRED PRACTICAL 10

A Leslie cube is a hollow, water-tight cube made of a metal, such as aluminium. The four vertical faces of the cube have different surfaces (for example, matt black paint, matt white paint, shiny metal and dull metal). You can use them to investigate IR emission by different surfaces (see below).

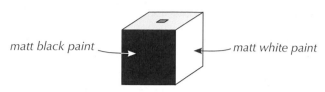

matt black paint ← → matt white paint

Figure 1: An example of a Leslie cube showing two vertical faces.

Figure 2: Runners often wrap themselves in space blankets after a race. They're light coloured and shiny to reduce heat loss by radiation.

Method

Place an empty Leslie cube on a heatproof mat and fill it with boiling water from a kettle. Wait a while for the cube to warm up, then hold a thermometer against each of the four vertical faces of the cube. You should find that all four faces are the same temperature.

Next hold an infrared detector a set distance (e.g. 10 cm) away from one of the cube's vertical faces, and record the amount of IR radiation it detects. Repeat this measurement for each of the cube's vertical faces. Make sure you position the detector at the same distance from the cube each time.

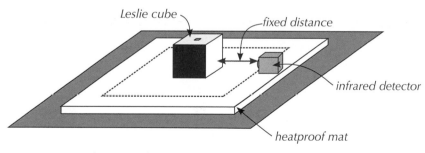

Leslie cube — — fixed distance

— infrared detector

heatproof mat

Figure 3: The setup of an investigation into the infrared radiation emitted by different surfaces.

It's important to be careful when you're doing this experiment. Don't try to move the cube when it's full of boiling water — you might burn your hands. Also, take care when carrying the kettle of boiling water and when pouring the water into the cube.

Tip: As with all practical investigations, make sure you do a risk assessment before you start.

Results

You should find that you detect more infrared radiation from darker surfaces on the Leslie cube than from lighter ones, and more from matt surfaces than from shiny ones. As always, you should do the experiment more than once, to make sure your results are repeatable (p. 9).

Tip: A bar chart (p. 16) is a good way to display your results from this experiment.

Investigating IR absorptions

REQUIRED PRACTICAL 10

Tip: Be very careful with the bunsen burner — make sure not to put anything which could catch fire near the flame. Take care with the metal plates too. Make sure you leave them to cool before you touch them once you've finished the experiment.

The amount of infrared radiation absorbed by different materials also depends on the material. You can do an experiment to show this, using a bunsen burner and some candle wax.

▪ Set up the equipment as shown in Figure 4. Two ball bearings are each stuck to one side of a metal plate with solid pieces of candle wax. The other sides of these plates are then faced towards the flame.

▪ The sides of the plates that are facing towards the flame each have a different surface colour — one is matt black and the other is silver.

▪ The ball bearing on the black plate will fall first as the black surface absorbs more infrared radiation — transferring more energy to the thermal energy store of the wax. This means the wax on the black plate melts before the wax on the silver plate.

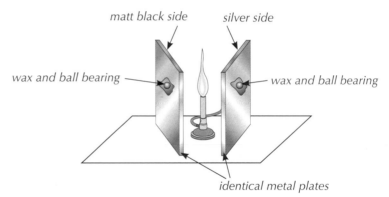

Figure 4: *The setup of an investigation into the amount of infrared radiation absorbed by different surfaces.*

Radiation and the Earth's temperature   `Higher`

Tip: 🄷 The atmosphere protects us from lots of the harmful wavelengths in solar radiation, such as gamma rays, by absorbing them.

Radiation from the sun (solar radiation) consists of wavelengths from across the EM spectrum. Some of the solar radiation transmitted to Earth is reflected back to space by the atmosphere, clouds or bright surfaces such as snow. The rest is absorbed by the atmosphere or by Earth's surface. The atmosphere and surface then emit the radiation as infrared.

The overall temperature of the Earth depends on the rates at which radiation is reflected, absorbed and emitted. During the day, more radiation is absorbed than is emitted. This causes an increase in local temperature (i.e. an increase in temperature at the points on Earth where it is currently daytime).

At night, less radiation is absorbed than is emitted, which causes a decrease in the local temperature.

Even though local temperatures vary, the overall temperatures of the Earth's surface and atmosphere stay fairly constant. That's because the overall amount of radiation absorbed from space is equal to the amount reflected or emitted into space.

You can show the flow of radiation for the Earth on a diagram:

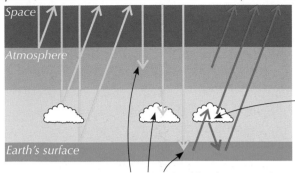

Some radiation is reflected by the atmosphere, clouds and the Earth's surface.

Some radiation is emitted by the atmosphere

Some of the radiation emitted by the surface is reflected or absorbed (and later emitted) by the clouds.

Some radiation is absorbed by the atmosphere, clouds and the Earth's surface.

Figure 5: *Diagram showing how radiation affects the temperature of the Earth's surface.*

Practice Questions — Fact Recall

Q1 How is the temperature of an object related to the amount of infrared radiation it emits in a given time?

Q2 Explain whether the following statement is true or false: 'Objects only absorb radiation if they're cooler than their surroundings.'

Q3 What is the relationship between the amount of IR radiation being emitted and the amount being absorbed over a certain time for an object at constant temperature?

Q4 What is a Leslie cube? Describe an investigation in which it is used.

Q5 Why does the temperature at a point on Earth decrease at night?

Practice Questions — Application

Q1 A security light comes on if the amount of infrared radiation that its sensor receives changes. Explain why a person walking in front of the sensor causes the light to switch on.

Q2 Explain, in terms of radiation emission and absorption, what happens to the temperature of a bowl of ice cream that is left on a counter in a warm room.

- Know that a perfect black body absorbs all the EM radiation that hits it and is also the best possible emitter.
- Understand how the temperature of an object determines the intensity and distribution of the wavelengths emitted by it.
- Know that when the temperature of an object rises, the intensity of all the wavelengths increases, but the intensity of the shorter wavelengths does so more quickly.

Specification References 4.6.3.1, 4.6.3.2

7. Black Bodies and Radiation

So you've seen that all objects emit and absorb IR radiation. But that's not the end of it — all sorts of different radiation can be emitted and absorbed, and some objects are better at it than others. That's where black bodies come in...

What are black bodies?

A perfect **black body** is an object that absorbs all of the electromagnetic radiation that hits it. No radiation is reflected or transmitted by it. A good absorber is also a good emitter, so perfect black bodies are the best possible emitters of radiation too. You won't come across many black bodies — they are just an ideal situation that physicists have devised to help them model the behaviour of stars, etc.

Radiation emitted by objects

Black bodies might be the best emitters of radiation, but all objects emit some radiation due to the energy in their thermal energy stores. They don't just emit infrared radiation — they emit radiation with a range of wavelengths and frequencies across the electromagnetic spectrum. The amount and type of radiation emitted depend on the object's temperature.

Intensity of wavelengths emitted

Intensity is power per unit area, i.e. how much energy is transferred to a given area in a certain amount of time. The intensity of radiation emitted by an object depends on its temperature:

> As the temperature of an object increases, the intensity of every emitted wavelength increases.

This basically means that a hotter object gives out more of every wavelength of radiation than a cooler object does. This is shown in Figure 1 on the next page — the curve for the hotter object is higher up than the curve for the cooler object.

Distribution of wavelengths emitted

The distribution of the wavelengths emitted by an object also depends on the object's temperature:

> As the temperature of an object increases, the peak wavelength decreases.

So as an object gets hotter, the intensity increases more rapidly for shorter wavelengths than for longer wavelengths. A bigger proportion of the radiation emitted by a hot object has a short wavelength, compared to that emitted by a cooler object. This is shown in Figure 1 by the peak of the curve for the hotter object being further to the left than the peak for the cooler object.

Tip: By 'distribution', we just mean the spread of intensities of different wavelengths. The general pattern is a peak wavelength (the wavelength that's emitted with the highest intensity), with a spread of intensities either side. The further you go from the peak wavelength, the lower the intensity of that wavelength (see Figure 1).

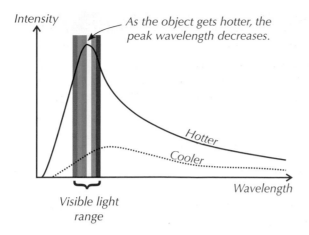

Intensity

As the object gets hotter, the peak wavelength decreases.

Hotter

Cooler

Wavelength

Visible light range

Figure 1: *Graph showing how the intensity and distribution of the wavelengths emitted by a black body depends on its temperature.*

Tip: The curves in Figure 1 are for a black body. However, this is a good approximation of how the intensity and distribution of wavelengths varies with temperature for any object.

Practice Questions — Fact Recall

Q1 What is a black body?

Q2 What types of objects emit radiation?

Q3 What does intensity mean?

Q4 Describe how the intensity and distribution of wavelengths emitted by an object changes as the object's temperature increases.

Practice Question — Application

Q1 The peak wavelength of light from the Sun is about 500 nm. The peak wavelength of light from a second star is about 850 nm. Which star is hotter? Explain your answer.

Topic 6b Checklist — Make sure you know...

What are Electromagnetic Waves?

☐ That electromagnetic waves are a group of transverse waves that all transfer energy from a source to an absorber and all travel at the same speed in a vacuum or in air.

☐ That the EM spectrum is a continuous spectrum of all the possible wavelengths (or frequencies) of EM waves, split into seven types: radio waves, microwaves, infrared, visible light, ultraviolet, X-rays and gamma rays (in order of decreasing wavelength/increasing frequency).

☐ That our eyes can only detect waves from the visible light part of the EM spectrum.

☐ That changes in atoms can generate EM waves and that gamma rays can be produced by changes in the nucleus.

☐ That different EM waves interact differently with different substances.

Radio Waves

☐ H That radio waves are produced by oscillating charges in an alternating electric current.

☐ H That when radio waves are absorbed by a receiver, they cause oscillations, which can lead to alternating currents of the same frequencies as the radio waves.

☐ How radio waves are used to transmit TV and radio.

☐ H Why radio waves are suitable for use in TV and radio communications.

More EM Waves and their Uses

☐ That microwaves are used for satellite communications and cooking food.

☐ How infrared radiation can be used in cooking, and how it allows electric heaters and infrared cameras to work.

☐ How fibre optic communications use visible light.

☐ How ultraviolet light allows energy efficient lamps and sun tanning lamps to work.

☐ How X-rays are used in medical imaging and treatments, e.g. to kill cancer cells.

☐ How gamma rays are used as medical tracers.

☐ H Why certain EM waves are suited to their uses.

Dangers of EM Radiation

☐ That UV waves can harm the body (e.g. through premature skin ageing and an increased risk of skin cancer).

☐ X-rays and gamma rays are forms of ionising radiation which can cause gene mutations and cancer.

☐ That radiation dose (measured in sieverts) is a measure of the risk of harm due to exposure to radiation, and that 1000 mSv = 1 Sv.

cont...

☐ That radiation dose depends on the radiation type, the amount absorbed and the area of the body exposed.

☐ Be able to understand the risks of exposure to radiation in given situations.

Visible Light and Colour

☐ That each colour of visible light relates to a narrow band of wavelengths/frequencies.

☐ That an opaque object's colour depends on the wavelengths of light that it reflects most strongly.

☐ That white objects reflect all wavelengths of visible light equally, and black objects absorb all wavelengths of visible light.

☐ That transparent and translucent objects transmit light.

☐ That colour filters transmit only certain wavelengths and how this affects the appearance of objects through them.

☐ How an object will appear when viewed through a primary colour filter, e.g. a blue object appears blue when viewed through a blue filter, but some other colour (e.g. red) objects will appear black.

Infrared Radiation

☐ That all objects are continually absorbing and emitting infrared radiation.

☐ That the hotter an object is, the more infrared radiation it will emit over a certain time period.

☐ **H** That an object's temperature is related to the balance between the amount of infrared radiation absorbed and emitted. For example, objects that are hotter than their surroundings emit more than they absorb, which leads to them cooling down.

☐ **H** That an object at a constant temperature emits and absorbs infrared radiation at the same rate.

☐ How to investigate which surfaces emit or absorb infrared radiation better than others.

☐ **H** That the Earth's temperature is dependent on the rates that radiation is absorbed, emitted and reflected back into space.

Black Bodies and Radiation

☐ That a perfect black body absorbs all the electromagnetic radiation that hits it and is also the best possible emitter.

☐ That hotter bodies emit a greater intensity of all wavelengths of radiation than cooler bodies do.

☐ That increasing the temperature of an object causes the intensity of the shorter wavelengths it emits to rise most rapidly.

Exam-style Questions

1   The diagram below shows the electromagnetic spectrum.

radio waves	microwaves	infrared	visible light	A	X-rays	gamma rays

1.1   Name a property of electromagnetic waves that decreases across the electromagnetic spectrum in the direction of the arrow shown.

(1 mark)

1.2   Which type of radiation is represented by the letter **A** in the diagram?

(1 mark)

1.3   Which two types of electromagnetic wave are used for transmitting TV signals?

(2 marks)

2   The table below shows the radiation dose of some X-ray scans.

	Abdomen	Spine	Teeth (Dental)
Radiation dose (mSv)	0.7	1.5	0.004

2.1   How many times greater is the risk of harm from having a single abdominal X-ray scan than from having a single dental X-ray?

(2 marks)

2.2   Explain why exposure to X-rays can be harmful to humans.

(2 marks)

2.3   Many dentists recommend regular dental X-rays to check for hidden problems. Suggest why it is not recommended that people have regular X-rays of the spine to screen for problems.

(2 marks)

3   A number of electromagnetic waves are used for communications purposes.

3.1   Describe how radio waves are transmitted and received, with reference to alternating current.

(4 marks)

X-ray telescopes detect X-ray emissions from astronomical objects.

3.2   There are no X-ray telescopes on the Earth's surface. They are all mounted on satellites in space. Suggest a reason for this.

(2 marks)

Microwaves are used to communicate the observations of X-ray telescopes through the Earth's watery atmosphere. Some wavelengths of microwaves are transmitted by water, but some are absorbed.

3.3   Explain why the wavelengths of microwaves used in a microwave oven could not be used to communicate with satellites.

(2 marks)

4 The setup below shows the apparatus for an investigation into the amount of radiation emitted by different surfaces. Half of a shiny aluminium can is painted matt black and the can is filled with boiling water. An infrared sensor, connected to a data logger, points towards the matt black side of the can and records the amount of infrared radiation it detects. The can is then turned around so that the detector points towards the shiny silver side of the can and another reading is taken.

Aluminium can filled with hot water

Infrared sensor connected to data logger

The results of the investigation are shown in the table below.

	Matt black	Shiny silver
Average infrared radiation reading	41 W/m²	5 W/m²

A thermometer is held so that it is touching the outside of the can. The temperature shown by the thermometer is the same when it is held against the shiny surface as when it is held against the black surface.

4.1 Explain why the results from the infrared sensor are different for each surface, while the thermometer readings are the same for each surface.

(2 marks)

The water was left in the can and the IR reading was taken again 15 minutes later.

4.2 Describe how you would expect this reading to differ from the earlier one, giving a reason for your answer.

(2 marks)

Another can, identical to the original, is painted matt silver and filled with boiling water. The IR radiation emitted is then measured in the same way as described above.

4.3 Suggest whether the IR radiation reading will be higher or lower than each of the readings above. Explain your answer.

(3 marks)

5* Some colour filters are set up and a white ball with a thick red stripe is viewed from different points, as shown.

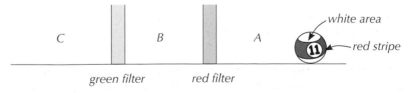

C   B   A   white area   red stripe

green filter   red filter

Describe how the red and white parts of the ball will appear when viewed from points A, B and C. Explain why this is, in terms of how the light is reflected and transmitted.

(6 marks)

Learning Objectives:
- Know that lenses refract light to form images.
- Know that convex lenses cause parallel rays of light passing through them to converge at the principal focus.
- Know that the focal length of a lens is the distance from the centre of the lens to the principal focus.
- Know that concave lenses cause parallel rays of light passing through them to diverge as if from the principal focus.
- Know that an image formed by a lens can be either real or virtual.
- Be able to describe an image formed by a lens.

Specification Reference 4.6.2.5

1. Lenses and Images

Lenses use refraction to bend light rays to form an image. They're used in things like glasses, magnifying glasses and cameras. This topic is all about the different types of lenses and the images they produce.

Lenses

Lenses form images by refracting light and changing its direction. There are two main types of lens — convex (converging) and concave (diverging). They have different shapes and have opposite effects on light rays.

Convex lenses

A **convex lens** is a lens which bulges outwards. It causes rays of light which are parallel to the axis of the lens (see Figure 1) to converge (come together) at the principal focus.

The **axis** of a lens is a line passing through the middle of the lens, perpendicular to the lens — see Figure 1. The **principal focus of a convex lens** is where rays hitting the lens parallel to the axis all meet.

The distance from the centre of the lens to the principal focus is called the **focal length**.

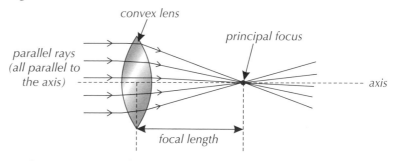

Figure 1: *A convex lens focusing parallel rays at the principal focus.*

There is a principal focus on each side of the lens. For a convex lens, rays parallel to the axis will always focus on the principal focus on the far side of the lens in relation to where the rays are coming from — see Figure 3.

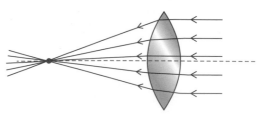

Figure 3: *The convex lens from Figure 1 viewed from the same angle. The rays this time are coming from the right of the lens and focusing on the left.*

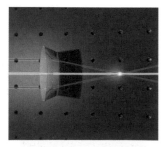

Figure 2: *A convex lens focusing laser beams at the principal focus (yellow). The light is coming from the left.*

Tip: You may also see convex lenses referred to as 'converging lenses'.

Three important rules for convex lenses

There are three rules you need to know about rays passing through convex lenses. Figure 4 shows these rules in action.

1. An incident ray travelling parallel to the axis refracts through the lens and passes through the principal focus on the other side.

2. An incident ray passing through the centre of the lens carries on in the same direction.

3. An incident ray passing through the principal focus before meeting the lens refracts through the lens and travels parallel to the axis.

Tip: In reality, the line passing through the centre of the lens will refract slightly as it enters and leaves the lens, but the deflection is so small that it can be ignored and just drawn as a straight line.

Tip: Diagrams that show the paths of rays are called ray diagrams. You'll learn how to draw them using these rules on pages 270-274.

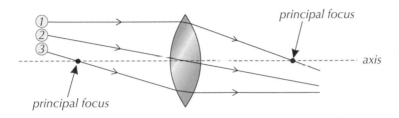

Figure 4: *Three rays passing through a convex lens.*

Concave lenses

A **concave lens** is a lens which caves inwards. It causes parallel rays of light to diverge (spread out).

Concave lenses have an axis just like convex ones, but the principal focus is slightly different. The **principal focus of a concave lens** is the point where rays hitting the lens parallel to the axis appear to all come from — you can trace them back until they all appear to meet up at a point behind the lens (see Figure 6).

Just like convex lenses, the focal length is the distance between the centre of the lens and the principal focus and there is a principal focus either side of the lens.

Tip: You may also see concave lenses referred to as 'diverging lenses'.

Figure 5: *A concave lens refracting beams of laser light. The light is coming from above.*

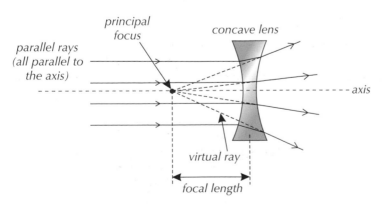

Figure 6: *A concave lens diverging parallel rays so that they appear to have come from the principal focus.*

Tip: Virtual rays are rays that aren't actually there. They show the path that it <u>looks like</u> the light has taken.

Three important rules for concave lenses

There are three rules for rays passing through concave lenses too. Figure 7 shows these.

1. An incident ray travelling parallel to the axis refracts through the lens and travels in line with the near-side principal focus (so it appears to have come from the principal focus).

2. An incident ray passing through the centre of the lens carries on in the same direction.

3. An incident ray passing through the lens towards the far-side principal focus refracts through the lens and travels parallel to the axis.

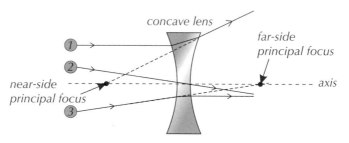

Figure 7: *Three rays passing through a concave lens.*

Images

Different lenses can produce different kinds of image. There are two types of image — **real images** and **virtual images**. It's all to do with whether two light rays coming from the same point on an object will eventually meet.

Real images

A real image is formed when the light rays from a point on an object come together to form an image. The light rays actually pass through the same point. A real image can be captured on a 'screen' — like the image formed on an eye's retina (the 'screen' at the back of an eye) or the image formed on a projector screen.

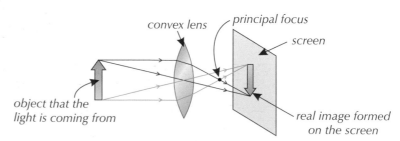

Figure 8: *A real image, formed by a convex lens, being projected onto a screen.*

Virtual images

A virtual image is formed when the light rays from a point on an object are diverging after they have left the lens, so the light from the point on the object appears to be coming from a completely different place. The light rays don't actually pass through that point — they just appear to.

Tip: Real and virtual images are a bit tricky to get your head around. Make sure you know the definitions and can explain the difference. If you ever need to work out what sort of image a lens will produce, a ray diagram will help you (see pages 270-274).

Tip: The screen has to be placed exactly where the rays meet (and the image is formed) otherwise it will be out of focus.

Exam Tip
Remember, "you can't project a virtual image onto a screen" — that's a useful phrase to use in the exam if they ask you about virtual images.

When you look in a mirror you see a virtual image of your face because the object (your face) appears to be behind the mirror (see Figure 9).

You can get a virtual image when looking at an object through a magnifying lens. The virtual image looks bigger than the object actually is (see Figure 10).

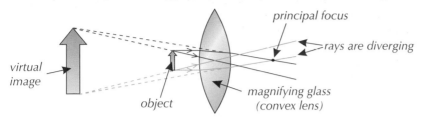

Figure 10: A virtual image formed by looking through a magnifying glass (a convex lens).

Figure 9: A man looking in the mirror at a virtual image of himself.

Describing an image

To describe an image properly, you need to say three things:

- How big it is compared to the object.
- Whether it's upright or inverted (upside down) relative to the object.
- Whether it's real or virtual.

Tip: There's more on how different images are produced on pages 272 and 274.

Example

In Figure 10, the image is larger than the object, the right way up and virtual.

Practice Questions — Fact Recall

Q1 What is meant by the focal length of a lens?

Q2 What sort of lens spreads out rays of light?

Q3 Draw a convex lens and a concave lens.

Q4 Define the principal focus of a convex lens and a concave lens.

Q5 Name the two types of images produced by lenses.

Q6 What three things do you need to say to describe an image formed by a lens?

Practice Question — Application

Q1 The eye contains a lens like the one shown in the diagram.

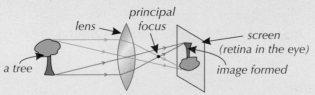

a) What kind of lens does the eye contain?

b) Describe the image formed in the diagram.

- Know that the images formed by concave or convex lenses can be shown using ray diagrams.
- Know the symbols for convex and concave lenses in ray diagrams.
- Be able to produce ray diagrams for convex lenses.
- Know that a convex lens can produce a real or virtual image, depending on the distance of the object from the lens.
- Be able to produce ray diagrams for concave lenses.
- Know that a concave lens always produces a virtual image, that is upright, smaller than the object and on the same side of the lens as the object.
- Be able to use ray diagrams to show the similarities and differences between convex and concave lenses.

Specification Reference
4.6.2.5

Tip: With this type of diagram you draw each ray only changing direction once as it passes through the lens. But the rays actually change direction twice, once on entering and once on leaving the lens. If you're given a ray diagram with a real lens, you either need to draw the ray changing once at the centre of the lens, or twice:

2. Ray Diagrams

Ray diagrams are really useful — they can be used to work out what sort of image is formed, its size, its position and its orientation.

What is a ray diagram?

A **ray diagram** is basically just a diagram showing the paths taken by light rays through a lens. It can help you to work out what an image formed by the lens will be like.

Convex lenses are drawn like this in a ray diagram:

And concave lenses are drawn like this:

Ray diagrams for convex lenses

Drawing ray diagrams is pretty straightforward — just follow these four steps:

1. Draw a ray through the lens centre

Draw a ray from the top of the object that passes straight through the centre of the lens without changing direction at all (see Figure 1).

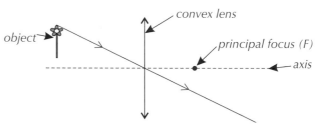

Figure 1: A diagram showing step 1 of the method for drawing a ray diagram for a convex lens.

2. Draw a ray parallel to the axis

Draw a second ray from the top of the object that travels parallel to the axis. The ray should refract at the lens so that it passes through the principal focus of the lens. There are two things that can happen when you do this. One option is that the two rays will meet on the far side of the lens — see Figure 2. If this happens, you're done with this step.

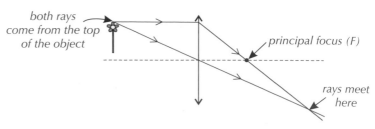

Figure 2: A diagram of step 2 of the method for drawing a ray diagram for a convex lens showing the refracted rays meeting.

The other option is that the rays don't meet (see Figure 3). In this case, you'll need to extend the refracted rays back on the near side of the lens as virtual rays (dotted lines). They will eventually meet.

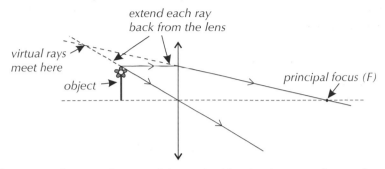

Figure 3: A diagram of step 2 of the method for drawing a ray diagram for a convex lens showing what to do if the refracted rays don't meet.

Tip: You should always draw arrows on real rays but not on virtual rays.

3. Repeat the process

Repeat steps 1 and 2 for the bottom of the object. If the rays met in step 2 the first time round, they'll meet this time too. If they didn't meet before, they won't this time and you'll need to extend them back again.

Exam Tip
It's sometimes possible to draw a third incident ray passing through the principal focus on the way to the lens (it will refract so that it goes parallel to the axis — see rule 3 on p. 267). In the exam, you can get away with two rays though, unless they ask for three.

4. Draw the image

Draw in the image. The point where the two rays from the top of the object meet is where the top (in this case, the petals) of the image is formed. The point where the two rays from the bottom of the object meet is where the bottom (in this case, the stalk) of the image is formed.

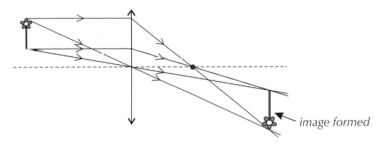

Figure 4: A diagram of steps 3 and 4 of the method for drawing a ray diagram for a convex lens.

Tip: Figure 4 only shows what to do when the rays meet. If the rays don't meet, you draw the image where the virtual rays meet.

When the bottom of the object is on the axis, the bottom of the image is also on the axis. In this case you only need to draw the top set of rays — see Figure 5.

Tip: The rules that describe how rays travel through convex lenses are on page 267.

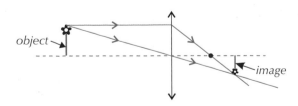

Figure 5: A ray diagram showing the image formed by a convex lens of an object that is 'sitting' on the axis of the lens.

Images formed by convex lenses

The type of image formed by a convex lens depends on where the object is placed in relation to the principal focus (F). Make sure you learn these four situations:

Tip: 2F just means twice the focal length away from the lens.

Beyond 2F

An object beyond 2F will produce a real, inverted (upside down) image that is smaller than the object. It will sit between F and 2F on the far side of the lens (see Figure 6).

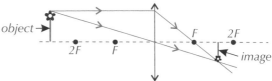

Figure 6: *A real, inverted and smaller image is formed by a convex lens when an object is placed beyond 2F.*

At 2F

An object placed at 2F will produce a real, inverted (upside down) image that is the same size as the object. It will sit at 2F on the far side of the lens (see Figure 8).

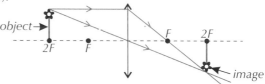

Figure 8: *A real, inverted image of the same size as the object is formed by a convex lens when an object is placed at 2F.*

Figure 7: *A convex lens forming a real and inverted, but smaller, image of the woman.*

Between F and 2F

An object placed between F and 2F will form a real, inverted image that's bigger than the object. It will sit beyond 2F on the far side of the lens (see Figure 9).

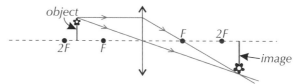

Figure 9: *A real, inverted and larger image is formed by a convex lens when an object is placed between F and 2F.*

Between the lens and F

An object placed closer to the lens than F will make a virtual image that is the right way up, but bigger than the object and on the same side of the lens as the object (see Figure 10).

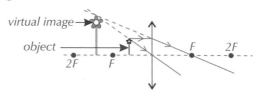

Figure 11: *A convex lens forming a virtual, upright and larger image of the eye.*

Figure 10: *A virtual, upright and larger image is formed by a convex lens when an object is placed closer to the lens than F.*

Ray diagrams for concave lenses

Drawing ray diagrams for concave lenses is similar to convex ones — it's just step 2 that works slightly differently.

1. Draw a ray through the lens centre

Draw a ray from the top of the object that passes straight through the centre of the lens without changing direction at all (see Figure 12).

Tip: The rules that describe how rays travel through concave lenses are on page 268.

2. Draw a ray parallel to the axis

Draw a second ray from the top of the object that travels parallel to the axis. The ray should refract at the lens so that it appears to have come from the near-side principal focus. Draw a virtual ray (dotted line) from that principal focus to where that ray meets the lens (see Figure 12).

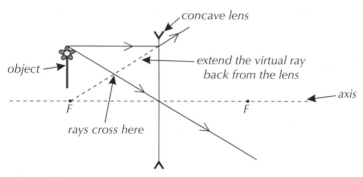

Figure 12: A diagram of steps 1 and 2 of the method for drawing a ray diagram for a concave lens.

Tip: You can draw a third incident ray in the direction of the principal focus on the far side of the lens if you want to. Just remember that it refracts to travel parallel to the axis.

3. Repeat the process

Repeat steps 1 and 2 for the bottom of the object. If the bottom of the object is on the axis, the bottom of the image will be too (see page 271).

Exam Tip
In the exam, you can get away with two rays from each point on the object. Just choose whichever two are easiest to draw — don't try to draw a ray that won't actually pass through the lens.

4. Draw the image

Draw in the image. Where the real and the virtual rays from the top of the object meet is where the top of the image is formed. Where the real and virtual rays from the bottom of the object meet is where the bottom of the image is formed.

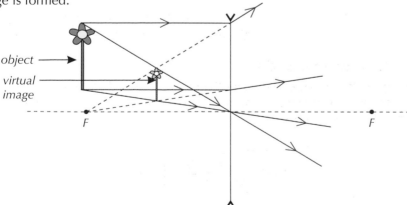

Figure 13: A diagram of steps 3 and 4 of the method for drawing a ray diagram for a concave lens.

Tip: It is sometimes difficult to work out where the image should be drawn. The top and bottom of the image should be where two rays from <u>the same point</u> on the object cross.

Images formed by a concave lens

A concave lens always produces a virtual image. The image is the right way up, smaller than the object, and on the same side of the lens as the object — no matter where the object is.

You can draw ray diagrams for both convex and concave lenses that show the similarities and differences between how the lenses refract light and form images. Remember:

- Convex lenses converge light whereas concave lenses diverge light.
- Rays parallel to the axis converge onto the principle focus of a convex lens, but diverge so that they appear to have come from the principal focus of a concave lens.
- Convex lenses can produce real and virtual images, whereas concave lenses can only produce virtual images.

Tip: Remember, a ray that goes straight through the centre of a lens doesn't change direction — this is true for both convex and concave lenses.

Practice Questions — Fact Recall

Q1 Draw and label the two symbols used for lenses in ray diagrams.

Q2 The principal focus of a convex lens is F. Describe the image produced by a convex lens if the object is placed:

a) beyond 2F,   b) at 2F,   c) between F and 2F,   d) closer than F.

Q3 The principal focus of a concave lens is F. Describe the image produced by the lens if the object is placed at 2F.

Practice Questions — Application

Q1 The principal focus of the lens in this diagram is F. A pencil is placed in front of the lens between F and 2F, as shown.

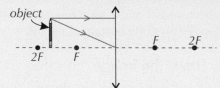

a) Complete the ray diagram to show the image being formed.

b) Describe the image produced by the lens.

c) What sort of lens is shown in the diagram?

Tip: Remember focal length is the distance between the centre of the lens and the principal focus.

Q2 A straight arrow is placed pointing upwards, perpendicular to the axis of a convex lens with a focal length of 3 cm, so that the bottom end of the arrow is touching the axis. The object is 1 cm from the lens.

a) Sketch a ray diagram to show the image being formed.

b) Describe the image produced by the lens.

Q3 A straight arrow is placed pointing upwards, perpendicular to the axis of a concave lens so that it is further away than the principal focus and the bottom end of the arrow is touching the axis.

a) Sketch a ray diagram showing the image that will be formed.

b) Describe the image produced by the lens.

c) The object is moved so that it is closer than the principal focus. Describe the image produced now.

3. Magnification

Convex lenses can produce images that are bigger than the object, which is useful if you need a magnifying glass.

Magnifying glasses

Magnifying glasses use convex lenses. They work by creating a virtual, upright image that is larger than the object and is on the same side of the lens as the object. For this to happen the object being magnified must be closer to the lens than the focal length (see page 272).

Since the image produced is a virtual image, the light rays don't actually come from the place where the image appears to be.

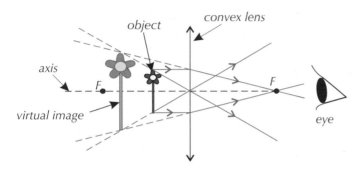

Figure 1: *A convex lens being used as a magnifying glass to produce a magnified image of a flower.*

Figure 2: *A musical score being magnified (region in the centre) by a convex lens.*

Magnification equation

You can use the **magnification** equation to work out the magnification produced by a lens at a given distance:

$$\text{magnification} = \frac{\text{image height}}{\text{object height}}$$

Magnification is a ratio, so it doesn't have any units. This means so long as the units are the same for both, you can measure the heights in whatever units of length you like. (Usually, you'll be using mm or cm.)

Tip: You can also use the ratio of distance to the image to distance to the object. It gives the same value for magnification, but it's always defined by image and object height.

Example 1

A coin with a diameter of 14 mm is placed a certain distance behind a magnifying lens. The virtual image produced has a diameter of 35 mm. What is the magnification of the lens at this distance?

The object height is 14 mm and the image height is 35 mm.
So just use the equation:
$$\text{magnification} = 35 \div 14 = 2.5$$

Tip: The magnification tells you how many times taller the image is than the object. In this example, the image height is 2.5 times the object height. So the lens makes the object appear 2.5 times taller.

Example 2

A 5 mm long insect is viewed through a magnifying glass, producing an image of magnification 4. Calculate the length of the image of the insect.

Use the magnification equation but rearranged:
image height = magnification × object height = 4 × 5 = 20 mm

Example 3

A 1 cm pin is placed perpendicular to the axis of a convex lens at a distance of 2 cm from the lens, so that one end of the pin is touching the axis. The focal length of the lens is 4 cm. Draw a ray diagram to scale to show where the image is formed and calculate the magnification of the lens at this distance.

To draw a ray diagram to scale, use a ruler to measure and mark the principal focus and the object position on the axis. Next, measure and draw the object. Then just draw the ray diagram as you usually would.

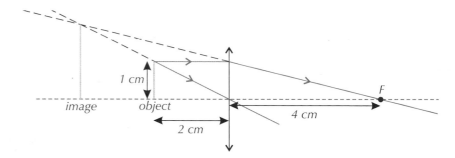

For the second part, you've drawn the diagram to scale, so you can measure the image height using a ruler — it's 2 cm high. Then just use the magnification equation:

$$\text{magnification} = \frac{\text{image height}}{\text{object height}} = \frac{2}{1} = 2$$

> **Tip:** See pages 270-274 for more on drawing ray diagrams.

> **Exam Tip**
> Make sure you draw the ray diagram very accurately if you're drawing it to scale in case you're asked to measure something from it.

> **Exam Tip**
> If you're given a ray diagram that is drawn to scale, you can just measure the image height and object height from it.

Practice Questions — Application

Q1 A 3 cm long paperclip is viewed through a magnifying glass. The image produced is 4.5 cm long.

    a) What kind of lens will be in the magnifying glass?

    b) What is the magnification of the lens at this distance?

Q2 An image of a pea as viewed through a magnifying glass has a diameter of 1.5 cm. The magnification at this distance is 3. What is the diameter of the actual pea?

> **Tip:** Remember that the units of object height and image height must be the same.

Topic 6c Checklist — Make sure you know...

Lenses and Images

☐ That lenses use refraction to form images.

☐ That convex lenses 'bulge out' and cause light rays parallel to the principal axis passing through them to converge (move together) towards a point, the principal focus.

☐ That the axis of a lens is the line perpendicular to the lens which passes through its centre.

☐ That lenses have a principal focus on each side and the distance between either principal focus and the lens is the focal length.

☐ That a ray parallel to the axis passing through a convex lens will bend to pass through the principal focus.

☐ That a ray that passes through the principal focus and into a convex lens will bend to be parallel to the axis.

☐ That concave lenses 'cave inwards' and cause light rays parallel to the axis passing through them to diverge (spread out) as if from a single point, the principal focus.

☐ That a ray parallel to the axis passing through a concave lens will bend so that it appears to have come from the principal focus on the side that the ray started from.

☐ That a ray in line with the principal focus on the opposite side of a concave lens will be bent by the lens so that it is parallel to the axis.

☐ That a ray which passes through the centre of a convex or concave lens will carry on in the same direction without being refracted.

☐ That a real image is formed when light rays from a point on an object come together at another point to form an image, and that this image can be projected onto a screen.

☐ That a virtual image is formed when light rays from a point on an object diverge, and so appear to have come from another point.

Ray Diagrams

☐ The symbols used in ray diagrams for a convex and concave lens.

☐ How to produce ray diagrams for convex lenses, and that convex lenses can produce real or virtual images, depending on the distance of the object from the lens.

☐ How to produce ray diagrams for concave lenses, and that concave lenses always produce virtual images that are upright, smaller than the object, and on the object's side of the lens.

☐ How to use ray diagrams to show the differences between convex and concave lenses.

Magnification

☐ That convex lenses can be used to created magnified images, e.g. in magnifying glasses.

☐ How to use the equation magnification = $\dfrac{\text{image height}}{\text{object height}}$.

☐ That magnification is a ratio with no units, but that the heights used to calculate it must both be in the same units.

Exam-style Questions

1   This incomplete ray diagram shows a convex lens producing an image of an object that is placed 4 cm away from the lens.

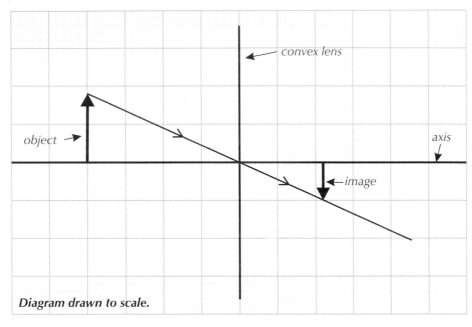

Diagram drawn to scale.

1.1   Copy and complete the ray diagram to show how the image is formed. Make sure the lens is correctly drawn and that you label its principal focus (F).

(3 marks)

1.2   Describe the image produced by the convex lens.

(3 marks)

1.3   Calculate the magnification of the image. Use the correct equation from those provided on page 394. Give your answer to 2 significant figures.

(2 marks)

1.4   The object is moved so that it is only 1 cm to the left of the lens.
Describe the new image that is formed.

(3 marks)

2   A 5 cm tall object is placed upright and perpendicular to the axis of a concave lens, with its base on the axis 10 cm from the lens. The lens has a focal length of 5 cm.

2.1   Describe the image of the object that is produced by the concave lens.

(3 marks)

2.2   Draw a ray diagram to show the formation of the image. Draw your diagram to scale.

(5 marks)

2.3   State the height of the image formed.

(1 mark)

1. Sound Waves

Sound waves are basically vibrations that pass through matter. If they get to your ear, they make bits inside it vibrate, causing the sensation of hearing.

What are sound waves?

Sound waves are longitudinal waves (see page 224) of vibrating particles caused by vibrating objects. These vibrations are passed through the surrounding matter as a series of compressions and rarefactions. The surrounding matter can be solid, liquid or gas. Sound generally travels faster in solids than in liquids, and faster in liquids than in gases.

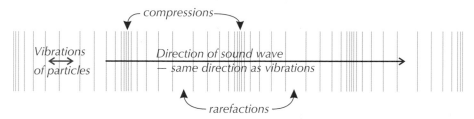

Figure 1: A sound wave made up of compressions and rarefactions of the particles in a medium.

Sound waves in different materials

Sound waves will refract (change direction — see page 232) as they enter different materials. However, since sound waves spread out a lot, the change in direction is hard to spot under normal circumstances.

When a sound wave enters a denser medium it speeds up. Its frequency stays the same, so for the wave equation speed = frequency × wavelength (see page 228) to stay true the wavelength of the sound wave must change too. If its speed increases, its wavelength also increases. If its speed decreases, then its wavelength decreases too.

Example

A guitar string produces a sound wave with a wavelength of 0.66 m and a speed of 330 m/s in air. The sound wave then enters water where its speed changes to 1500 m/s. Calculate its new wavelength.

In air, $v = 330$ m/s and $\lambda = 0.66$ m.
Rearrange the wave equation, $v = f\lambda$ to find frequency: $f = v \div \lambda$

So $f = 330 \div 0.66 = 500$ Hz

The frequency stays the same in the water, so $f = 500$ Hz and $v = 1500$ m/s.
Rearrange the wave equation to find wavelength: $\lambda = v \div f$

So $\lambda = 1500 \div 500 = 3$ m

Learning Objectives:

- Understand how, when sound waves move from one medium to another, changes in speed, wavelength and frequency are linked.

- **H** Be able to describe how sound waves travel through a solid as a series of vibrations, and how they are passed between a solid and air.

- **H** Know that sound waves cause structures inside the ear to vibrate, leading to the sensation of sound.

- **H** Know that the range of human hearing is from 20 Hz to 20 kHz.

- **H** Be able to explain why human hearing is restricted to a limited range of frequencies.

Specification References
4.6.1.2, 4.6.1.4

Tip: Remember — the word 'medium' just refers to the material that a wave is travelling through.

Tip: The speed of sound in air is about 330 m/s.

Hearing Higher

To understand how sound waves are heard by the ears, you need to understand how sound travels through solids. Figure 2 shows how sound travels through air and then through a solid. The paper diaphragm in a speaker vibrates backwards and forwards, which causes the surrounding air to vibrate. This creates compressions and rarefactions — a sound wave.

Tip: Compressions are areas of high pressure and rarefactions are areas of low pressure.

When the sound wave meets a solid object, the air particles hitting the object (and producing changes in pressure — p. 114) cause the closest particles in the solid to move back and forth (vibrate). These particles hit the next particles along and so on. This series of vibrations passes the sound wave through the object.

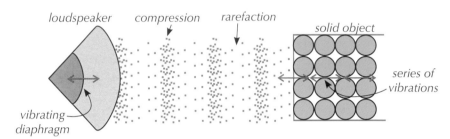

Figure 2: The conversion of sound waves to vibrations in a solid.

Sometimes, sound waves can cause the entire solid object to vibrate. Not all sound waves will be converted to vibrations of a solid in this way. It depends on the type of solid material and the frequency of the sound wave. Most materials will only convert sound waves within a certain frequency range into vibrations.

Tip: H Microphones work in a similar way to the ear. In a microphone, sound waves cause a diaphragm to vibrate and this movement is transferred into an electrical signal.

Sound waves of certain frequencies that reach your **ear drum** cause it to vibrate. These vibrations are passed on to tiny bones in your ear called ossicles, through the semicircular canals and to the cochlea. The cochlea turns these vibrations into electrical signals which get sent to your brain and allow you to sense (i.e. hear) the sound.

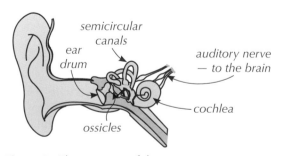

Figure 3: The structure of the ear.

Tip: H The exact range of frequencies that can be heard varies a lot between individuals. Also, as we age, the upper limit decreases.

Humans can only hear sounds with frequencies between 20 Hz and 20 kHz. The limits are caused by the size and shape of our ear drum, as well as the structure of all the parts within the ear that vibrate to transfer the energy from the sound wave.

Practice Questions — Fact Recall

Q1 How does the frequency, speed and wavelength of a sound wave change when the sound wave enters a denser medium?

Q2 How do sound waves pass into and travel through a solid?

Q3 Explain how sound waves reaching the ear drum cause the sensation of hearing.

Q4 What frequency range can humans hear?

Q5 Why can humans only hear a range of frequencies of sound waves?

Practice Questions — Application

Q1 The density of lead is 11 g/cm^3 and the density of aluminium is 2.7 g/cm^3 at room temperature. Predict what will happen to a sound wave's frequency, wavelength and speed when it moves from the lead to the aluminium.

Q2 Dog whistles produce high-frequency sound waves that dogs can hear, but humans cannot. Suggest why humans cannot hear sound produced by these whistles.

- **H** Know that ultrasound waves are sound waves with a frequency of more than 20 kHz.
- **H** Know that at a boundary between substances, ultrasound waves are partially reflected — some of the wave is reflected, some is transmitted.
- **H** Know that the distance to a boundary can be calculated from the time taken for reflected ultrasound waves to reach a detector.
- **H** Understand and be able to explain how the interaction of waves with boundaries between different materials can help us explore structures which are hidden from view.
- **H** Know how ultrasound is used in medical and industrial imaging.
- **H** Know how echo sounding uses high-frequency sound waves (including ultrasound) to explore deep water.

Specification Reference 4.6.1.5

2. Ultrasound   `Higher`

Ultrasound waves are high frequency sound waves. How they reflect at boundaries makes them a useful tool for imaging things we can't see directly.

What is ultrasound?

Ultrasound waves are sound waves with frequencies greater than 20 kHz. Ultrasound cannot be heard by people, as it is outside the normal range of human hearing.

Ultrasound waves are fairly easy to produce. Electrical devices can be made which produce electrical oscillations of a large range of frequencies. These can easily be converted into mechanical vibrations to produce sound waves with frequencies above 20 kHz.

Properties of ultrasound

When waves pass from one medium into another, some of the waves are reflected by the boundary between the two media, and some are transmitted (and possibly refracted). This is **partial reflection**.

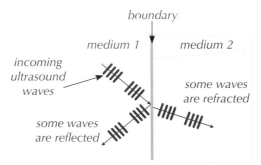

Figure 1: *A diagram of ultrasound waves undergoing partial reflection at a boundary.*

What this means is that you can point a pulse of ultrasound at an object, and wherever there are boundaries between one substance and another, some of the ultrasound gets reflected back.

When sound waves are reflected by surfaces, there is a delay between you hearing the original sound and the reflected sound. This is because the reflected sound waves have to travel further, taking longer to reach your ears. The reflected wave is known as an echo.

The same goes for ultrasound, the pulse takes time to return. The time it takes for the reflections to reach a detector can be used to measure how far away the boundary is.

Exploring structures with waves

As you've seen on p. 233, waves have different properties (e.g. speed) depending on the material they're travelling through.

When a wave arrives at a boundary between materials, a number of things can happen. It can be completely reflected or partially reflected (as above), continue travelling in the same direction but at a different speed, or be refracted or absorbed (p. 232).

Studying the properties and paths of waves through structures can give you clues to some of the properties of the structure that you can't see by eye. You can do this with lots of different waves — ultrasound and seismic waves (page 285) are two good, well-known examples.

Exploring structures with ultrasound

Partial reflection means that ultrasound has a number of useful applications. These chiefly involve imaging or locating objects that we can't directly see. Ultrasound can be used for medical and industrial purposes, as well as for echo sounding, which can be used in the military or in commercial fishing.

Medical imaging

Ultrasound is used in medicine to "see" inside the body. This allows doctors to diagnose medical conditions and check how the body is functioning.

> **Example** — **Higher**
>
> Ultrasound is used in prenatal scanning (the imaging of an unborn baby).
>
> Ultrasound waves can pass through the body, but whenever they reach a boundary between two different media (like fluid in the womb and the skin of the foetus) some of the wave is reflected back and detected.
>
>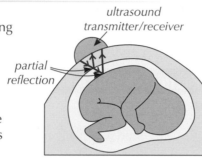
> *ultrasound transmitter/receiver*
> *partial reflection*
>
> The exact timing and distribution of these echoes are processed by a computer to produce a video image of the foetus.

Industrial imaging

Ultrasound is used in industry to see inside materials and products to check that they are suitable for use.

> **Example** — **Higher**
>
> Ultrasound can be used to find flaws in objects such as pipes, or in materials such as wood or metal.
>
> Ultrasound waves entering a material will usually be reflected by the far side of the material. If there is a flaw such as a crack inside the object, the wave will be reflected from it, and so a reflected wave will be detected sooner.
>
>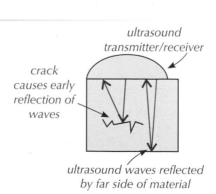
> *ultrasound transmitter/receiver*
> *crack causes early reflection of waves*
> *ultrasound waves reflected by far side of material*

Echo sounding

Echo sounding is a type of sonar used by boats and submarines, where high-frequency sound waves (including ultrasound) are used to find out the depth of the water they are in, or to locate objects in deep water.

Tip: H Waves which have been transmitted could go on to be partially reflected and transmitted at another boundary. So if you have a set of boundaries one after the other, the wave will undergo partial reflection at each boundary. This allows you to work out the distances between boundaries, based on when you detect the reflected waves.

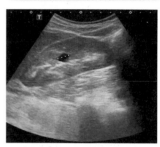

Figure 2: *An ultrasound image of a kidney. The black area marked with yellow dots is a cyst.*

Figure 3: *An image of the Mariana Trench, produced by echo sounding. The darker the blue, the lower the sea bed.*

A pulse of sound waves is emitted by the boat or submarine, and then the time taken for the echo to return is measured. It can be used to calculate the distance to the object that reflected the sound, if you know the speed of sound in the water.

Example — Higher

A pulse of ultrasound takes 4.50 seconds to travel from a submarine to the sea bed and back again. If the speed of sound in seawater is 1520 m/s, how far away is the submarine from the sea bed?

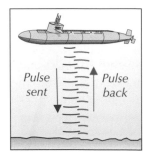

You need to use the equation $s = vt$ (distance = speed × time) from page 179:

$$s = vt = 1520 \times 4.50 = 6840 \text{ m}$$

But, this is a reflection question — the pulse travelled to the sea bed AND back in 4.50 s. So you need to divide your distance by 2.

distance to seabed = 6840 ÷ 2 = 3420 m

Pulse sent / *Pulse back*

Practice Questions — Fact Recall

Q1 What is ultrasound?

Q2 Ultrasound is partially reflected at a boundary. What does this mean?

Q3 State a use of ultrasound in medicine.

Q4 Explain how ultrasound echo sounding is used by a submarine to determine how far it is above the ocean floor.

Practice Questions — Application

Q1 A food manufacturer discovers that some of their tins of syrup have been contaminated with shards of plastic. Suggest how ultrasound might be used to detect which tins contain shards of plastic.

Q2 An ultrasound pulse is directed into a solid block of material, as shown in the diagram. A reflected pulse is detected 15.0 μs later.

Tip: μs means microseconds. One microsecond is equal to 1×10^{-6} s. (See p. 18 for more on prefixes like this.)

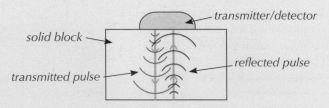

transmitter/detector

solid block

reflected pulse

transmitted pulse

The ultrasound pulse travels at 1720 m/s through the material. What is the distance (in mm) between the transmitter and the boundary from which the pulse is reflected?

3. Seismic Waves Higher

You've just seen how ultrasound waves can help us create an image of things we can't see on a fairly small scale. But what about structures as big as the Earth itself? That's where seismic waves come in.

What are seismic waves?

When there's an earthquake somewhere, it produces **seismic waves** which travel out through (or across the surface of) the Earth.

There are two different types of seismic waves you need to know about — P-waves and S-waves. P-waves are longitudinal waves. They can travel through both solids and liquids. Like sound waves, P-waves travel much faster in solids than in liquids. S-waves are transverse waves. They aren't able to pass through liquid materials.

P-waves always travel faster than S-waves, and both will travel faster through denser materials than they will through less dense ones.

Exploring the Earth's structure

When there's an earthquake, the seismic waves that are produced can be detected at different points on the surface of the planet using detectors called seismometers. Seismologists (scientists who study earthquakes) work out the time it takes for the shock waves to reach each seismometer. They also note which parts of the Earth don't receive the shock waves at all. They can use the data they gather to form an understanding of the inner structure of the Earth.

Like all waves, when seismic waves reach a boundary between two different materials, they can be absorbed, reflected, transmitted and refracted. Most of the time, as seismic waves travel through the Earth, they change speed gradually (as the density of the material they are moving through is also gradually changing). This causes the waves to follow a slightly curved path.

But when the medium the waves are travelling through changes suddenly (like at the boundary between two different materials), the direction changes abruptly, and the path has a kink (see Figure 2).

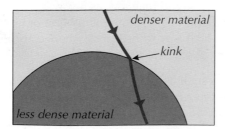

Figure 2: A diagram showing the 'kink' in the path when a seismic wave refracts at a boundary between two materials.

By measuring the seismic waves detected at various points on Earth, seismologists can build up a picture of the paths the waves have travelled along from the point of origin (the earthquake) to the points of detection and work out where any boundaries between different materials occur within the Earth.

Learning Objectives:

- **H** Know that seismic waves are produced by earthquakes.
- **H** Know that P-waves are a type of seismic wave that are longitudinal and travel faster in solids than in liquids.
- **H** Know that S-waves are a type of seismic wave that are transverse, and can't travel through liquids.
- **H** Understand how the detection of seismic waves has led to our current understanding of the Earth's interior, including the size and structure of its core.

Specification Reference 4.6.1.5

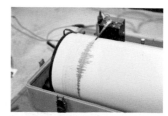

Figure 1: A seismograph, attached to a seismometer. Seismic waves cause the needle to move, and draw out a graph of the waves.

Tip: **H** P-waves and S-waves are 'body waves' — they travel through the Earth, not across its surface.

An earthquake produces seismic waves which travel through the inside of the Earth. Figure 3 shows the path of the P-waves.

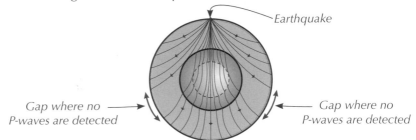

Figure 3: A diagram showing how P-waves travel through the Earth.

Some P-waves are detected on the other side of the Earth to the earthquake, which means they have been able to pass through the Earth's interior. But there are some gaps where no waves are detected. At these points, the waves must have abruptly changed direction (i.e. refracted) at a boundary somewhere along the way. This suggests that the Earth has a 'core' inside it, made of a different material to the rest of the Earth.

Figure 4 shows the path of the S-waves generated by the earthquake.

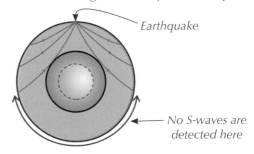

Figure 4: A diagram showing how S-waves travel through the Earth.

There's a large area on the opposite side of the Earth to the earthquake where no S-waves are detected. This indicates that, whatever the 'core' is made of, the S-waves cannot pass through it. So the core must be (at least partially) made from a liquid material, as S-waves cannot pass through liquids. This would also help explain the motion of P-waves — they travel slower in liquids, which is why their paths change abruptly at the boundary.

Different observations such as the one described above have led seismologists to develop a model (p. 3) of the internal structure of the Earth — see Figure 5.

Practice Questions — Fact Recall

Q1 State two types of seismic waves.

Q2 Which type of seismic wave cannot travel through liquids?

Q3 Which type of seismic wave can pass through the Earth's core?

Tip: [H] The P-waves travelling directly downwards from the source to the opposite side of the Earth don't change direction. This is because they hit the core at 90° — and waves that hit a boundary at 90° don't refract (see p. 233).

Tip: [H] P-waves are 'primary waves' — they arrive at the detectors first. S-waves are 'secondary waves' — they travel slower, so are detected second. It's a good way to remember the difference between the two.

Tip: [H] The mantle (see Figure 5) isn't quite solid — but it's close enough to solid that S-waves can still travel through it.

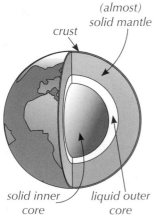

Figure 5: The internal structure of the Earth.

Q1 A seismometer is placed at the bottom of a lake to detect S-waves and P-waves created by earthquakes. Suggest why the seismometer must be placed at the bottom of the lake.

Topic 6d Checklist — Make sure you know...

Sound Waves

☐ That when a sound wave moves from one medium to another, its frequency stays the same but its wavelength and speed change.

☐ H How sound waves travel through solids as a series of vibrations.

☐ H How sound waves causing your ear drum to vibrate leads to the sensation of hearing.

☐ H That whether a certain frequency of sound will cause a solid to vibrate depends on the solid object's size, shape and what it's made of.

☐ H That the normal range of human hearing is 20 Hz to 20 kHz.

☐ H That human hearing is limited to the range of frequencies that will cause the ear drum to vibrate.

Ultrasound

☐ H That sound waves with frequencies greater than 20 kHz (the upper limit of human hearing) are known as ultrasound.

☐ H That when an ultrasound waves meets a boundary between substances, some of the wave is reflected and some is transmitted (and possibly refracted) — this is partial reflection.

☐ H How interpreting the time between the emission of ultrasound waves and the detection of partially reflected ultrasound waves can be used to determine the locations of boundaries and form images of structures which are hidden from view.

☐ H Some examples of how ultrasound is used in medicine (e.g. prenatal scanning), in industry (e.g. finding faults in materials) and in echo sounding.

Seismic Waves

☐ H That seismic waves are created by earthquakes, and that two examples are S-waves and P-waves.

☐ H That P-waves are longitudinal waves which can pass through solids and liquids, and travel faster through solids than liquids.

☐ H That S-waves are transverse waves which can only pass through solids.

☐ H That P-waves travel faster through substances than S-waves.

☐ H How observations of S-waves and P-waves have provided evidence which has improved our knowledge of the structure of the Earth, including the size and structure of the Earth's core.

Exam-style Questions

1   An ultrasonic scanner is a device that emits and detects ultrasound waves. It can be used in medicine to diagnose and treat illnesses and to monitor pregnancies.

1.1   Use the correct words from the box below to complete the following sentences.

higher than	**Hz**	**within**
MHz	**less than**	**kHz**

Ultrasound is sound with a frequency .... the range of human hearing.

20 to 20 000 .... is the range of frequencies of human hearing.

(2 marks)

A pregnant patient has an ultrasound scan to produce an image of the foetus. A pulse of ultrasound is transmitted into her abdomen. A reflected pulse is detected by the scanner 0.000020 s later.

1.2   Calculate the distance between the scanner and the boundary from which the pulse is reflected. You can assume that ultrasound travels at 1500 m/s in the abdomen.
Give your answer in metres.

(3 marks)

2   Here is a picture of the Earth. An earthquake occurs at point A. It produces P-waves and S-waves, which travel through the Earth.

Explain why, at point B, P-waves are detected but no S-waves are detected.

(3 marks)

3   Sound waves can travel through solids, liquids and gases.

3.1   A sound wave with a wavelength of 3.3 m travels through a concrete block at 3300 m/s.
It passes into a steel beam, and its wavelength changes to 6.0 m.
Calculate the speed of the sound wave in the steel beam.
Give your answer in m/s.

(3 marks)

3.2   The sound wave then passes into air. State how the speed of the sound wave changes.
Explain your answer.

(2 marks)

1. Magnetic Fields

Magnetic fields are responsible for the non-contact force experienced by magnetic materials. All magnets have one.

What is a magnetic field?

A magnet has a **magnetic field** around it. It's the magnetic field that causes magnetic objects to be attracted or repelled by a magnet, and they don't just exist around magnets.

> A magnetic field is a region where magnets, magnetic materials (like iron and steel), and also wires carrying currents, experience a force acting on them.

This magnetic force is a non-contact force — it can act between objects that are not physically touching each other. It is similar to the force on charges in an electric field, like you saw on page 103.

A magnetic field can be represented by a field diagram (see Figure 1). A field diagram is just a series of lines that show where a magnetic field exists and its direction. All magnets have two poles, where the field is strongest — a north (or north seeking) pole and south (or south seeking) pole. Magnetic field lines have arrows on them that always point from north to south. The direction of the field lines show the direction of the force a north pole would feel if it was placed in that location. The stronger the magnetic field at any point, the closer together the field lines are. The further away from a magnet you get, the weaker the field.

Example

This is the magnetic field around a bar magnet. It's strongest at the north and south poles, where the field lines are closest together.

A north pole placed here will feel a force to the right

Figure 1: *A field diagram showing the magnetic field around a bar magnet.*

The force between a magnet and a **magnetic material** is always attractive, no matter the pole (see next page).

The force between two magnets placed close to each other can be attractive or repulsive.

- Two poles that are the same (these are called 'like poles') will repel each other (see Figure 2).

- Two different poles ('unlike poles') will attract each other (see Figure 3).

Learning Objectives:

- Know that magnets have a magnetic field — a region where magnets and magnetic materials experience a non-contact force.
- Know that all magnets have two poles, where the magnetic field is strongest.
- Know that field lines point from the north pole to the south pole and give the direction of the force that another north pole would feel.
- Know that the field is stronger closer to the magnet.
- Be able to draw the magnetic field of a bar magnet.
- Know that the force on a magnetic material in a magnetic field (or an induced magnet) is always attractive.
- Know that magnets exert a force on each other, and that like poles repel and unlike poles attract.
- Know how to plot a magnetic field using a compass.
- Understand how the motion of a compass needle indicates that the Earth has a magnetic core.
- Know that an induced magnet is different to a permanent magnet, as it demagnetises when removed from a magnetic field.

Specification References 4.7.1.1, 4.7.1.2

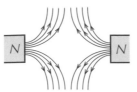
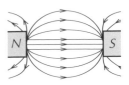

Figure 2: The magnetic field between two like poles repelling each other.

Figure 3: The magnetic field between two unlike poles attracting each other.

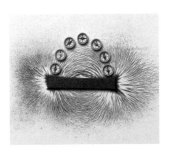

Figure 4: A series of compasses pointing along the magnetic field lines of a bar magnet, which are also shown by iron filings.

Compasses

Inside a compass is a tiny bar magnet. The north pole of this magnet is attracted to the south pole of any other magnet it is near. So the compass points in the direction of the magnetic field it is in. You can move a compass around a magnet and trace its position on some paper to build up a picture of what the magnetic field looks like.

When they're not near a magnet, compasses always point north. This is because the Earth generates its own magnetic field, which shows that the inside (core) of the Earth must be magnetic.

Permanent and induced magnets

There are two types of magnet — **permanent magnets** and **induced magnets**. Permanent magnets produce their own magnetic fields. Induced magnets are magnetic materials that turn into magnets when they're in a magnetic field.

The force between permanent and induced magnets is always attractive (as with all magnetic materials). When you take away the magnetic field, induced magnets quickly lose their magnetism (or most of it).

Example

The magnetic material becomes magnetised when it is brought near the bar magnet. It has its own poles and magnetic field.

permanent magnet   *induced magnet*

Figure 5: A diagram of magnetic poles induced in a magnetic material when it is brought near to a bar magnet.

Practice Questions — Fact Recall

Q1   What is a magnetic field?

Q2   Draw the magnetic field lines around a bar magnet.

Q3   How do we know that the Earth's core is magnetic?

Q4   Describe what is meant by an 'induced magnet'.

2. Electromagnetism

Magnets aren't the only things with magnetic fields — anything that's carrying an electric current also has one. This can have some clever applications.

The magnetic field around a current-carrying wire

When a current (see p. 63) flows through a wire, a magnetic field is created around the wire. The field is made up of concentric circles (circles which share the same centre) perpendicular to the wire, with the wire in the centre (see Figure 1).

You can see this by placing a compass near a wire which is carrying a current. As you move the compass, it will trace the direction of the magnetic field.

Changing the direction of the current changes the direction of the magnetic field — use the **right-hand thumb rule** to work out which way it goes:

> **The right-hand thumb rule**
>
> Using your right hand, point your thumb in the direction of the current, and curl your fingers. The direction of your fingers is the direction of the field.

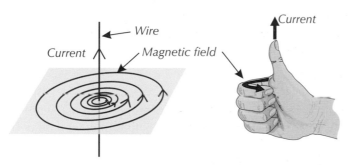

Figure 1: A diagram showing how a magnetic field is created around a current-carrying wire, alongside a demonstration of the right-hand thumb rule.

The strength of the magnetic field produced changes with the current and the distance from the wire. The larger the current through the wire, or the closer to the wire you are, the stronger the field is.

Solenoids

You can increase the strength of the magnetic field that a wire produces by wrapping the wire into a coil called a **solenoid**.

This happens because the field lines around each loop of wire line up with each other and form the magnetic field shown in Figure 2.

Learning Objectives:

- Know that a current-carrying wire has a magnetic field.
- Know how to demonstrate the magnetic properties of a current-carrying wire, e.g. by using a compass.
- Be able to draw the magnetic field around a straight current-carrying wire.
- Know how the strength of this magnetic field varies with distance from the wire and size of current in the wire.
- Understand why coiling a current-carrying wire into a solenoid increases the strength of the magnetic field.
- Be able to draw the magnetic field of a solenoid.
- Know that the magnetic field inside a solenoid is strong and uniform, and the field outside a solenoid is similar to a bar magnet.
- Know that adding an iron core to a solenoid increases the strength of the magnetic field.
- Know that a solenoid with an iron core is known as an electromagnet.
- Be able to interpret diagrams of devices which use electromagnets and understand how they work.

Specification Reference 4.7.2.1

The result is lots of field lines pointing in the same direction that are very close to each other. As you saw on page 289, the closer together the field lines are, the stronger the field is.

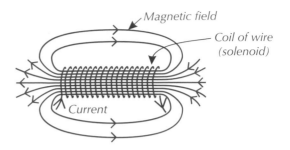

Magnetic field

Coil of wire (solenoid)

Current

Figure 2: A field diagram showing the magnetic field created around a current-carrying solenoid.

Figure 3: Iron filings showing the magnetic field lines around a current-carrying solenoid.

The magnetic field inside a solenoid is strong and uniform (it has the same strength and direction at every point in that region). Outside the coil, the magnetic field is just like the one round a bar magnet.

You can increase the strength of the magnetic field of a solenoid even more by putting a block of iron in the centre of the coil. The iron core becomes an induced magnet whenever current is flowing. The magnetic field of the core and the coil combine, making a stronger magnet overall.

If you stop the current, the magnetic field disappears. A solenoid with an iron core (a magnet whose magnetic field can be turned on and off by an electric current) is called an **electromagnet**.

Figure 4: An electromagnet used in the Large Hadron Collider at CERN. Solenoids can be used to make incredibly large and powerful magnets.

Uses of electromagnets

Magnets you can switch on and off (electromagnets) are really useful. They're usually used because they're so quick to turn on and off or because they can create a varying force.

Example 1

Magnets can be used to attract and pick up things made from magnetic materials like iron and steel. Electromagnets are used in some cranes, e.g. in scrap yards and steel works.

If an ordinary magnet was used, the crane would be able to pick up the scrap metal etc., but then wouldn't let it go, which isn't very helpful.

Using an electromagnet means the magnet can be switched on when you want to pick stuff up, then switched off when you want to drop it, which is far more useful.

Figure 5: A scrap yard crane that uses an electromagnet. The electromagnet is switched off when the crane needs to drop the metal.

Example 2

Electromagnets can be used in relay switches. Relay switches link together two circuits, so that turning on one circuit causes the other to turn on too. They are used to turn on very high-current circuits using a lower-current circuit, as they help to stop the user from coming into contact with the high current.

A relay switch is used to turn on a car's starter motor. A car's starter motor needs a very high current, but the part you control (when you're turning the key) is in the low-current circuit, making it safer for the person driving the car.

Here's how it works (see Figure 6 for a diagram). When the switch in the low-current circuit is closed, current flows through the electromagnet, producing a magnetic field. The electromagnet attracts the iron contact on the rocker, which pivots and closes the contacts in the high-current circuit. Current flows through the starter motor and the motor spins.

When the low-current switch is opened, the electromagnet stops pulling, the rocker returns, and the high-current circuit is broken again.

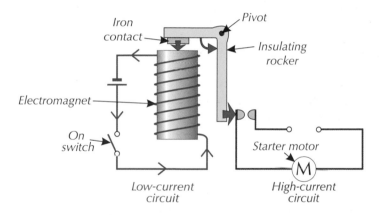

Figure 6: *A relay switch in a starter motor. An electromagnet in a low-current circuit is used to close a switch in the higher-current starter motor circuit.*

Practice Questions — Fact Recall

Q1 In the right-hand thumb rule, what does your thumb represent? What do your fingers represent?

Q2 Draw the magnetic field of a solenoid.

Q3 State how you can increase the strength of the magnetic field produced by a solenoid.

Q4 Describe what an electromagnet is and explain how it works.

Q5 Give one use of an electromagnet.

Practice Questions — Application

Q1 A current-carrying wire (shown by the red dot) produces the following magnetic field pattern when viewed from above.

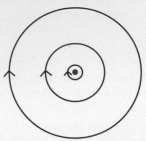

Tip: You'll need to use the right-hand thumb rule to find the direction of the current.

a) What direction is the current flowing in?

b) In what direction would the field lines point if the current began flowing in the opposite direction?

Q2 An electromagnet is used in an electric door lock, as shown. Normally, the circuit has a current flowing through it and the iron bolt is pulled across the doorway, stopping the door from being opened. When the correct code is typed into the keypad, the circuit is broken.

a) Explain why the bolt is pulled across the door when current flows through the circuit.

b) Explain why breaking the circuit allows the door to be opened.

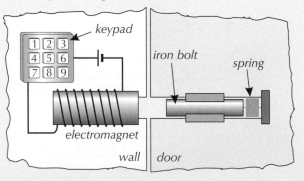

3. The Motor Effect Higher

If the magnetic field produced by a current-carrying wire is within a magnetic field in the first place, the interacting fields can make things start to move...

What is the motor effect?

As you know from page 291, passing an electric current through a wire (or other conductor) produces a magnetic field around the wire. If you put that wire into another magnetic field, you end up with two magnetic fields combining. The result is that the wire and the magnet exert a force on each other.

When a current-carrying wire in a magnetic field experiences a force it is known as the **motor effect**.

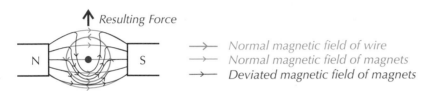

Figure 1: *A diagram of the magnetic field interactions which lead to the motor effect. The red dot represents a wire carrying current out of the page (towards you).*

Increasing the strength of the magnetic field, or the size of the current flowing through the wire, will increase the size of the force.

To experience the full force, the wire has to be at 90° to the magnetic field (see Figure 2). If the wire runs along parallel to the magnetic field it won't experience any force at all. At angles in between it'll feel some force.

Example Higher

A straight current-carrying wire will experience a force if it is placed at right angles to a magnetic field. The force will be at 90° to both the wire and the magnetic field.

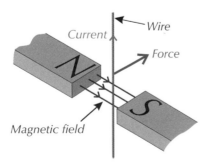

Figure 2: *A diagram showing the force on a current-carrying wire placed at 90° to a magnetic field.*

Learning Objectives:

- **H** Know that a current-carrying conductor and a magnet exert a force on one another while the conductor is inside the magnet's magnetic field. This is the motor effect.
- **H** Know that the size of the force due to the motor effect depends on the current, the length of conductor inside the field and the magnetic flux density.
- **H** Know how to use the equation $F = BIl$.
- **H** Know and be able to use Fleming's left-hand rule.

Specification Reference 4.7.2.2

Tip: **H** You'll see how to work out the size and direction of the force on the next few pages.

Example **Higher**

A horseshoe magnet is used to provide a magnetic field. A conducting
bar is placed on two conducting rails fixed at 90° to the magnetic field
(see Figure 4).

If a current is applied to the rails, the current will flow through the bar
(at 90° to the field) and the motor effect will cause the bar to experience
a force.

The bar is free to move, so it will roll in the direction of the force. As in
the previous example, the force on the conductor is at 90° to both the
conductor and the magnetic field.

Figure 3: *An experiment
showing the motor effect in
action. The coil of yellow
current-carrying wire leaps
into the air due to interaction
between its magnetic field,
and that of the magnet.*

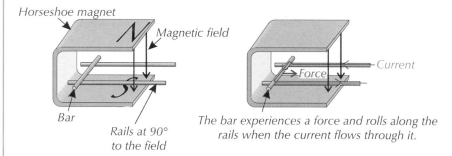

*The bar experiences a force and rolls along the
rails when the current flows through it.*

Figure 4: *A diagram showing how the motor effect
can be used to produce movement.*

Calculating the size of the force

The force acting on a conductor in a magnetic field depends on three things:

- The **magnetic flux density** — how many field (flux) lines there are in a
  region. This shows the strength of the magnetic field. It is represented by
  the symbol B and is measured in tesla (T).

- The size of the current through the conductor, I.

- The length of the conductor that's in the magnetic field, l.

When the current is at 90° to the magnetic field it is in, then you can calculate
the size of the force using the equation:

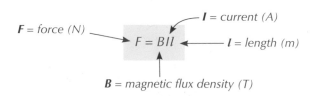

F = force (N)
I = current (A)
l = length (m)
$F = BIl$
B = magnetic flux density (T)

A 10 cm length of wire carrying a current of 3 A sits inside a magnetic field. The current flows at 90° to the direction of the magnetic field. It experiences a force of 0.12 N from the motor effect.

Calculate the magnetic flux density of the magnet.

First, convert the length into metres.

$$l = 10 \text{ cm} = 0.1 \text{ m}$$

You're looking for the magnetic flux density, so rearrange $F = BIl$ to find B, then substitute in the values you're given.

$$F = BIl, \text{ so } B = F \div (I \times l)$$
$$= 0.12 \div (3 \times 0.1) = 0.4 \text{ T}$$

Fleming's left-hand rule

If a current is flowing at 90° to a magnetic field, you can tell which way the force acting on the conductor due to the motor effect will act using **Fleming's left-hand rule**.

Here's what you do:

- Using your left hand, point your **F**irst finger in the direction of the **F**ield.

- Point your se**C**ond finger in the direction of the **C**urrent.

- Stick your thu**M**b out so it's at 90° to the other two fingers (see Figure 5). It will then point in the direction of the force (**M**otion).

If the direction of the current or magnetic field is reversed, then the direction of the force is reversed too (try it).

Tip: H You can use Fleming's left-hand rule to find the missing direction of any of these three quantities — as long as you know the other two.

Exam Tip H
You need to practise this before you get into the exam — you might need to use it to answer an exam question. It's no good pointing all over the place and ending up with the wrong answer because you can't remember which hand you're supposed to be using or what each finger represents.

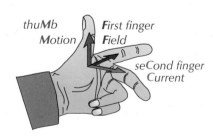

Figure 5: *A diagram showing how Fleming's left-hand rule can be used to find the direction of the force experienced by a current-carrying wire at 90° to a magnetic field.*

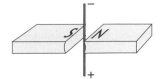

Example — **Higher**

Draw in the direction of the force acting on this wire.

- Start by drawing in the current arrows and the magnetic field lines. Current goes from positive to negative and magnetic fields go from north to south. Then use Fleming's left-hand rule to work out the direction of the force (motion).

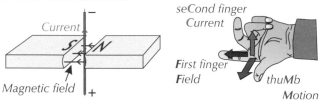

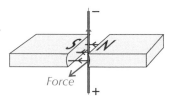

- Finally, draw in the direction of the force.

Practice Questions — Fact Recall

Q1 For the motor effect, give two ways the force on a current-carrying wire at 90° to a magnetic field can be increased.

Q2 What is the size of the force due to the motor effect on a current-carrying wire if it's parallel to a magnetic field?

Q3 State the equation which gives the size of the force due to the motor effect, and give the units for each quantity. State the condition that is required for this equation to work.

Q4 In Fleming's left-hand rule, what do the middle finger, index finger and thumb each represent?

Practice Questions — Application

Q1 In each of the situations below, the current-carrying wire is at 90° to the magnetic field. Give the direction of the force on the wire.

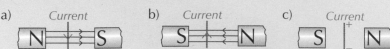

Q2 Give the direction of the force on the wire in Q1 c) if the direction of the current is swapped and the poles of the magnets are swapped.

Q3 A wire carrying a current of 5.0 A through a 0.20 T magnetic field (at 90° to the field) experiences a force of 0.25 N. Calculate the length of the wire, in cm, inside the magnetic field.

Tip: [H] You'll have to work out the direction of the current and magnetic field yourself for part c).

298   Topic 7 Magnetism and Electromagnetism

4. Using the Motor Effect Higher

The motor effect has many useful applications. It becomes really handy once you start using it to turn an axle.

The simple dc electric motor

You know from page 295 that the motor effect is when a current-carrying wire experiences a force in a magnetic field. Electric motors use the motor effect to produce rotation — and that rotation is the basis of an awful lot of appliances.

How it works

A loop of wire that's free to rotate about an axis is placed in a magnetic field (see Figure 1).

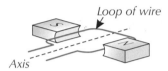
Loop of wire
Axis

Figure 1: *A loop of wire in a magnetic field free to rotate about its axis.*

When a direct current flows through the loop, the two side arms, which are at 90° to the field, each experience a force due to the motor effect. They experience forces in opposite directions because the direction of the current in each arm is opposite (see Figure 2). The loop will start to rotate around its axis because the forces act one up and one down.

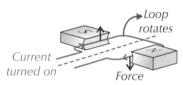
Loop rotates
Current turned on
Force

Figure 2: *A loop of wire in a magnetic field rotating when current flows through it.*

When the wire loop reaches a vertical position the forces will still be acting one up and one down on the same arms of the loop, so the loop gets stuck (see Figure 3).

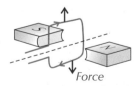
Force
Force

Figure 3: *A loop of wire getting stuck in the same position due to the forces acting on it.*

For the motor to keep rotating in the same direction, the forces acting on the arms of the loop need to swap direction.

Learning Objectives:

- H Understand how the motor effect can be used to make a coil of conductor rotate in a magnetic field.
- H Know that the rotation of a current-carrying coil in a magnetic field is the basis for many simple motors.
- H Understand how the motor effect can be used to create sound waves in a moving-coil loudspeaker and headphones.

Specification Reference 4.7.2.3, 4.7.2.4

Tip: Remember, direct current (dc) is current that only flows in one direction, see page 89.

Tip: H Fleming's left-hand rule can be used to find the direction of each force — see page 297 if you need a reminder.

Tip: H The loop will actually vibrate back and forth slightly as it'll have some momentum, but you don't need to worry about that.

Reversing the direction of the current reverses the direction of the force (see page 297). A **split-ring commutator** is a clever way of doing this — it swaps the contacts of the loop every half turn, reversing the current — see Figures 4 and 5.

A split-ring commutator is just a conducting ring with a gap between the two halves. As it rotates, the part of the commutator that is touching each contact changes every half turn.

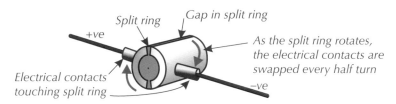

Figure 4: A split-ring commutator.

By linking each end of the loop to one half of a split-ring commutator, you change the electrical contacts of the loop (and so the direction of the current) every half turn. This means that the force acting on each arm of the loop will swap every half turn, allowing rotation to continue in the same direction.

Tip: H You can have an ac (alternating current) electric motor too, in which case you don't need a split-ring commutator — it just has to rotate at or close to the frequency of the current.

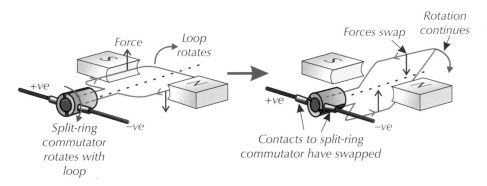

Figure 5: A dc electric motor using a split-ring commutator to swap the direction of the current, and so the forces, every half turn.

And that's it — a simple dc electric motor. Make sure you understand how it uses the motor effect to work.

Speed of a simple electric motor

The speed of an electric motor can be increased in two main ways:

- by increasing the current.

- by increasing the strength of the magnetic field.

Both of these factors increase the force experienced by the wire due to the motor effect, so it rotates faster.

Using a coil of wire instead of a single loop will also increase the force caused by the motor effect. It's simple really — the sides of each individual loop experience a force, so the more loops you have, the larger the total force on the sides of the coil. So, if you're calculating the force on a coil of wire in a magnetic field, you need to first calculate the force on a single loop (using the equation on p. 296) and then multiply it by the number of loops in the coil.

Direction of a simple electric motor

The direction that a motor turns in can be found using Fleming's left-hand rule, and can be reversed either by:

- swapping the polarity of the direct current (dc) supply, or

- swapping the magnetic poles over.

Example — **Higher**

Is the electric motor turning clockwise or anticlockwise?

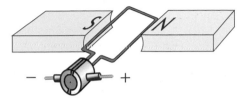

- Start by drawing in the magnetic field lines from north to south, and the direction of the current from positive to negative.

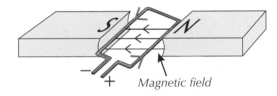

Magnetic field

- Then use Fleming's left-hand rule on one side of the loop to work out the direction of the force (motion). Let's use the right side of the loop.

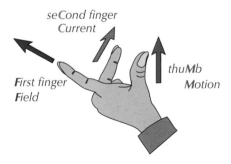

seCond finger
Current

thuMb
Motion

First finger
Field

- Finally, draw in the direction of the force.

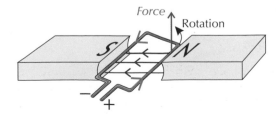

Force
Rotation

- So — the motor is turning anticlockwise.

> **Tip:** **H** The first step of this method should look familiar — it's the same one that was used on page 298 to find the direction of the force on a straight wire in a magnetic field.

Loudspeakers

Loudspeakers work due to the motor effect. Loudspeakers and headphones (which are just tiny loudspeakers) both use electromagnets.

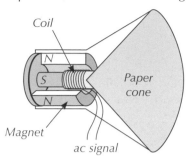

Figure 6: A diagram of the internal workings of a loudspeaker.

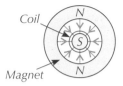

Figure 7: A diagram of a cylindrical magnet, as used in a loudspeaker, showing the magnetic field.

An alternating current is sent through a coil of wire attached to the base of a paper cone (see Figure 6). The coil surrounds one pole of a permanent magnet, and is surrounded by the other pole, so the current causes a force on the coil (which causes the cone to move).

When the current reverses, the force acts in the opposite direction, which causes the cone to move in the opposite direction too. So variations in the current make the cone vibrate, which makes the air around the cone vibrate and creates the variations in air pressure that form a sound wave (see page 279).

The frequency of the sound wave produced is the same as the frequency of the ac, so by controlling the frequency of the ac you can alter the sound wave.

Figure 8: A cross-section of a loudspeaker, showing the outer ring of the magnet, and the copper coil, attached to a paper cone, surrounding the inner pole of the magnet.

Practice Questions — Fact Recall

Q1 Why is a split-ring commutator needed in an electric motor that uses direct current?

Q2 Give two ways the direction of rotation of a simple electric motor can be reversed.

Q3 How do loudspeakers use the motor effect to produce sound waves?

Practice Question — Application

Q1 Look at this diagram of a simple electric motor.

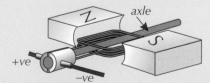

a) Which direction is the axle turning — clockwise or anticlockwise?

b) What will happen to the speed of the axle if the current is increased?

c) What will happen if the magnetic poles are reversed?

5. The Generator Effect `Higher`

The generator effect is quite a tricky idea to get your head around. It's really important though — it's how we generate most of our electricity.

What is the generator effect?

You've already seen on page 295 that a magnetic field can create a force that causes a current-carrying wire to move.

Well, it works the other way round too. An electrical conductor that is moving relative to a magnetic field can induce (create) a potential difference across the conductor. Similarly, a pd can also be induced if there is a change in an external magnetic field (i.e. one other than the conductor's) around a conductor. If the conductor is part of a complete circuit, current will flow.

When a potential difference is created like this it's called the **generator effect** (or electromagnetic induction).

> The generator effect is the induction of a potential difference (and current if there's a complete circuit) across a conductor which is experiencing a change in an external magnetic field.

The generator effect happens when a conductor 'cuts' through the magnetic field lines. So a potential difference across the ends of a conductor can be induced in two ways:

- By moving the electrical conductor in a magnetic field.

- By moving or changing a magnetic field (e.g. moving a magnet) relative to the electrical conductor.

You need to be able to use the generator effect to explain why a pd is induced in certain situations. Here are some examples:

Examples — `Higher`

- Moving a wire in a magnetic field will cause a potential difference to be induced across the ends of the wire.

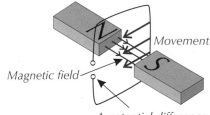

Movement

Magnetic field

A potential difference is induced across the ends of the wire.

- Moving a bar magnet through a coil of wire will induce a potential difference across the ends of the wire.

  If you connect each end of the coil to a bulb, you'll see the bulb light up as the magnet moves.

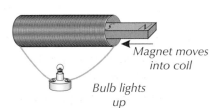

Magnet moves into coil

Bulb lights up

Tip: The 'polarity' of a magnet is which way round its north and south poles are.

Tip: **H** There's more about ac on page 89.

If you move the magnet (or conductor) in the opposite direction, then the potential difference/current will be reversed. Likewise, if the polarity of the magnet is reversed, then the potential difference/current will be reversed too.

So, if you keep the magnetic field (or the coil) moving backwards and forwards, you produce a potential difference/current that keeps swapping direction. This is an alternating current (ac).

Figure 1: *A portable electricity generator. It uses a combustion engine to turn a coil in a magnetic field, and so generate electricity.*

Example **Higher**

You can create the same effect by turning a magnet end to end in a coil, or turning a coil inside a magnetic field.

- As you turn the magnet, the magnetic field through the coil changes. This change in the magnetic field induces a potential difference, which can make a current flow in the wire.

- When you've turned the magnet through half a turn, the direction of the magnetic field through the coil reverses. When this happens, the potential difference reverses, so the current flows in the opposite direction around the coil of wire.

- If you keep turning the magnet in the same direction — e.g. always clockwise — then the potential difference will keep on reversing every half turn and you'll get an alternating current.

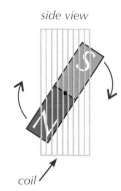

side view

coil

If you want to change the size of the induced pd, you have to change the rate that the magnetic field is changing. Induced potential difference (and so induced current) can be increased by:

- Increasing the speed of the movement — cutting more magnetic field lines in a given time.

- Increasing the strength of the magnetic field (so there are more field lines that can be cut).

Induced current

So, a change in magnetic field can induce a current in a wire. But, as you saw on page 291, when a current flows through a wire, a magnetic field is created around the wire. So you get a second magnetic field — different to the one whose field lines were being cut in the first place.

The magnetic field created by an induced current always acts against the change that made it (whether that's the movement of a wire or a change in the field it's in). Basically, it's trying to return things to the way they were.

This means that the induced current always opposes the change that made it.

Generating electricity

Generators make use of the generator effect to induce a current. They rotate a coil in a magnetic field (or a magnet in a coil). As the coil (or magnet) spins, a current is induced in the coil. Whether they generate an alternating or direct current depends on the device.

Alternators

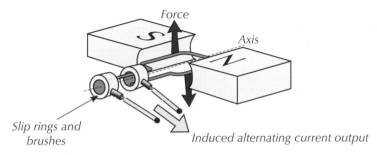

Figure 2: A diagram of a simple alternator.

Alternators generate alternating current. Their construction is similar to a motor (see page 299). As the coil spins, a current is induced in the coil. Every half-turn, this current changes direction. Alternators use slip rings and brushes (see Figure 2) so that the contacts don't swap every half-turn. This maintains the alternating potential difference, generating an alternating current.

Dynamos

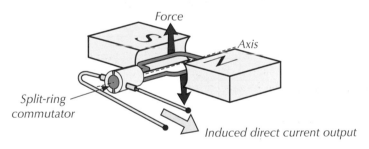

Figure 4: A diagram of a simple dynamo.

Dynamos generate direct current. They're similar to alternators, with one important difference — they use a split-ring commutator as their electrical contact (see Figure 4). This swaps the connection every half turn, to keep the current flowing in the same direction and generate direct current.

Oscilloscopes

You can use an oscilloscope to see the generated potential difference. Oscilloscopes show how the potential difference generated in the coil changes over time.

Tip: ⓗ Remember — the generator effect works whether the coil or the field is moving. It's how most of our electricity is generated, whether it's in a coal-fired power station or a wind turbine.

Figure 3: Dynamos are used in wind-up torches. The crank turns a coil, which induces a pd and powers the bulb.

Tip: ⓗ Split-ring commutators are also used in electric motors — see page 300.

Tip: ⓗ You've met oscilloscopes before, on page 229.

Alternating pd

Tip: H The curve on an oscilloscope is known as a trace.

Alternating current comes from an alternating potential difference, where the positive and negative ends keep alternating. So the trace cycles between positive and negative peaks, crossing the horizontal axis — see Figure 5.

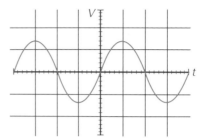

Figure 5: *An oscilloscope trace of the potential difference (V) generated by an alternator.*

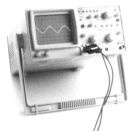

Figure 6: *An example of an oscilloscope displaying a trace for an alternating potential difference.*

Direct pd

For direct current, the potential difference doesn't alternate and the current always flows in the same direction. The trace never crosses the horizontal axis — see Figure 7. You might expect the direct pd trace to be a straight line because it doesn't change direction. But because the dynamo reverses direction every half turn, you still get peaks like in the alternating pd trace — they're just all above the horizontal axis.

Tip: H If you see a series of troughs which are all below the horizontal axis, you have probably connected your oscilloscope incorrectly, so the current is flowing in the opposite direction.

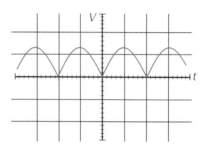

Figure 7: *An oscilloscope trace of the potential difference (V) generated by a dynamo.*

Understanding traces

Tip: H The frequency of rotations is just the number of full rotations of the coil (or magnet) per second.

The height of the line at a given point is the generated pd at that time. Increasing the frequency of rotations in the generator increases the overall pd — so you get higher peaks, but you also get more of them. The wave will be stretched vertically but squashed horizontally, as shown in Figure 8.

Tip: H Increasing the frequency means more field lines are cut in a given time, so the induced pd is larger (see page 304).

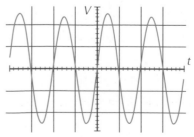

Figure 8: *An oscilloscope trace of an alternating pd generated by a greater frequency of rotation than in Figure 5.*

Microphones

Microphones generate current from sound waves.

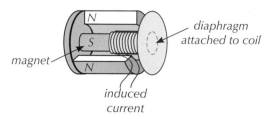

Figure 9: A diagram of the internal workings of a microphone.

Tip: H Microphones are basically loudspeakers (see page 302) in reverse.

Sound waves hit a flexible diaphragm that is attached to a coil of wire, wrapped around a magnet (see Figure 9). This causes the coil of wire to move back and forth in the magnetic field, which generates a current.

The movement of the coil (and so the generated current) depends on the properties of the sound wave (louder sounds make the diaphragm move further). This is how microphones can convert the pressure variations of a sound wave into variations in current in an electric circuit.

Figure 10: A microphone connected to an oscilloscope, displaying the generated pd trace.

Practice Questions — Fact Recall

Q1 What is the generator effect?

Q2 Give two ways the generator effect can be used to generate a potential difference across a coil of wire.

Q3 Sketch an oscilloscope trace of an alternating potential difference.

Q4 Describe how a microphone converts sound waves into electrical signals.

Practice Question — Application

Q1 The diagram shows how a windmill can be used to power a light bulb. As the axle turns, the slip rings make sure that each end of the coil of wire stays connected to the same end of the bulb circuit.

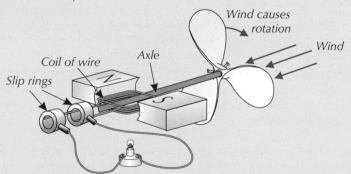

a) Explain how the windmill rotating causes a current to flow in the bulb circuit.

b) What sort of current will flow in the bulb circuit?

Learning Objectives:

- **H** Know the basic structure of a transformer.
- **H** Know that iron is used as a core because it can magnetise and demagnetise quickly.
- **H** Understand how a transformer uses an alternating current in the primary coil to induce an alternating current in the secondary coil.
- **H** Know that the ratio of the potential differences across the coils depends on the ratio of the number of turns on each coil.
- **H** Know that the pd across the secondary coil is greater than the pd across the primary coil in a step-up transformer.
- **H** Know that the pd across the primary coil is greater than the pd across the secondary coil in a step-down transformer.
- **H** Understand that if transformers are assumed to be 100% efficient, their power output is equal to their power input.
- **H** Be able to use the equations $\frac{V_p}{V_s} = \frac{n_p}{n_s}$ and $V_s I_s = V_p I_p$ and be able to apply them to the scenarios where transformers are used to transmit power.

Specification Reference 4.7.3.4

Another nifty use of electromagnetic induction is in transformers. They're handy for changing the potential difference of an ac supply.

What do transformers do?

Transformers are devices that can change the potential difference of an electrical supply using the generator effect (see page 303). It's important to remember that they will only work on alternating current, not direct current. Keep reading to see why...

The structure of a transformer

Figure 1 shows the structure of a transformer — it consists of two coils, the primary and the secondary, wrapped around an iron core. Iron is used for the core because it can magnetise and demagnetise quickly.

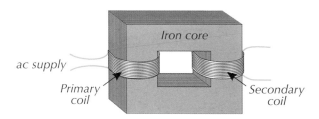

Figure 1: *The structure of a transformer.*

How a transformer works

Here's how a transformer works:

- As you know from page 291, passing an electric current through a wire produces a magnetic field around it. So, when an alternating current flows through the primary coil in a transformer it produces a magnetic field which magnetises the iron core — see Figure 2 on the next page.

- Because there is alternating current (ac) in the primary coil, the magnetic field in the iron core is alternating too — i.e. it is a constantly changing magnetic field.

- This constantly changing magnetic field cuts through the secondary coil.

- The changing field induces an alternating potential difference across the ends of the secondary coil by the generator effect.

- If the secondary coil is part of a complete circuit, this potential difference causes an alternating current to flow — and it has the same frequency as the alternating current in the primary coil.

- The size of the potential difference induced across the secondary coil depends on the size of the potential difference across the primary coil and the number of turns on each coil (see page 310).

Remember that the iron core is purely for transferring the changing magnetic field from the primary coil to the secondary. No electricity flows round the iron core.

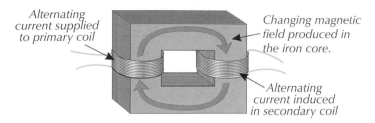

Alternating current supplied to primary coil

Changing magnetic field produced in the iron core.

Alternating current induced in secondary coil

Figure 2: *A transformer using electromagnetic induction to induce an alternating current in the secondary coil.*

If you supplied dc to the primary, you'd get nothing out of the secondary at all. Sure, there'd still be a magnetic field in the iron core, but it wouldn't be constantly changing, so there'd be no induction in the secondary because you need a changing field to induce a potential difference.

Step-up and step-down transformers

There are a few different types of transformer. Two that you need to know about are **step-up transformers** and **step-down transformers**. They both have two coils, the primary and the secondary, wrapped around an iron core.

These transformers change an electricity supply so that the size of the output potential difference is not the same as the input potential difference.

The ratio between the primary and secondary potential differences is the same as the ratio between the number of turns on the primary coil and the number of turns on the secondary coil — see the equation on the next page.

- In a step-up transformer, the number of turns on the secondary coil and the size of the pd across it are greater than across the primary coil.

- In a step-down transformer, the number of turns on the secondary coil and the size of the pd across it are smaller than across the primary coil.

> **Tip:** **H** Step-up and step-down transformers change the size of the input potential difference, but not the frequency.

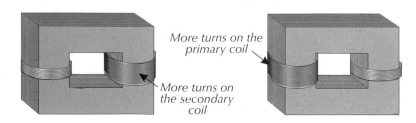

More turns on the primary coil

More turns on the secondary coil

Figure 3: *A step-up transformer (left) and a step-down transformer (right).*

Step-up and step-down transformers are used in the national grid — see page 95. Step-up transformers increase the potential difference of the supply for transmission around the country via power lines. This is useful because for a given power, increasing the pd decreases the current (see the equation on page 312), and a lower current means less energy is lost by heating (see page 91).

Step-down transformers then reduce the potential difference of the supply before it reaches our homes, making it safer to use.

| **Example** | **Higher** |

A step-down transformer can be used on the national grid (see page 95) to bring a potential difference of around 400 000 V in the power lines down to the 230 V that is needed in the home. It has more turns on the primary coil than it does on the secondary coil.

Transformers are also used in many household appliances. Many appliances don't need the 230 V supplied by the mains, so they have transformers in them to reduce (or increase) the potential difference to a suitable level.

The transformer equation

You can calculate the output potential difference from a transformer if you know the input potential difference and the number of turns on each coil.

This is the equation to use:

Tip: H The potential differences can be in any units, as long as they are both in the same units.

$$\frac{\text{Potential difference across primary coil}}{\text{Potential difference across secondary coil}} = \frac{\text{Number of turns on primary coil}}{\text{Number of turns on secondary coil}}$$

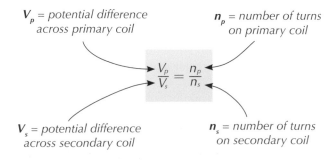

V_p = potential difference across primary coil

n_p = number of turns on primary coil

$$\frac{V_p}{V_s} = \frac{n_p}{n_s}$$

V_s = potential difference across secondary coil

n_s = number of turns on secondary coil

Tip: H There's less rearranging to do if you put whichever variable you're trying to find on the top of the fraction.

Handily, you can write it either way up. So:

$$\frac{V_s}{V_p} = \frac{n_s}{n_p}$$

So, for a step-up transformer, $V_s > V_p$ and for a step-down transformer, $V_s < V_p$.

Example 1 — **Higher**

A transformer has a potential difference of 12 V across the primary
coil and a potential difference of 6 V across the secondary coil.

The primary coil has 20 turns. How many does the secondary coil have?

The pd across the primary coil is twice as big as across the secondary coil.
So, the primary coil must have twice as many turns as the secondary coil.

Therefore, the secondary coil has $20 \div 2 = 10$ turns.

Example 2 — **Higher**

A transformer has 40 turns on the primary coil and 800 on the secondary
coil. If the input potential difference is 1000 V, find the output potential
difference. What type of transformer is this?

$V_p = 1000$ V, $n_s = 800$, $n_p = 40$

You're looking to find the potential difference across the secondary coil, so
use the version of the formula with V_s at the top.

$$\frac{V_s}{V_p} = \frac{n_s}{n_p}$$
$$\frac{V_s}{1000} = \frac{800}{40}$$
$$\Rightarrow \; V_s = \frac{800}{40} \times 1000$$
$$= 20\,000 \text{ V}$$

This is what you'd expect — there are 20 times as many turns on the
secondary coil as on the primary coil, so the output voltage is 20 times the
input voltage.

This is a step-up transformer.

Transformers and efficiency

Transformers are almost 100% efficient. If we assume that they are, this
means their electrical power output is equal to their electrical power input.

You know from page 93 that the formula for power is:

power = potential difference × current (or $P = VI$)

This means you can write "electrical power output = electrical power input"
as an equation in terms of potential difference and current.

Tip: H There's more
about efficiency on
pages 40-41.

Tip: Remember, power
is measured in watts, W.

So, for a transformer:

Tip: H Don't be put off by the subscripts (the little *s* and *p* symbols). Just remember when you see two terms written next to each other in an equation, it means they're multiplied. So you could also write this equation as:
$V_s \times I_s = V_p \times I_p$

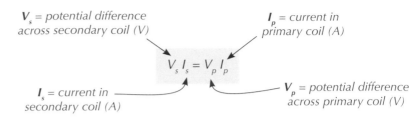

V_s = potential difference across secondary coil (V)

I_p = current in primary coil (A)

$$V_s I_s = V_p I_p$$

I_s = current in secondary coil (A)

V_p = potential difference across primary coil (V)

$V_s I_s$ is the power output at the secondary coil, and $V_p I_p$ is the power input at the primary coil. This equation shows why, for a given power, a high pd is needed for a low current, which is useful in the national grid (see page 95). The equation on page 310 can be used to work out the number of turns needed to increase the pd (and decrease the current) to the right levels.

Example 1 — Higher

The current through the primary coil of a step-down transformer is 0.5 A. The potential difference across the primary coil is 200 V. The transformer can be assumed to be 100% efficient.

a) **Find the power output of the transformer.**

The transformer can be considered 100% efficient, which means:

power output = power input

And the power input can be found using the formula $P = VI$:

power output = power input
= $V_p I_p$ = 200 × 0.5 = 100 W

b) **The potential difference across the primary coil is doubled, but the current is kept the same.**
What effect will this have on the output power?

If the potential difference across the primary coil is doubled, the input power will also be doubled. Because the transformer can be considered 100% efficient, the output power will be double its original value (i.e. 200 W).

Example 2 — Higher

A transformer in a travel adaptor steps up a 115 V ac mains electricity supply to the 230 V needed for a hair dryer. The current through the hair dryer is 5 A.

Exam Tip H
You might be given the power output at the secondary coil and asked to find the current through the primary coil. You'll just need to do the same calculation as here, but replace $V_s I_s$ with the power value you're given.

Assume the transformer is 100% efficient and calculate how much current is drawn by the transformer from the mains supply.

V_p = 115 V, V_s = 230 V, I_s = 5 A

Rearranging $V_s I_s = V_p I_p$:

$I_p = (V_s \times I_s) \div V_p = (230 \times 5) \div 115 = 10$ A

Practice Questions — Fact Recall

Q1 What does a transformer do?

Q2 Look at this diagram of a transformer. What do labels A, B and C represent?

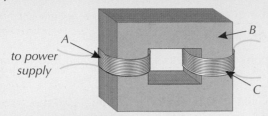

Q3 Explain how a transformer works.

Q4 What is a step-up transformer?

Q5 What equation should be used to calculate the output potential difference of a transformer if you know the input potential difference and the number of turns on each coil?

Q6 What can you say about the electrical power input and electrical power output of a transformer if it's assumed to be 100% efficient?

Practice Questions — Application

Q1 A transformer has 15 turns on the primary coil and 30 turns on the secondary coil.

   a) Is it a step-up transformer or a step-down transformer? How do you know?

   b) Will the output potential difference be greater or smaller than the input potential difference?

Q2 A transformer, assumed to be 100% efficient, has an output potential difference of 12 V and an output current of 10 A. What is the electrical power input? How do you know?

Q3 A step-up transformer has a potential difference across the primary coil of 4 V. It has 8 turns on the primary coil and 12 turns on the secondary coil. Calculate the potential difference across the secondary coil.

Q4 A transformer has a pd of 400 V across the primary coil and a pd of 20 V across the secondary coil. The current in the secondary coil is 100 A. What is the current in the primary coil? Assume the transformer is 100% efficient.

Topic 7 Checklist — Make sure you know...

Magnetic Fields

☐ That magnets have a magnetic field — a region in which other magnets, magnetic materials and wires carrying a current experience a non-contact force.

☐ That all magnets have two poles, north and south, where the magnetic field is strongest.

☐ That the strength of a magnetic field decreases with distance from the magnet.

☐ That magnetic field lines point away from north poles and towards south poles, and how to draw the magnetic field pattern of a bar magnet.

☐ That like poles repel each other and unlike poles attract each other.

☐ That compasses work due to magnetism, and provide evidence that the Earth's core is magnetic.

☐ How to use a compass to plot a magnetic field.

☐ What an induced magnet is, and the difference between permanent and induced magnets.

☐ That the force on induced magnets (like all magnetic materials) in a magnetic field is always attractive.

Electromagnetism

☐ That a current flowing through a conductor generates a magnetic field.

☐ That you can demonstrate the existence of the magnetic field around a current-carrying conductor using a compass.

☐ That the direction of the magnetic field generated by a current can be found using the right-hand thumb rule.

☐ That the strength of the magnetic field generated by a current increases with the size of the current, and decreases with distance from the wire.

☐ Why bending a current-carrying conductor into a coil increases the magnetic field strength.

☐ How to draw the magnetic fields of a current-carrying wire and a solenoid.

☐ That the magnetic field inside a solenoid is strong and uniform, and the field outside resembles that of a bar magnet.

☐ That the strength of the magnetic field of a solenoid can be increased by adding an iron core, and that this forms a simple electromagnet.

☐ How electromagnets can have useful applications in a number of devices, and be able to explain how they work in a given situation.

The Motor Effect

☐ **H** That a current flowing through a conductor in a magnetic field causes a force between the magnet producing that field and the conductor — this is the motor effect.

☐ **H** That the size of the force depends on the size of the current, the length of conductor inside the magnetic field, and the magnetic flux density of the magnetic field.

☐ **H** How to calculate the size of the force when the current is at 90° to the magnetic field using $F = BIl$.

☐ **H** That the direction of the force depends on the direction of the current and the direction of the magnetic field, and how to use Fleming's left-hand rule.

cont...

Using the Motor Effect

☐ **H** How the motor effect can cause a coil of current-carrying wire to rotate inside a magnetic field.

☐ **H** How this is the basis of a simple electric motor, and how the speed and direction of an electric motor can be changed.

☐ **H** How the motor effect is used in a loudspeaker to generate sound waves from alternating current.

The Generator Effect

☐ **H** That a change in an external magnetic field around a conductor, or relative motion of the conductor to the field, induces a potential difference in that conductor. This is called the generator effect.

☐ **H** That if a pd is induced in a conductor that is part of a complete circuit, a current will flow.

☐ **H** How the direction of the pd induced depends on the direction of the magnetic field and the direction of motion of the conductor.

☐ **H** How the size of the pd induced depends on the speed of motion of the conductor and the strength of the magnetic field.

☐ **H** That a current induced by the generator effect will have a magnetic field which opposes the change which induced the current.

☐ **H** How the generator effect is applied in different situations, including in alternators to generate ac and in dynamos to generate dc.

☐ **H** How to interpret oscilloscope traces of induced alternating and direct potential differences.

☐ **H** How the generator effect is used in a microphone to convert sound waves into electrical signals.

Transformers

☐ **H** That a transformer is a device which changes the potential difference of an electrical supply using the generator effect.

☐ **H** That a transformer consists of two coils (a primary coil and a secondary coil) wrapped around an easily magnetised core (usually iron).

☐ **H** That if the secondary coil is part of a complete circuit, a current will flow when a pd is induced.

☐ **H** How a transformer changes the pd of an ac source using the generator effect, and why it will not work for a dc source.

☐ **H** That the ratio of potential differences across the coils is proportional to the ratio of the number of turns of wire on the coils, and how to use $\frac{V_p}{V_s} = \frac{n_p}{n_s}$.

☐ **H** That a step-up transformer increases pd and has more turns on the secondary than the primary coil.

☐ **H** That a step-down transformer decreases pd and has more turns on the primary than the secondary coil.

☐ **H** That transformers are assumed to be 100% efficient and so the power input equals the power output.

☐ **H** How to use the equation $V_s I_s = V_p I_p$.

☐ **H** How to apply the two transformer equations to how transformers are used to achieve low currents and transmit power efficiently (e.g. in the national grid).

Exam-style Questions

1   A student designs a simple battery-powered screwdriver, as shown. The coil of wire is placed in a uniform magnetic field between two permanent bar magnets. The coil is supplied with direct current (dc), and a split-ring commutator is used to reverse the direction of the current flowing through the coil of wire every half turn.

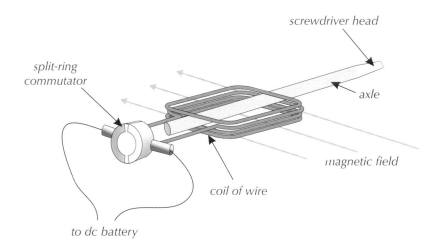

split-ring commutator

screwdriver head

axle

magnetic field

coil of wire

to dc battery

1.1   Describe what is meant by the motor effect.

(1 mark)

1.2   The screwdriver uses the motor effect to produce rotation of the axle. Usually, the screwdriver head rotates clockwise. A reverse switch rotates the position of the magnets so that the direction of the magnetic field is reversed.

Explain why this causes the screwdriver head to rotate anticlockwise.

(2 marks)

1.3   The current flowing through the wire is 3 A. The coil forms a square with sides 0.5 cm long and four wires tall. The magnetic flux density of the magnet is 0.2 T.
Calculate the total force on one of the two sides perpendicular to the magnetic field. Give your answer in newtons.
Use the correct equation from the equations listed on page 394.

(2 marks)

The bar magnets used by the drill lose their magnetism over time, so the drill becomes gradually less powerful. One way of avoiding this is using electromagnets in place of the bar magnets. The electromagnets are made of a coil of wire, connected to a dc supply, that is wrapped around an iron core.

1.4   Explain how turning on the dc supply to the electromagnets produces a magnetic field around them.

(2 marks)

1.5   Explain why an ac supply would not be suitable to power an electromagnet in the screwdriver.

(3 marks)

2 A basic transformer consists of a primary coil and a secondary coil wrapped around an iron core, as shown below.

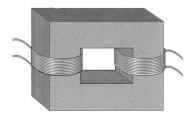

2.1 Select the correct option from the phrases below to complete the sentence.

| an alternating current | an alternating pd | a changing magnetic field |

When an alternating current flows in the primary coil,
it produces .... in the iron core.

(1 mark)

2.2 Explain why the power supply to a transformer must supply alternating current, and why it won't work as it should if it is supplied with direct current.

(2 marks)

2.3 One type of transformer is a step-down transformer. Explain what a step-down transformer does, and compare the number of turns on its primary and secondary coils.

(2 marks)

2.4 A transformer has 50 turns on the primary coil and 60 turns on the secondary coil. The input potential difference is 100 V. Calculate the output potential difference. Give your answer in volts. Use the correct equation from those listed on page 394.

(3 marks)

2.5 The transformer is assumed to be 100% efficient. The output current is 10 A. Calculate the input current. Give your answer in amps. Use the correct equation from those listed on page 394.

(3 marks)

3 A dynamo is attached to the wheel of an exercise bike. When the wheel turns, it causes a coil in the dynamo to rotate in a magnetic field, and a current is induced in the coil. The dynamo generates direct current.

3.1 Explain why a split-ring commutator needs to be used in a dynamo.

(3 marks)

The speed of a cyclist using the bike is monitored by reading the potential difference produced by the dynamo through an oscilloscope.

3.2 Sketch the oscilloscope trace of the potential difference generated by the dynamo.

(2 marks)

3.3 The cyclist increases the speed at which they are pedalling. State **two** ways in which the oscilloscope trace changes.

(2 marks)

1. The Life Cycle of Stars

Stars give out a huge amount of heat and light — this is fuelled by nuclear fusion inside the star. These pages are all about how stars are formed and what happens to stars once all of their fuel runs out.

The formation of stars

Stars initially form from a cloud of dust and gas called a **nebula**. The force of gravity makes the gas and dust spiral in together to form a **protostar** — see Figure 3. Gravitational attraction causes the density of the protostar to increase and particles within the protostar to collide with each other more frequently, so the temperature rises.

When the temperature gets high enough, hydrogen nuclei begin to undergo nuclear fusion (see pages 140-141) to form helium nuclei. This gives out massive amounts of energy which keeps the core of the star hot. At this point, a star is born. Smaller masses of gas and dust around the star may also be pulled together to make planets that orbit the star.

Main sequence stars

Once a star has been formed, it immediately enters a long stable period. The energy released by the nuclear fusion provides an outward pressure that tries to expand the star, which balances the force of gravity pulling everything inwards. It is in equilibrium. In this stable period it's called a **main sequence star** and it lasts several billion years. The Sun is a stable main sequence star in the middle of this stable period.

Eventually the hydrogen begins to run out. Fusion of helium (and other elements) occurs and the star ceases to be a main sequence star. Heavier elements (up to iron) are created in the core of the star. Stars (and their life cycles) produce all naturally occurring elements in the universe.

The death of stars

Exactly what happens to a star as it starts to run out of fuel and dies depends on how big the star is.

Stars about the size of the Sun

A small star that is about the size of the Sun will expand into a **red giant** when it starts to run out of hydrogen. It becomes red because the surface cools.

It will then become unstable and eject its outer layer of dust and gas as a planetary nebula. This leaves behind a hot, dense solid core — a **white dwarf**. As a white dwarf cools down, it emits less and less energy. When it no longer emits a significant amount, it is called a **black dwarf** and eventually disappears from sight (see Figure 3).

Figure 1: *A star cluster — the orange star in the centre is a red giant.*

Stars much larger than the Sun

Stars that are larger than the Sun expand into **red super giants** when they start to run out of hydrogen. Red super giants are much bigger and burn for much longer than regular red giants. They expand and contract several times, forming elements as heavy as iron in various nuclear reactions.

Eventually they run out of elements to fuse and become unstable. They explode in a **supernova**, forming elements heavier than iron and ejecting them into the universe to form new planets and stars.

The exploding supernova throws the outer layers of dust and gas into space, leaving a very dense core called a **neutron star**. If the star is big enough, it will become a **black hole** instead — a super dense point in space that not even light can escape from (see Figure 3).

Figure 2: *The remnants of a supernova.*

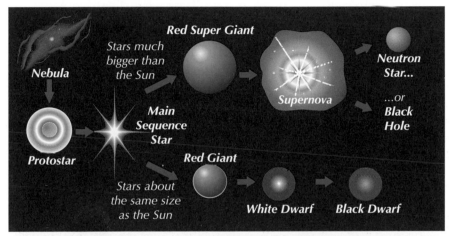

Figure 3: *A flow diagram to show the life cycles of stars that are about the same size as the Sun and stars much bigger than the Sun.*

Tip: Black holes don't emit any light, so you can't actually see them.

Practice Questions — Fact Recall

Q1 How are stars and planets formed?

Q2 By what process is energy released in stars?

Q3 Explain why main sequence stars are stable.

Q4 Describe the stages that a star the same size as the Sun will go through when it stops being a main sequence star.

Q5 Hydrogen was the only element present in the early universe. How did other elements form and get spread through the universe?

Practice Question — Application

Q1 Melnick-34 is a star that is much larger than the Sun. State whether it will form a white dwarf, or a black hole.

2. The Solar System and Orbits

The solar system is made up of the Sun at the centre, everything that orbits around it — and everything else which orbits around them. These pages are all about our solar system, and how objects are kept in orbits.

The solar system

Our solar system contains one star — the Sun. The rest is made up of everything that orbits the Sun. An **orbit** is the path on which one object moves around another. The Sun is orbited by:

- **Planets** — these are large objects which orbit a star. There are eight of them orbiting the sun. They have to be large enough to have "cleared their neighbourhood". This means that their gravity is strong enough to have pulled in any nearby objects apart from their natural satellites.

- **Dwarf planets** — e.g. Pluto. These are planet-like objects that orbit stars, but are too small to meet all of the rules for being a planet.

- **Satellites** — these are objects that orbit a second more massive object. For example:

   1. **Moons** — these orbit planets. They're a type of natural satellite (i.e. they're not man-made).

   2. **Artificial satellites** are satellites that humans have built. There are lots orbiting the Earth and some orbiting the Sun and other planets.

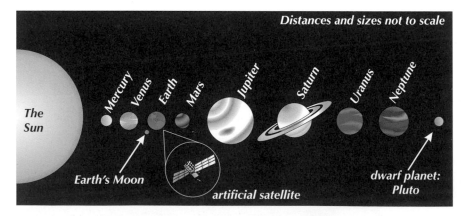

Figure 1: A diagram of our solar system, showing the eight planets in order and examples of a dwarf planet, a moon and an artificial satellite.

Our solar system is a tiny part of the Milky Way galaxy. This is a massive collection of billions of stars that are all held together by gravity.

Orbits — the basics

The planets move around the Sun in almost circular orbits. The same is true for moons and artificial satellites orbiting planets. Gravity provides the force which keeps an object within this orbit.

Circular motion in orbits Higher

If an object is travelling in a circle, it is constantly changing direction, which means it is constantly accelerating (just like a car going round a roundabout, page 179). This also means that it is constantly changing velocity (but not changing speed).

For an object to accelerate, there must be a force acting on it (page 194). For circular motion, this force is directed towards the centre of the circle. In the solar system, the force that is acting towards the centre of the circle is the gravitational force (gravity) between a planet and the Sun (or a planet and its satellites).

An orbit is a balance between the force providing the acceleration and the forward motion (**instantaneous velocity**) of the object. The object keeps accelerating towards what it's orbiting, but the instantaneous velocity (which is at right angles to the acceleration and to the force of gravity) keeps it travelling in a circle — see Figure 2.

Tip: Velocity is a vector quantity — it has a magnitude and a direction.

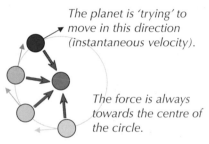

The planet is 'trying' to move in this direction (instantaneous velocity).

The force is always towards the centre of the circle.

Figure 2: A diagram of a planet orbiting around the Sun, showing its instantaneous velocity (black arrows) and the gravitational force (blue arrows).

Orbital speed and radius Higher

The size of an orbit depends on the object's speed. The closer you get to a star or planet, the stronger the gravitational force. The stronger the force, the faster the orbiting object needs to travel to remain in orbit. For an object in a stable orbit, if the speed of the object changes, the size (radius) of its orbit must do so too. If the object moves faster, the radius of its orbit must be smaller. If it moves slower, the radius must be larger — see Figure 3.

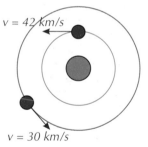

$v = 42$ km/s

$v = 30$ km/s

Figure 3: A diagram of two different orbits around a body, showing how distance (radius) affects speed.

Practice Questions — Fact Recall

Q1 State what is meant by a 'natural satellite', and give one example.

Q2 State the force which keeps objects in orbit around the Sun.

Q3 In what direction does the force that keeps the Earth orbiting the Sun act on Earth?

Practice Question — Application

Q1 If the Earth moved towards the Sun, what would happen to the speed of its orbit, provided it were to remain in a stable orbit? Why?

- Know that the light from most distant galaxies is red-shifted — the wavelength of the observed light is greater than it should be.

- Understand that red-shift of light from distant galaxies suggests they are moving away from us.

- Know that generally, the further away the galaxy, the greater the red-shift and the faster they're moving away.

- Know that red-shift of light from galaxies is evidence that the universe (space itself) is expanding.

- Know the Big Bang theory of the beginning of the universe.

- Know how red-shift supports the Big Bang theory of how the universe began.

- Understand how scientists use observations to form theories like the Big Bang theory.

- Know that recent observations suggest that the expansion of the universe is accelerating.

- Know that there are many features of the universe we cannot yet explain, such as dark matter and dark energy.

Specification Reference
4.8.2

3. Red-shift and the Big Bang

The universe is getting bigger as time goes on — it's expanding. Scientists now think it has been expanding since the day it began, as described by the Big Bang Theory.

Red-shift

When we look at light from most distant galaxies we find that its wavelength (page 226) has increased. The wavelengths are all longer than they should be — they're shifted towards the red end of the spectrum (see page 242 for the order of the electromagnetic spectrum). This effect is called **red-shift**.

Red-shift occurs when the source of the light is moving away from the observer. You can think of it as the light wave 'stretching out' as the source moves away.

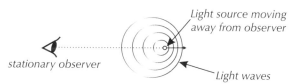

Light source moving away from observer

stationary observer

Light waves

Figure 1: *Diagram showing the effect of red-shift. The observer receives the wavefronts further apart, so sees a redder light.*

The expansion of the universe

As the light from most distant galaxies has been red-shifted, this suggests that the galaxies are moving away from us. Measurements of the red-shift indicate that most distant galaxies are moving away from us (receding) very quickly — and it's the same result whatever direction you look in. More distant galaxies have a greater red-shift than nearer ones. This means that more distant galaxies are moving away faster than the nearer ones. This suggests that all galaxies are moving away from each other. The conclusion of these results is that the whole universe (space itself) is expanding.

Example

To understand the expansion of the universe, imagine a balloon covered with pompoms. As you blow into the balloon, it stretches. The pompoms move further away from each other, but each pompom stays the same size.

The balloon represents the universe, and each pompom is a galaxy. As time goes on, space stretches and expands, moving galaxies away from each other. The galaxies are held together by gravity and so don't stretch themselves.

This is a simple model (balloons only stretch so far, and there would be galaxies 'inside' the balloon too) but it shows how the expansion of space makes it look like galaxies are moving away from us.

The Big Bang theory

All galaxies are moving away from each other at great speed — suggesting something must have got them going. That something was probably a big explosion — the Big Bang. The **Big Bang theory** says:

- Initially, all matter in the universe occupied a very small space. This tiny space was very dense (p. 106) and so was very hot.

- Then it 'exploded' — space started expanding, and the expansion is still going on.

The Big Bang theory is supported by red-shift measurements, and is the generally accepted theory of how the universe began — but it's only the best guess we have so far. Scientists use observations to come up with theories like the Big Bang theory. But whenever scientists discover new evidence that isn't explained by their theory, they have to either make a new theory, or change a current one to explain what they have observed.

Exam Tip
You may be asked to explain how a piece of evidence supports a theory, or how it could lead to other theories being abandoned.

Unexplained observations

There are a number of observations of the universe which science does not yet have an explanation for. Observations of supernovae from 1998 to the present day appear to show that distant galaxies are moving away from us faster and faster (the speed at which they're receding is increasing). This indicates that the expansion of the universe is accelerating, but we don't yet know how or why.

Scientists currently think that the universe is mostly made up of dark matter and dark energy. Dark matter is the name given to an unknown substance which holds galaxies together, but does not emit any electromagnetic radiation. Dark energy is thought to be responsible for the accelerated expansion of the universe. But no one knows what these things are, so there are many different theories about them.

Tip: You might hear dark matter called dark mass instead.

Practice Questions — Fact Recall

Q1 a) How is the light observed from a distant galaxy different from the light that the galaxy actually emits?

   b) What is the name given to this effect?

Q2 The 'steady state' theory of the universe supposes that the universe has always existed and has remained at the same size forever.

   a) State one observation which contradicts the steady state theory, and explain why it contradicts the theory.

   b) State the Big Bang theory and explain whether the observation in part a) supports it.

Q3 What have observations of supernovae since 1998 shown scientists about the expansion of the universe?

Topic 8 Checklist — Make sure you know...

The Life Cycle of Stars

☐ That all stars, including the Sun, begin as a cloud of dust and gas called a nebula.

☐ That nebulae are pulled together by gravity to form a protostar.

☐ That gravitational attraction causes the density of a protostar to increase, causing more collisions between particles to occur, which causes the temperature to rise enough for nuclear fusion to begin.

☐ That in main sequence stars hydrogen is fusing to form helium.

☐ That main sequence stars are stable due to a balance between the energy released through nuclear fusion, which tries to expand the star, and gravitational attraction, which tries to compress it.

☐ That the Sun is a main sequence star, and so is in a stable period.

☐ That stars begin to fuse heavier elements as they run out of hydrogen.

☐ That fusion in stars is responsible for producing all naturally occurring elements up to iron.

☐ That the life cycle of a star after the main sequence depends on its mass.

☐ That stars about the same size as the Sun become red giants after the main sequence.

☐ That red giants eventually eject their outer layer and become white dwarfs, which cool to become black dwarfs.

☐ That stars much bigger than the Sun become red super giants.

☐ That red super giants explode in supernovae, forming heavier elements than iron and ejecting them into the universe.

☐ That fusion in stars and supernovae has produced all naturally occurring elements and distributed them throughout the universe.

☐ That supernova explosions leave behind either a neutron star or, for stars with a high enough mass, a black hole.

The Solar System and Orbits

☐ That our solar system contains a single star, the Sun, and eight planets: Mercury, Venus, Earth, Mars, Jupiter, Saturn, Uranus and Neptune.

☐ That the solar system also includes dwarf planets (e.g. Pluto) and satellites of the planets, both natural satellites (e.g. moons) and artificial satellites.

☐ That an orbit is the path on which one object moves around another.

☐ That planets and dwarf planets orbit stars, and that dwarf planets are too small to be considered planets.

☐ That satellites orbit another more massive object.

☐ That our solar system is located in the Milky Way galaxy and is only a very small part of it.

☐ That the orbits of planets around the Sun, and the orbits of moons and artificial satellites around planets, are roughly circular.

☐ That objects travel in orbits due to the force of gravity between the two objects.

☐ **H** That gravity acts towards the centre of the orbit, causing an acceleration which changes the velocity, but not the speed, of the orbiting object.

cont...

☐ 🅗 That the gravitational force is stronger the closer you are to an object.

☐ 🅗 That a faster orbital speed must mean a smaller orbital radius for the object to be in a stable orbit.

Red-shift and the Big Bang

☐ That light from distant galaxies has longer wavelengths than expected, an effect known as red-shift.

☐ That red-shift suggests that all galaxies are moving away from us.

☐ That more distant galaxies appear to be moving away faster than nearer galaxies.

☐ That this suggests the universe is expanding — all galaxies are moving away from each other.

☐ That observations of red-shift support the Big Bang theory.

☐ That the Big Bang theory says that the universe was once confined to a very small area that was extremely hot and dense, which exploded and has been expanding ever since.

☐ That observations of distant supernovae since 1998 have shown that the expansion of the universe is accelerating.

☐ That there is a lot about the universe that is not understood, e.g. dark matter and dark energy.

Exam-style Questions

1   The Sun is a main sequence star in the middle of its stable period.

1.1   State the element which undergoes nuclear fusion in a main sequence star and the element produced in the fusion reaction.

(2 marks)

1.2   Explain why a main sequence star does not collapse due to gravitational attraction.

(2 marks)

1.3   A main sequence star evolves and eventually becomes a black hole. Describe this evolution from a main sequence star to the point of becoming a black hole.

(4 marks)

1.4   Will the Sun ever become a black hole? Explain your answer.

(1 mark)

1.5   Nuclear fusion in stars can only produce elements which are no heavier than iron. Describe how elements heavier than iron are produced and spread through the universe.

(2 marks)

2*   A planet is orbiting around a star in a stable, circular orbit.
Explain why the planet orbits the star in this way and how the speed of the planet's orbit relates to the radius of its orbit.

(6 marks)

3   Our solar system is located within a larger structure of stars called a galaxy.

3.1   State the name of the galaxy in which our solar system is located.

(1 mark)

3.2   Light from most distant galaxies is observed to have been red-shifted.
Explain what is meant by the term red-shift, and state when it occurs.

(2 marks)

3.3   Explain what this observation suggests about our universe.

(2 marks)

Until 1998, scientists believed that the effect described in **3.3** was slowing down.

3.4   Describe the observations that caused scientists to change their minds after 1998.

(2 marks)

3.5   Scientists aren't agreed on what could be causing the observations made since 1998, but there are several different theories. Suggest why none of the theories have been accepted yet.

(1 mark)

1. Apparatus and Techniques

As part of GCSE Physics, you'll have to do at least ten practicals, called Required Practical Activities. You'll need to know how to use various pieces of apparatus and carry out different scientific techniques. And not only do you need to carry out the practicals, you could also be asked about them in the exams. Luckily, all the Required Practical Activities are covered in this book, and the next few pages cover some of the techniques that you'll need to know about.

Measuring mass

Mass should be measured using a balance. For a solid, set the balance to zero, place your object onto the scale and read off the mass.

If you're measuring the mass of a liquid (or a granular solid, like sand) start by putting an empty container onto the balance. Next, reset the balance to zero, so you don't include the mass of the container in your measurement. Then just pour the substance you want to measure the mass of into the container and record the mass displayed. Easy peasy.

Measuring weight

Remember not to get weight and mass confused. Mass is the amount of 'stuff' in an object. Weight is the force acting on the object due to gravity (p. 149).

You could calculate the weight of an object by measuring its mass (see above) and then multiplying by the gravitational field strength (p. 150). But to measure the weight of an object directly, you should use a newtonmeter (see Figure 1). Make sure that whatever you're measuring is securely attached to the hook of the newtonmeter and can hang freely. Remember to wait until it's stopped swinging or bouncing before you read off the value.

Measuring length

In most cases a standard centimetre ruler can be used to measure length. It depends on what you're measuring though — metre rulers are handy for large distances, while micrometers (which have smaller divisions than a standard ruler) are used for measuring tiny things like the diameter of a wire. If you're dealing with something where it's tricky to measure just one accurately (e.g. water ripples, p. 230), you can measure the length of ten of them and then divide by ten to find the length of one.

If you're taking multiple measurements of the same object (e.g. to measure changes in length) then make sure you always measure from the same point on the object. It can help to draw or stick small markers onto the object to line up your ruler against — see Figure 2.

The ruler should always be parallel to what you want to measure. You should also make sure the ruler and the object are always at eye level when you take a reading. This stops parallax affecting your results (see next page).

Tip: The Required Practical Activities in this book are marked with a big stamp like this...

The practicals that you do in class might be slightly different to the ones in this book (as it's up to your teacher exactly what method you use), but they'll cover the same principles and techniques.

***Figure 1:** Measuring weight using a newtonmeter.*

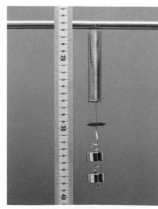

***Figure 2:** A red marker being used to help read a length measurement from a ruler.*

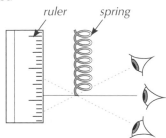

Example

When measuring the length of a spring, you need to make sure you avoid the parallax effect.

Parallax is where a measurement appears to change based on where you're looking from.

In this example, looking from above gives you a longer measurement, while looking from below gives you a shorter one.

The blue line is the measurement taken when the spring is at eye level. It shows the correct length of the spring.

ruler   *spring*

Figure 3: *The parallax effect.*

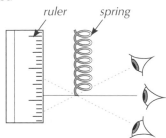

Tip: Whenever you're reading off a scale, use the value of the nearest mark on the scale (the nearest graduation).

Measuring angles

You should use a protractor to measure angles. First align the vertex (point) of the angle with the mark in the centre of the protractor. Line up the base line of the protractor with one line that forms the angle and then measure the angle of the other line using the scale on the protractor (see Figure 4).

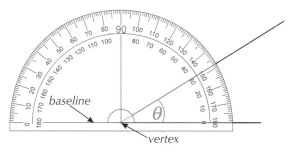

Figure 4: *A protractor, correctly aligned to measure angle θ. The angle is 32°.*

If the lines creating the angle are very thick, align the protractor and measure the angle using the centre of the lines. Using a sharp pencil to trace light rays or draw ray diagrams (page 270) helps to reduce errors when measuring angles.

If the lines are too short to measure easily, you may have to extend them. Again, make sure you use a sharp pencil and a ruler to do this.

Figure 5: *The bulb of a thermometer fully submerged in the liquid it is measuring.*

Measuring temperature

You measure temperature with a thermometer. To ensure you measure the temperature accurately, make sure the bulb of your thermometer is completely submerged in any substance you're measuring if it's a fluid (see Figure 5), or held directly against what you're measuring if it's a solid.

If you're taking a single measurement, wait for the temperature to stabilise before you take your reading. If you're measuring how the temperature of a substance is changing, take readings at regular time intervals (e.g. every 10 s).

Again, read your measurement off the scale on a thermometer at eye level.

Measuring time

You should use a stopwatch to measure time in most experiments — they're more accurate than regular watches. You can start and stop the timer whenever you need, or you can set an alarm so you know exactly when to stop an experiment or take a reading.

For some time measurements, you might be able to use a light gate. This will reduce the errors in your experiment. Have a look at page 199 for an example of a light gate being used.

A light gate sends a beam of light from one side of the gate to a detector on the other side. When something passes through the gate, the beam of light is interrupted. The light gate measures how long the beam was interrupted for. Light gates can use their measurements of time to calculate speed and acceleration, given the right information.

Tip: You can input data into light gate software, for example, the length of the thing that interrupts the beam. This allows it to calculate other quantities such as speed.

Measuring volume

Measuring the volume of a liquid

Measuring cylinders are the most common way to measure the volume of a liquid. They come in all different sizes. Make sure you choose one that's the right size for the measurement you want to make. It's no good using a huge 1 dm³ (1000 cm³) cylinder to measure out 2 cm³ of a liquid — the graduations (markings for scale) will be too big and you'll end up with massive errors. It'd be much better to use one that measures up to 10 cm³.

You can also use a pipette to measure volume. Pipettes are used to suck up and transfer volumes of liquid between containers. Graduated pipettes are used to transfer accurate volumes. A pipette filler is attached to the end of a graduated pipette, to control the amount of liquid being drawn up.

Whichever method you use, always read the volume from the bottom of the meniscus (the curved upper surface of the liquid) when it's at eye level — see Figure 6.

Figure 6: The meniscus of a fluid in a measuring cylinder, viewed at eye level. This one reads 32.

Measuring the volume of a solid

Eureka cans (or displacement cans) are used in combination with measuring cylinders to find the volumes of solids. A eureka can is essentially a beaker with a downward spout. A solid object will displace an amount of water equal to its volume, which can then be measured using a measuring cylinder.

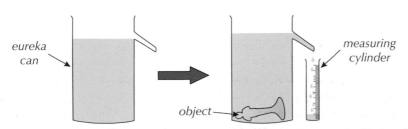

Figure 7: A eureka can being used to measure volume.

This method is particularly useful for irregularly shaped objects, where you can't simply measure its dimensions and calculate the volume mathematically. However, this method will only work for objects that sink. If the object floats, it will only displace a volume of water equal to the volume of the object that is submerged. You can see more on how to use a eureka can on page 108.

Tip: There's more on what causes objects to float and sink on pages 172-173.

Good laboratory practice

Tip: For lots more info on doing experiments, see the Working Scientifically section at the start of this book.

When it comes to actually doing an experiment, it's important that you use good laboratory practice. This means working safely and accurately. To ensure you get good results, make sure you do the following:

- Measure all your quantities carefully — the more accurately you measure things the more accurate your results will be.

- Try to be consistent — for example, if you're using a piece of apparatus, make sure you use the same one throughout the experiment.

- Don't let yourself get distracted by other people — if you're distracted by what other people are doing you're more likely to make a mistake or miss a reading.

- As you're going along, make sure you remember to fill in your table of results — it's no good doing a perfect experiment if you forget to record the data.

Working safely

There are always hazards in any experiment, so before you start a practical you should do a risk assessment first. A risk assessment identifies all possible hazards in an experiment and lists the precautions you'll take to deal with them. You should also read and follow any safety precautions provided with the apparatus or by your teacher to do with your method or the apparatus you're using.

The hazards will depend on the experiment and the apparatus you're using, but the examples given here should give you some ideas of things to think about.

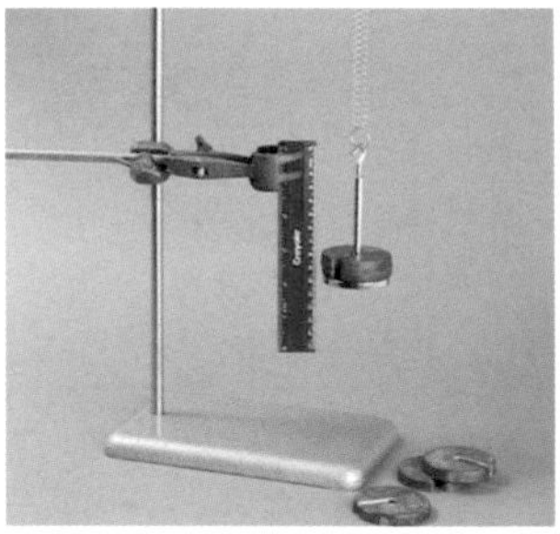

Figure 8: A clamp stand being used in a hanging mass experiment.

Figure 9: A scientist wearing laser safety goggles while working with lasers.

Examples

- Stop masses and equipment falling by using clamp stands. Make sure masses are of a sensible weight so they don't break the equipment they're used with, and use pulleys of a sensible length. That way, any hanging masses won't hit the floor during the experiment.

- When heating materials, make sure to let them cool before moving them, or wear insulated gloves while handling them.

- When working with water, clean up any spillages immediately to avoid a slip hazard, and be extra careful when using water around electricity.

- If you're using a laser, there are a few safety rules you must follow. Always wear laser safety goggles and never look directly into the laser or shine it towards another person. Make sure you turn the laser off if it's not needed to avoid any accidents.

- When working with electronics, make sure you use a low enough voltage and current to prevent wires overheating (and potentially melting) and avoid damage to components, like blowing a filament bulb.

You also need to be aware of general safety in the lab — handle glassware carefully so it doesn't break, don't stick your fingers in sockets and avoid touching frayed wires. That kind of thing.

2. Working with Electronics

You'll have to do a lot of experiments using electrical circuits. Here's a run down of some of the measurements you'll need to make most often and the equipment you'll need to use to do so.

Circuit diagrams

You need to be able to interpret circuit diagrams. Before you get cracking on an experiment involving any kind of electrical devices, you have to plan and build your circuit using a circuit diagram. Make sure you know all of the circuit symbols on page 62 so you're not stumped before you've even started.

Tip: Make sure you draw wires as straight lines, and don't let them cross each other unless there's a connection. This makes your circuit diagrams a lot clearer.

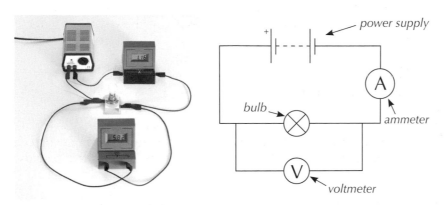

Figure 1: A photo of a simple electrical circuit, alongside a circuit diagram of the same electrical circuit.

Making measurements in circuits

Using voltmeters

A voltmeter measures potential difference (see page 64). If you're using an analogue voltmeter, choose the voltmeter with the most appropriate unit for what you're measuring (e.g. V or mV). If you're using a digital voltmeter, you'll most likely be able to switch between them.

Connect the voltmeter in parallel (p. 74) across the component you want to test. The voltmeter will usually have red (positive) and black (negative) ports, to help you connect them to a circuit correctly. If you get them backwards, your voltage reading will be negative. Once everything's set up, simply read the potential difference from the scale (or from the screen if it's digital).

Figure 2: An example of an analogue voltmeter (left) and a digital voltmeter (right).

Using ammeters

Ammeters measure electrical current (see page 63). Just like with voltmeters, choose the ammeter with the most appropriate unit (usually A or mA), unless it's a digital one.

Connect the ammeter in series (p. 70) with the component you want to test, making sure they're both on the same branch. Again, they usually have red and black ports to show you where to connect your wires. Then simply read off the current shown on the scale or by the screen.

Tip: You should turn your circuit off between readings to prevent wires overheating and affecting your results or damaging the equipment.

Figure 3: *An example of a digital multimeter.*

Using multimeters

Instead of having a separate ammeter and voltmeter, many circuits use multimeters (see Figure 3). These are devices that measure a range of properties — usually potential difference, current and resistance.

One wire should always be plugged into the black (negative) port. If you want to find potential difference, make sure the other wire is plugged into the red (positive) port that says 'V' (for volts), and connect it like a voltmeter. To find the current, use the red (positive) port labelled 'A' or 'mA' (for amps), and connect it like an ammeter.

The dial on the multimeter should then be turned to the relevant section, e.g. to 'A' to measure current in amps (on Figure 3, the 'DCA' section measures direct current, dc). The screen will display the value you're measuring in the units you've chosen with the dial.

Oscilloscopes

An oscilloscope is basically a snazzy voltmeter. You can use one to 'see' how the potential difference of an electricity supply changes over time. Figure 4 shows an oscilloscope and how it can be used.

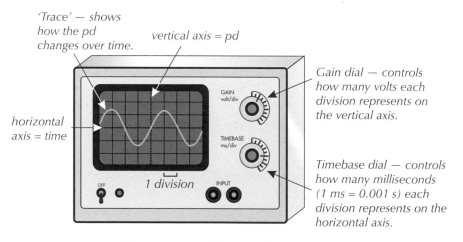

Figure 4: *An oscilloscope showing a trace.*

Tip: You can find examples of oscilloscopes being used on page 229 and page 306.

To use an oscilloscope, simply connect it up as you would any voltmeter. You can then read values of the potential difference from the trace displayed.

Each main division (or square) on the screen is usually divided into five smaller divisions so it can be read to a higher resolution. So each minor division is 0.2 of a major division. You'll probably want to adjust the 'gain' and 'timebase' dials to make sure each division represents a sensible value for what you want to measure.

Maths skills for GCSE Physics

Maths crops up quite a lot in GCSE Physics so it's really important that you've mastered all the maths skills needed before sitting your exams. Maths skills are covered throughout this book but here's an extra little section, just on maths, to help you out.

Exam Tip
At least 30% of the marks in your physics exams will test mathematical skills. That's a lot of marks, so it's definitely worth making sure you're up to speed.

1. Calculations

Sometimes the numbers you use in Physics are just plain awkward — they're either too big, too small or go on forever. The next few pages show how calculations can be made a lot easier.

Standard form

You need to be able to work with numbers that are written in **standard form**. Standard form is used for writing very big or very small numbers in a more convenient way. Standard form must always look like this:

This number must always be between 1 and 10. → $A \times 10^n$ ← *This number is the number of places the decimal point moves.*

You can write a standard form number out in full by moving the decimal point. Which direction to move the decimal point, and how many places to move it depends on 'n'. If 'n' is positive, the decimal point needs to move to the right. If 'n' is negative the decimal point needs to move to the left.

> **Tip:** If a number isn't written in standard form, it's said to be in decimal form — e.g. 0.00012 or 34 500.

Example

Here's how to write out 9.3×10^4 in full.

- Work out which way the decimal point needs to move, by looking at the 'n' number. Here it's a positive number (4) so the decimal point needs to move four places to the right:

$$9.3 \times 10^4 = 9\,3\,0\,0\,0.$$

- So 9.3×10^4 is the same as 93 000.

Here's how to write out 5.6×10^{-5} in full.

- 'n' is a negative number (–5) so the decimal point needs to move five places to the left.

$$5.6 \times 10^{-5} = .0\,0\,0\,0\,5\,6$$

- So 5.6×10^{-5} is the same as 0.000056.

> **Tip:** You need to add a zero into any space left by the decimal point moving.

Figure 1: *The 'Exp' or '×10ˣ' button is used to input standard form on calculators.*

The key things to remember with numbers in standard form are...

- When 'n' is positive the number is big. The bigger 'n' is, the bigger the number is.

- When 'n' is negative the number is small. The smaller 'n' is (the more negative), the smaller the number is.

- When 'n' is the same for two or more numbers, you need to look at the start of each number to work out which is bigger. For example, 4.5 is bigger than 3.0, so 4.5×10^5 is bigger than 3.0×10^5.

There's a special button on your calculator for using standard form in a calculation — it's the 'Exp' button. So if, for example, you wanted to type in 2×10^7, you'd only need to type in: '2' 'Exp' '7'. Some calculators may have a different button that does the same job, for example it could say 'EE' or '×10ˣ' instead of 'Exp' — see Figure 1.

Ratios, fractions and percentages

You need to be able to use ratios, fractions and percentages, and know what they mean.

Fractions and ratios

Fractions and ratios are two different things in maths, but they are used to express relationships between quantities in physics in very similar ways.

A fraction is just one number divided by another number, written as $\frac{x}{y}$. They're used all over the place in physics:

Tip: When using the equation $E_k = \frac{1}{2}mv^2$, you could type 0.5 into your calculator instead. Either will work, as $\frac{1}{2} = 0.5$.

Tip: Calculating power is covered on page 33.

Example 1

- $E_k = \frac{1}{2}mv^2$ uses the fraction $\frac{1}{2}$, or 1 divided by 2.

- $\text{Power} = \dfrac{\text{Energy transferred}}{\text{time}}$ uses a fraction to express power as energy transferred over time.

A ratio is a proportional relationship between two quantities. It tells you how the two quantities are related to each other.

Tip: For a given magnification, object height and image height are proportional to each other.

Example 2

Magnification is a ratio. It is the relationship between two heights, image height and object height:

$$\text{Magnification} = \dfrac{\text{image height}}{\text{object height}}$$

But it is also a fraction. This is an example of 'ratio' and 'fraction' being used to describe similar things.

Example 3 — Higher

A step-down transformer has 960 turns on its primary coil and 640 turns on its secondary coil. Calculate the ratio of the number of turns on the primary coil to the number of turns on the secondary coil.

$$\frac{\text{turns on primary coil}}{\text{turns on secondary coil}} = \frac{960}{640} = \frac{3}{2}$$

Tip: H This ratio means that for every 3 turns on the primary coil, there are 2 turns on the secondary coil.

The ratio of the input voltage to the output voltage is the same as the ratio of turns on the primary coil to turns on the secondary. If the output voltage from the transformer is 20 V, what is the input voltage?

Because the ratios are the same, this gives $\frac{\text{input voltage}}{\text{output voltage}} = \frac{3}{2}$.

To find the input voltage, multiply both sides by the output voltage:

Input voltage = output voltage × $\frac{3}{2}$ = 20 × $\frac{3}{2}$ = 30 V.

Usually in maths, ratios are expressed in a special form when comparing two quantities.

A colon separates one quantity from the other.   x and y stand for the two quantities.

To write a ratio in this form, first write down the numbers you have of each thing, separated by a colon. Then divide the numbers by the same amount until they're the smallest they can be whilst still being whole numbers — this is called simplifying the ratio. You can also find what a ratio simplifies to using your calculator.

Tip: H So in the example above, the ratio of the number of turns on the primary coil to the number of turns on the secondary coil would be 960 : 640. This ratio simplifies to 3 : 2.

Example 4

To find the ratio 120 : 150 in its simplest form using your calculator, just type in $\frac{120}{150}$ as a fraction and press equals. Your calculator will give you the most simplified version, which in this case is $\frac{4}{5}$.

So the ratio in its simplest form is 4 : 5.

If your calculator gives a decimal, use the button on your calculator that swaps between fractions and decimals — it'll probably look like one of these:

Percentages

Percent means 'out of 100'. A percentage is really just a fraction out of 100, so 62% means 62 out of 100. In physics, percentages are mostly used to express one number as a percentage of another number. For example, useful output energy transfer as a percentage of total input energy transfer is efficiency (see page 40). To find one number as a percentage of another, divide the first number by the second and multiply by 100.

Example

1092 J of energy is transferred to a motor. 819 J of this transferred usefully by the motor. Calculate the efficiency of the motor as a percentage.

To find 819 as a percentage of 1092, first divide 819 by 1092:
$819 \div 1092 = 0.75$

> **Tip:** Take a look at page 40 for more on calculating efficiencies.

Then multiply this by 100:
$0.75 \times 100 = 75\%$

So efficiency of motor = 75 %

Estimating

Estimating can be a really useful tool in physics. You've already seen that you can use typical values (e.g. speed and mass, see pages 179 and 197) in calculations to give estimates.

You can also use estimating to check if your final answer is sensible or not.

Example 1

A spring with spring constant 12.05 N/m is extended by 0.98 m. Calculate the energy transferred to its elastic potential energy store. Use the equation: $E_e = \frac{1}{2}ke^2$.

First estimate what the answer should be:
elastic potential energy $= 0.5 \times 12.05 \times 0.98^2 \approx 0.5 \times 10 \times 1^2 \approx 5$ J

The actual answer is 5.8 J (to 2 s.f.). From the estimated calculation, 5.8 J is a sensible answer. If your answer was 5800 J, you would know that your calculation had gone wrong somewhere.

Estimating can also be useful for choosing apparatus to use in an experiment.

Example 2

> **Tip:** You need to be able to find the volume of an irregularly shaped object in order to find its density. See page 108.

The apparatus for measuring the volume of an awards statue is shown on the right.

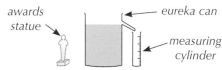

When the statue is put in the eureka can, the water level rises, which causes water to flow into the measuring cylinder. The volume of the statue is equal to the volume of water displaced. To decide what measuring cylinder to use, you should estimate the volume of the object.

An estimate for the volume of the statue can be found if the statue is considered to be a cuboid with a height of 20 cm and a base of 7 cm by 7 cm.

Volume of a cuboid $= w \times h \times d$
$= 7 \times 20 \times 7$
$= 980$ cm³

The measuring cylinder should be able to hold about 1000 cm³.

---

2. Algebra

Physics involves a lot of rearranging equations and substituting values into equations. It can be easy to make simple mistakes, so here's a few things to remember...

Algebra symbols

Here's a reminder of some of the symbols that you may come across:

Symbol	Meaning
=	is equal to
<	is less than
<<	is much less than
>	is greater than
>>	is much greater than
∝	is directly proportional to
~	is approximately
Δ	change in (a quantity)
≈	is approximately equal to

Tip: An example of using ∝ can be found on page 150.

Tip: Δ is the Greek capital letter 'delta'. An example of using Δ can be found on page 181.

Figure 1: It can be easy to make a mistake rearranging equations when you're stressed in an exam. It's a good idea to double check rearrangements, especially if it's a tricky one where you've had to combine and rearrange equations.

Rearranging equations

Being able to rearrange equations is a must in physics — you'll often need to change the subject of an equation. The subject of an equation is just the value that the rest of the equation is equal to (usually a single letter on the left-hand side of the equals sign). For rearranging equations, remember the golden rule — whatever you do to one side of the equation, you must do to the other side.

Tip: Remember — two letters written next to each other means they're being multiplied, so $mv = m \times v$.

Example 1

Rearrange the equation for momentum, $p = mv$, to make v the subject.

$p = mv$

$\dfrac{p}{m} = v$ — Divide by m to get v by itself.

So $v = \dfrac{p}{m}$.

Example 2

For an object travelling with a uniform acceleration, the equation that links the initial velocity, final velocity, acceleration and distance travelled is: $v^2 - u^2 = 2as$. Rearrange the equation to make v the subject.

$v^2 - u^2 = 2as$

$v^2 = 2as + u^2$ — Add u^2 to both sides to get v^2 on its own.

$v = \sqrt{2as + u^2}$ — Take the square root

Tip: There's an example of substituting into this equation on the next page.

Substituting into equations

Tip: Your values you put into an equation should be in the units given throughout the book, unless you're asked to give your answer in different units.

Once you've rearranged your equation, you'll probably need to substitute values into it to find your answer. Pretty easy stuff — make sure your values are in the right units — getting this wrong is a common mistake. Take a look at pages 18-19 for how to convert between different units.

Example

A train pulls out of a station and is initially travelling at 2.0 m/s. The train accelerates with a constant acceleration over a distance of 3.4 km. At this distance, the train reaches a final velocity of 150 km/h. Calculate the acceleration of the train. Use the equation $v^2 - u^2 = 2as$. Give your answer in m/s².

Your answer needs to be in m/s², so you need to be using metres and seconds. Some of the given values are in different units, so you need to convert them.

$u = 2.0$ m/s

$s = 3.4$ km $= 3.4 \times 1 \times 10^3$ m $= 3400$ m

$v = 150$ km/h $= 150 \times 1 \times 10^3$ m/h $= 150\ 000$ m/h
$\qquad\qquad\qquad\qquad\qquad\qquad = 150\ 000 \div 3600$ m/s $= 41.66...$ m/s

Rearrange the equation $v^2 - u^2 = 2as$ to make a the subject:

$a = \dfrac{v^2 - u^2}{2s} = \dfrac{41.66...^2 - 2.0^2}{2 \times 3400} = 0.254...$ m/s² $= 0.25$ m/s² (to 2 s.f.)

Tip: Converting all the values into the correct units <u>before</u> putting them into the equation stops you making silly mistakes.

Tip: Rounding to the correct number of significant figures is covered on page 15.

Formula triangles

Tip: The word formula is sometimes used instead of the word equation, e.g. the formula for power is $\dfrac{\text{energy transferred}}{\text{time}}$.

If three terms are related by a formula that looks like $v = f\lambda$ or $f = \dfrac{v}{\lambda}$, then you can put them into a formula triangle like this:

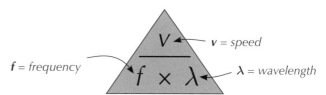

Figure 2: *The formula triangle for $v = f\lambda$ — v on the top and f × λ on the bottom.*

Tip: For an equation with two terms multiplied together, these go on the bottom of the formula triangle (and so the other must go on the top). For an equation with one term divided by another, the one on top of the division goes on top in the formula triangle and the others go on the bottom (in any order).

- To use the triangle, put your thumb over the term you want to find and write down what's left showing. This gives you your formula (for example $f = \dfrac{v}{\lambda}$).

- Then put in the values for the other terms and work out the term you want.

Example

Give the equation for power (*P*) in terms of energy transferred (*E*) and time (*t*) given the formula $E = Pt$.

Tip: Alternatively, you could work this out using the method for rearranging equations given on the previous page.

- As *P* is multiplied by *t*, *E* goes on top, leaving *P* × *t* on the bottom.

- Covering *P* leaves $\dfrac{E}{t}$.

- So, power = energy transferred ÷ time.

---

3. Graphs

Results are often presented using graphs, as you've seen on pages 16-17. They make it easier to work out relations between variables and can also be used to calculate other quantities.

Linear graphs

You will most often come across **linear graphs**. A linear graph means that the two variables plotted on the axes produce a straight line. If the line goes through the origin (0,0), then the two variables are directly proportional. An example of a linear graph is shown in Figure 1.

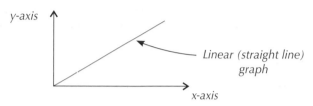

Figure 1: *A linear graph, which passes through the origin.*

Tip: The origin of a graph is the point (0,0) on the graph, i.e. it's the point at which the x-axis and y-axis meet.

Tip: A non-linear graph is just any graph that's curved.

Finding the gradient

The **gradient** (slope) of a graph tells you how quickly the variable on the y-axis changes if you change the variable on the x-axis. It is calculated using:

$$\text{gradient} = \frac{\text{change in } y}{\text{change in } x}$$

Tip: If the value on the x-axis is time, then the gradient will be equal to the rate of change of the value on the y-axis. For example, on a distance-time graph, the gradient is the rate of change of distance, which is speed.

Example

This linear graph shows the force acting on a spring against its extension. The gradient of the line is equal to the spring constant of the spring.

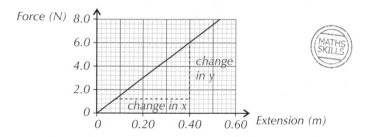

Tip: There's more on force-extension graphs on page 161.

To calculate the gradient, pick two points on the line that are easy to read and a good distance apart. Draw a line down from one of the points and a line across from the other to make a triangle. The line drawn down the side of the triangle is the change in y and the line across the bottom is the change in x.

Change in y = 6.0 − 1.2 = 4.8 N   Change in x = 0.40 − 0.08 = 0.32 m

Spring constant = gradient = $\dfrac{\text{change in } y}{\text{change in } x} = \dfrac{4.8}{0.32}$ = 15 N/m

Tip: You can use this method to calculate the rate of change from a graph, so long as the x-axis is time.

y = mx + c

The equation of a straight line is given by:

y = y-axis variable ⟶   **c** = y-intercept

$$y = mx + c$$

m = gradient ⟶   **x** = x-axis variable

The y-intercept is the point at which the line crosses the y-axis. If the straight line passes through the origin of the graph, then the y-intercept is just zero. You can use this equation to work out what the gradient and y-intercept values of a graph represent.

Example

A student is heating a block of aluminium with an electric heater in order to find its specific heat capacity. The expected shape of the graph from his experiment is shown. The change in temperature is the y-axis variable and the energy transferred to the aluminium is the x-axis variable.

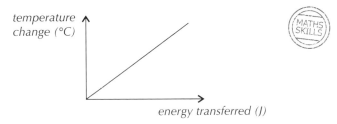

temperature change (°C) (y-axis)

energy transferred (J) (x-axis)

MATHS SKILLS

The equation that relates temperature change and energy transferred is $\Delta E = mc\Delta\theta$, where ΔE = energy transferred, m = mass of the aluminium, c = specific heat capacity of the aluminium and $\Delta\theta$ = temperature change.

This equation can't be compared to the equation of a straight line yet — the subject of the equation (see page 337) needs to be the y-axis variable, which in this case is $\Delta\theta$. So divide both sides of the equation by mc, to get:

$$\Delta\theta = \frac{1}{mc}\Delta E$$

When you compare this to the equation of the straight line, you can see that the gradient is equal to $\frac{1}{mc}$:

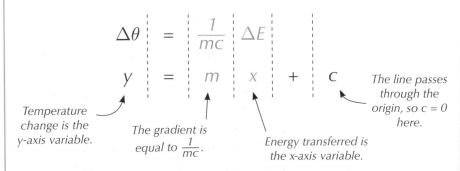

$$\Delta\theta \quad = \quad \frac{1}{mc} \quad \Delta E$$
$$y \quad = \quad m \quad x \quad + \quad c$$

Temperature change is the y-axis variable.

The gradient is equal to $\frac{1}{mc}$.

Energy transferred is the x-axis variable.

The line passes through the origin, so c = 0 here.

Once the student has plotted his data on a graph he can find the gradient, which can then be used to calculate the unknown value of c, as gradient = $\frac{1}{mc}$.

Tip: Take a look at pages 31-32 for more about this experiment and the graph.

Tip: You won't always need to rearrange an equation before you compare it to the equation of a straight line. For example, the gradient of a distance-time graph is equal to speed, which can be seen from the equation $s = vt$ (see page 183).

Tip: You'll often find that the y-intercept is equal to zero, and that the graph passes through the origin. If c wasn't equal to zero, then the straight line would be shifted up or down, to pass through the y-axis at the value of c.

Curved graphs

For a curved graph, the gradient is always changing. So you can't use the same method as the one on page 339 to calculate the gradient — you would end up with an average gradient between the two points chosen.

To find the gradient of a curve at a point, you need to draw a **tangent** to the curve at that point. A tangent is a straight line that touches the curve at that point, but doesn't cross it. So to draw one, you position a ruler so that it just touches the curve at the point you're interested in, and draw a straight line. Then you just find the gradient of the tangent using the method for straight lines on page 339.

Figure 2: *Make sure you use a really sharp pencil and a ruler whenever you're drawing graphs and tangents.*

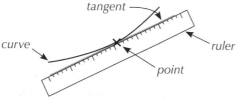

Figure 3: *Drawing the tangent to a curve.*

Example — **Higher**

**The velocity-time graph of a cyclist is shown below.
Find the acceleration of the cyclist at 70 s.**

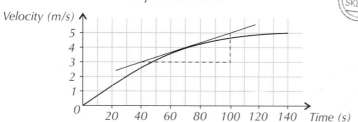

Tip: **H** Another example of finding the gradient of a tangent is shown on page 184.

A tangent to the curve at 70 s is drawn on the graph. Its gradient is:

$$\frac{\text{change in } y}{\text{change in } x} = \frac{5-3}{100-40} = \frac{2}{60} = 0.03333... = 0.03 \text{ (to 1 s.f.)}$$

The rate of change of velocity is acceleration.
So the gradient of a velocity-time graph is acceleration.

So the acceleration at 70 s = 0.03 m/s² (to 1 s.f.)

Tip: Just like with linear graphs, if the variable of the x-axis is time, the gradient at a point is rate of change of the quantity on the y-axis.

Area under a graph

Sometimes the area between the curve or line and the horizontal axis of a graph represents a quantity. For example:

- **H** The area under a velocity-time graph of an object is equal to the distance travelled by the object (page 187).

- The area under a linear force-extension graph for a stretched spring is equal to the energy transferred to the spring's elastic potential energy store.

To find an area under a graph, you'll either need to work it out exactly by breaking it up into triangles and rectangles or estimate the area by counting squares. Which method you use depends on the graph's shape. You'll have to count squares for a curved graph, but you can work it out exactly for a straight line graph. There are examples of both methods on pages 187-188.

Tip: The formulas for the area of a triangle and a rectangle are on the next page.

4. Geometry

You'll be expected to be comfortable with working out measurements of 2D and 3D shapes, such as areas, surface areas and volumes, in physics contexts.

Area

Make sure you remember how to calculate the areas of triangles and rectangles.

Tip: These come in handy when calculating the area under a graph.

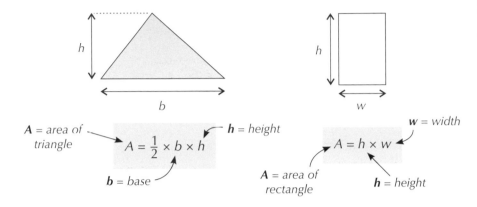

A = area of triangle

$$A = \frac{1}{2} \times b \times h$$

h = height

b = base

w = width

$$A = h \times w$$

A = area of rectangle

h = height

Surface area and volume

Tip: Surface area crops up when you're working with pressure (page 169).

If you need to work out the surface area of a 3D shape, you just need to add up the areas of all the 2D faces of the shape. So, for example, if you need to work out the surface area of a cuboid, you just find the area of each rectangular face and then add them together.

Tip: Make sure you don't forget any sides when finding the surface area. For a cube or cuboid, there should be 6.

Make sure you remember how to calculate the volume of a cuboid:

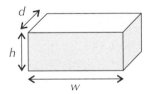

V = volume of cuboid

$$V = w \times h \times d$$

d = depth

w = width

h = height

Tip: You need to be able to calculate volumes to calculate densities (see page 106).

Example

A block of copper is shown. Calculate the volume and surface area of the copper.

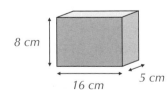

8 cm

16 cm

5 cm

Volume = $w \times h \times d$ = 16 × 8 × 5 = 640 cm³.

Surface area of front face = $h \times w$ = 8 × 16 = 128 cm².
Therefore the surface area of back face = 128 cm².

Surface area of right face = $h \times w$ = 8 × 5 = 40 cm².
Therefore the surface area of left face = 40 cm².

Surface area of top face = $h \times w$ = 5 × 16 = 80 cm².
Therefore the surface area of bottom face = 80 cm².

Total surface area = 128 + 128 + 40 + 40 + 80 + 80 = 496 cm².

1. The Exams

Unfortunately, to get your GCSE you'll need to sit some exams. And that's what this page is about — what to expect in your exams.

Assessment for GCSE Physics

To get your GCSE in physics you'll have to do some exams that test your knowledge of physics, your understanding of the Required Practical Activity experiments and how comfortable you are with Working Scientifically. You'll also be tested on your maths skills in at least 30% of the marks.

All the content that you need to know is in this book. All the Required Practical Activities are also covered in detail and clearly labelled, examples that use maths skills are marked up, and there are even dedicated sections on Working Scientifically (p. 2-21), Maths Skills (p. 333-342) and Practical Skills (p. 327-332).

Grading

When you sit your exams, you'll be given a grade between 1 and 9 based on your results. 9 is the highest grade, and 1 is the lowest. Which grades you can get will depend on which exams you sit — Foundation Tier, or Higher Tier.

If you take the Foundation Tier exams, you can get a grade between 1 and 5, with 5 being the maximum grade you can get. If you sit the Higher Tier exam, you can get a grade between 3 and 9.

The exams

You'll sit two separate exams at the end of Year 11. You will be tested on maths skills in both exams, and could be asked questions on the Required Practical Activities and Working Scientifically requirements in either of them. You're allowed to use a calculator in both of your GCSE Physics exams, so make sure you've got one.

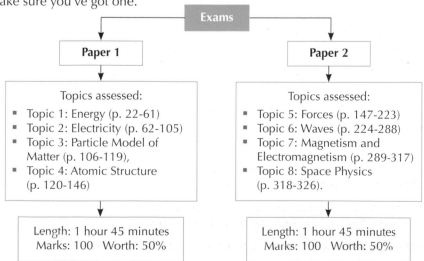

Exams

Paper 1

Topics assessed:
- Topic 1: Energy (p. 22-61)
- Topic 2: Electricity (p. 62-105)
- Topic 3: Particle Model of Matter (p. 106-119),
- Topic 4: Atomic Structure (p. 120-146)

Length: 1 hour 45 minutes
Marks: 100   Worth: 50%

Paper 2

Topics assessed:
- Topic 5: Forces (p. 147-223)
- Topic 6: Waves (p. 224-288)
- Topic 7: Magnetism and Electromagnetism (p. 289-317)
- Topic 8: Space Physics (p. 318-326).

Length: 1 hour 45 minutes
Marks: 100   Worth: 50%

> **Exam Tip**
> Make sure you have a good read through these pages. It might not seem all that important now but you don't want to get any surprises just before an exam.

> **Exam Tip**
> As well as a calculator, you should make sure you've got a ruler for both exams. In Paper 2, where you might have to deal with ray diagrams, you'll need a protractor too. And don't forget the basics — a couple of black pens and sharp pencils.

> **Exam Tip**
> Use these topic lists as a guide, but be aware that they may expect you to use basic physics knowledge from any part of the course in either exam. In particular, you'll probably need to apply your knowledge of energy stores and transfers (from Topics 1 and 2) in Paper 2.

2. Exam Technique

Knowing the science is vitally important when it comes to passing your exams. But having good exam technique will also help. So here are some handy hints on how to squeeze every mark you possibly can out of those examiners.

Time management

Good time management is one of the most important exam skills to have — you need to think about how much time to spend on each question. Check out the length of your exams (you'll find them on page 343 and on the front of your exam papers). These timings give you about 1 minute per mark, and 5 minutes spare to check your answers. Try to stick to this to give yourself the best chance to get as many marks as possible.

Don't spend ages struggling with a question if you're finding it hard to answer — move on. You can come back to it later when you've bagged loads of other marks elsewhere. Also, you might find that some questions need a lot of work for only a few marks, while others are much quicker — so if you're short of time, answer the quick and easy questions first.

> **Exam Tip**
> You shouldn't really be spending more time on a 1 mark question than on a 4 mark question. Use the marks available as a rough guide for how long each question should take to answer.

> ### Example
>
> The questions below are both worth the same number of marks but require different amounts of work.
>
> **1.1**   Name **two** renewable energy resources.
>
> *(2 marks)*
>
> **2.1**   Sketch a graph of activity against time for a radioactive isotope.
>
> *(2 marks)*
>
> Question 1.1 only asks you to write down the names of two energy resources — if you can remember them this shouldn't take you too long.
>
> Question 2.1 asks you to sketch a graph — this may take you a bit longer than writing down a couple of names, and it might take a couple of attempts to get right.
>
> So if you're running out of time, it makes sense to do questions like 1.1 first and come back to 2.1 if you've got time at the end.

> **Exam Tip**
> Don't forget to go back and do any questions that you left the first time round — you don't want to miss out on marks because you forgot to do the question.

Reading the question

You've probably heard it a million times before, but make sure you always read the whole question carefully. It can be easy to look at a question and read what you're expecting to see rather than what it's actually asking you. Read it through before you start answering, and read it again when you've finished, to make sure your answer is sensible and matches up to what the question is asking.

Remember to pay attention to the marks available too. They can often give you a sense of how much work is needed to answer the question. If it's just a 1 mark question, it'll often only need a single word or phrase as an answer, or a very simple calculation. Questions with 4, 5 or 6 marks are likely to be longer questions, which need to be clearly structured and may involve writing a short paragraph or a more complicated calculation.

Exam Tip
The amount of space given for your answer should also give you an idea about how much you need to write.

Making educated guesses

Make sure you answer all the questions that you can — don't leave any blank if you can avoid it. If a question asks you to tick a box, circle a word or draw lines between boxes, you should never, ever leave it blank, even if you're short on time. It only takes a second or two to answer these questions, and even if you're not absolutely sure what the answer is you can have a good guess.

> **Example**
>
> Look at the question below.
>
> **1.1** Which of the following are types of ionising radiation?
> Tick **two** boxes.
>
> Ultrasound ☐   Gamma rays ☐
>
> Microwaves ☐   Alpha particles ☐
> *(2 marks)*
>
> Say you know that ultrasound isn't a type of ionising radiation and gamma rays definitely are, but aren't sure about the other two answers.
>
> You can tick gamma rays — you know it's a type of ionising radiation. You know ultrasound is wrong, so leave that box blank. That leaves you with microwaves and alpha particles. If you're not absolutely sure which is ionising and which isn't, just have a guess. You won't lose any marks if you get it wrong and there's a 50% chance that you'll get it right.

Exam Tip
If you're asked, for example, to tick two boxes, make sure you only tick two. If you tick more than two, you won't get the marks even if some of your answers are correct.

Calculations

Calculations can seem daunting if you're not so keen on maths, but they're often where a load of marks are hiding. And there are some dead easy ways to make sure you don't miss out on them.

You'll be given an Equation Sheet in the exam that tells you some of the equations you might need. There are plenty of other equations you won't be given though — you'll be expected to remember them. Make sure you always write down any equation you use before you put any numbers into it.

Make sure you write down all your working. Getting the final answer is only one part of the question — if it's all you write down, and you get it wrong, that's all the marks gone. But if you write down all the steps that lead to the answer, and you've used the correct method, you'll be awarded marks for your working. Then you'll be able to pick up most of the marks, regardless of whether your final answer's correct or not.

Exam Tip
Take a look at p. 394 to see which equations will be given on the Equation Sheet in the exam.

Exam Tip
You can do an estimate to check that your answers to calculations are a sensible number. Check out page 336 for more on estimates.

3. Question Types

If all questions were the same, exams would be mightily boring. So really, it's quite handy that there are lots of different question types. Here are just a few...

Command words

Command words are just the bits of a question that tell you what to do. You'll find answering exam questions much easier if you understand exactly what they mean, so here's a brief summary of the most common ones:

Exam Tip
When you're reading an exam question, you might find it helpful to underline the command words. It can help you work out what type of answer to give.

Command word:	What to do:
Give / Name / Identify / State / Write	Give a brief one or two word answer, or a short sentence.
Choose	Select your answer from a range of options.
Complete	Write your answer in the space given. This could be a gap in a sentence or table, or you might have to finish a diagram.
Describe	Write about what something's like, e.g. describe the trend in a set of results.
Suggest	Use your scientific knowledge to work out what the answer might be.
Determine	Use the data or information you've been given to reach your answer.
Explain	Make something clear, or give the reasons why something happens. The points in your answer need to be linked together, so you should include words like because, so, therefore, due to, etc.
Calculate / Work out	Use the numbers in the question to work out an answer.
Show	Give clear evidence, and state a conclusion which this evidence supports.
Compare	Give the similarities and differences between two things.
Evaluate	Give the arguments both for and against an issue, or the advantages and disadvantages of something. You may also need to give an overall judgement.
Sketch	Draw without a lot of detail, e.g. for a graph you just need the general shape and correct axes.

Exam Tip
It's easy to get describe and explain mixed up, but they're quite different. For example, if you're asked to describe some data, just state the overall pattern or trend. If you're asked to explain data, you'll need to give reasons for the trend.

Some questions will also ask you to answer 'using the information provided' (e.g. a graph, table or passage of text) — if so, you must refer to the information you've been given or you won't get the marks. You'll often need to use information or diagrams that are provided for you when answering questions with command words such as 'measure' or 'plot'.

Required Practical Activities

The Required Practical Activities are ten specific experiments that you need to cover during your lessons. You'll be asked about them in the exams too. At least 15% of the total marks in your exams will test your understanding of these experiments and the apparatus and techniques involved. There are a lot of different types of question you could be asked on these experiments. Here are some basic areas they might ask you about:

- Carrying out the experiment — e.g. planning or describing a method, describing how to take measurements or use apparatus.

- Risk assessment — e.g. identifying or explaining hazards associated with the experiment, or safety precautions which should be taken.

- Understanding variables — e.g. identifying control, dependent and independent variables.

- Data handling — e.g. plotting graphs or doing calculations using some sample results provided.

- Analysing results — e.g. making conclusions based on sample results.

- Evaluating the experiment — e.g. making judgements on the quality of results, identifying where mistakes may have been made in the method, suggesting improvements to the experiment.

Required Practical Activity questions won't be pointed out to you in the exam, so you'll need to make sure you know the practicals inside out, and can recognise them easily. For an example of a question testing your understanding of a practical, see page 176.

Exam Tip
The Required Practical Activity questions are likely to have some overlap with Working Scientifically, so make sure you've brushed up on pages 2-21.

Levels of response questions

Some questions are designed to assess your ability to present and explain scientific ideas in a logical and coherent way, as well as your scientific knowledge. These questions often link together different topics, and are worth more marks than most other question types. You'll be told which questions these are on the front of your exam paper.

This type of question is marked using a 'levels of response' mark scheme. Your answer is given a level depending on the number of marks available and its overall quality and scientific content. Here's an idea of how the levels may work out for a 6 mark question:

Example

Level 0

A Level 0 answer has no relevant information, and makes no attempt to answer the question. It receives no marks.

Level 1

A Level 1 answer usually makes one or two correct statements, but does not fully answer the question. For instance, when asked to describe and explain the differences between two materials, it might state one or two correct properties of the materials, but not explain them or attempt to compare the two. These answers receive 1 or 2 marks.

Exam Tip
Make sure your writing is legible, and be careful with your spelling, punctuation and grammar.

Exam Tip
It might be useful to write a quick plan of your answer in some spare space on your paper. This can help you get your thoughts in order, so you can write a logical, coherent answer. But remember to cross your plan out after you've written your answer so it doesn't get marked.

Exam Tip
Make sure your writing style is appropriate for an exam. You need to write in full sentences and use fairly formal language.

Level 2

A Level 2 answer usually makes a number of correct statements, with explanation, but falls short of fully answering the question. It may miss a step, omit an important fact, or not be organised as logically as it should be. These answers receive 3 or 4 marks.

Level 3

A Level 3 answer will answer the question fully, in a logical fashion. It will make a number of points that are explained and related back to the question. Any conclusions it makes will be supported by evidence in the answer. These answers receive 5 or 6 marks.

Make sure you answer the question fully, and cover all points indicated in the question. You also need to organise your answer clearly — the points you make need to be in a logical order. Use specialist scientific vocabulary whenever you can. For example, if you're talking about the structure of the atom, you need to use scientific terms like 'the nuclear model'. Obviously you need to use these terms correctly — it's no good knowing the words if you don't know what they actually mean.

There are some exam-style questions that use this type of mark scheme in this book (marked up with an asterisk, *). You can use them to practise writing logical and coherent answers. Use the worked answers given at the back of this book to mark what you've written. The answers will tell you the relevant points you could've included, but it'll be down to you to put everything together into a full, well-structured answer.

Answers

Topic 1 — Energy

Topic 1a — Energy Transfers

1. Energy Stores and Transfers

Page 25 — Fact Recall Questions

Q1 Thermal (or internal) energy store, kinetic energy store, gravitational potential energy store, elastic potential energy store, chemical energy store, magnetic energy store, electrostatic energy store and nuclear energy store.

Q2 A closed system is a system where neither matter nor energy can enter or leave.

Q3 Energy can be transferred usefully, stored or dissipated, but can never be created or destroyed.

Page 25 — Application Question

Q1 a) Energy is transferred mechanically from the elastic potential energy store of the bow to the kinetic energy store of the arrow.

b) Energy is transferred by heating from the chemical energy store of the gas to the thermal energy stores of the soup and the surroundings.

c) Energy is transferred electrically from the chemical energy store of the battery to the kinetic energy store of the motor / the blades of the fan.

2. Kinetic and Potential Energy Stores

Page 29 — Fact Recall Questions

Q1 The large dog has more energy in its kinetic energy store, because the energy in this store is related to mass and speed by $E_k = \frac{1}{2}mv^2$.

Q2 $E_k = \frac{1}{2}mv^2$. E_k is the energy in the kinetic energy store in J, m is mass in kg and v is speed in m/s.

Q3 $E_p = mgh$. E_p is the energy transferred to the gravitational potential energy store in J, m is mass in kg, g is gravitational field strength in N/kg and h is height in m.

Q4 $E_e = \frac{1}{2}ke^2$. E_e is the energy in the elastic potential energy store in J, k is the spring constant in N/m and e is extension (or compression) in m.

Page 29 — Application Questions

Q1 $E_p = mgh = 25\,000 \times 9.8 \times 12\,000 = \textbf{2 940 000 000 J}$

Q2 $E_e = \frac{1}{2}ke^2$ so,
$k = 2E_e \div e^2 = (2 \times 18) \div 0.6^2 = \textbf{100 N/m}$

Q3 Rearranging $E_k = \frac{1}{2}mv^2$,
$$v = \sqrt{\frac{2E_k}{m}} = \sqrt{\frac{2 \times 40}{0.0125}} = \textbf{80 m/s}$$

Q4 Due to conservation of energy, the energy lost from the potato's g.p.e. store is all transferred to its kinetic energy store, so $E_p = E_k = 450$ J
Rearranging $E_k = \frac{1}{2}mv^2$,
$$v = \sqrt{\frac{2E_k}{m}} = \sqrt{\frac{2 \times 450}{1}} = \textbf{30 m/s}$$

If there had been air resistance, some energy would have been transferred to other stores, such as the thermal energy store of the air.

3. Specific Heat Capacity

Page 32 — Fact Recall Question

Q1 The amount of energy needed to change the temperature of 1 kg of a substance by 1°C.

Page 32 — Application Questions

Q1 $\Delta\theta = 100.0\,°C - 20.0\,°C = 80.0\,°C$
$\Delta E = mc\Delta\theta$
$\quad = 0.20 \times 4200 \times 80.0 = \textbf{67 200 J}$

Q2 $m = 400$ g $= 0.4$ kg
$\Delta\theta = 113\,°C - 25\,°C = 88\,°C$
$\Delta E = mc\Delta\theta$
$c = \Delta E \div (m\Delta\theta) = 70\,400 \div (0.4 \times 88)$
$\quad = \textbf{2000 J/kg°C}$

4. Power

Page 34 — Fact Recall Questions

Q1 Power is the rate of doing work — i.e. how much work is done per second.
Remember, you can also define power in terms of energy transferred. Energy transferred and work done are the same thing.

Q2 $P = \dfrac{E}{t}$ and $P = \dfrac{W}{t}$. P is power in watts, E is energy transferred in joules, W is work done in joules and t is time in seconds.

Page 34 — Application Questions

Q1 a) $P = E \div t = 150 \div 37.5 = \textbf{4 W}$

b) 79.8 kJ = 79 800 J
$P = E \div t = 79\,800 \div 42 = \textbf{1900 W}$

c) 6 840 kJ = 6 840 000 J
9.5 minutes = 570 s
$P = E \div t = 6\,840\,000 \div 570 = \textbf{12 000 W}$

Q2 Lift B will lift the load in the least time. Since both lifts need to supply the same amount of energy to perform the task, the lift with the larger power, B, will perform the task in less time, as it supplies more energy per second.

Q3 a) $P = W \div t$, so $t = W \div P$
$t = 1344 \div 525 = \textbf{2.56 s}$

b) 2.86 kW = 2860 W
$P = E \div t$, so $t = E \div P$
$t = 1430 \div 2860 = \textbf{0.5 s}$

Q4 a) $P = E \div t$, so $E = Pt$
$E = 1240 \times 35 = \mathbf{43\ 400\ J}$
   b) 17 minutes = 1020 s
$E = Pt = 1500 \times 1020 = \mathbf{1\ 530\ 000\ J}$

5. Conduction and Convection
Page 36 — Fact Recall Questions
Q1 The transfer of energy by heating through a substance by vibrating particles colliding. The particles of the part of the substance that is heated have more energy in their kinetic energy stores and vibrate more. The particles collide with their neighbouring particles and pass on some of this energy, causing energy eventually to be spread out through the substance.
Q2 Convection is the transfer of energy by the movement of more energetic particles in a gas or liquid from a hotter to a cooler region.
Q3 Solids
The particles in a solid can only vibrate about their fixed positions — they can't move from one place to another and take their energy with them.

6. Reducing Unwanted Energy Transfers
Page 39 — Fact Recall Questions
Q1 a) Foam squirted into the gap between the bricks of a cavity wall stops convection currents being set up in the gap.
   b) An insulating material laid on a loft floor reduces conduction through the loft floor and prevents convection currents forming in the loft space, which helps reduce energy transferred out of thermal energy stores in the house.
   c) Double-glazed windows have a double layer of glass separated by an air gap. Because air conducts poorly it will reduce energy loss by conduction through the window.
Q2 E.g. A lubricant (e.g. oil) can be used on an axle in a fan to decrease the friction between the axle and its support and so decreases the unwanted energy transfers to thermal energy stores due to friction.

Page 39 — Application Question
Q1 Any two of: e.g.
The cups were uncovered and the insulation was only wrapped around in a strip, so energy will still be transferred by heating directly to the surroundings from the top and uninsulated sides. This would mean the effect of the insulation would be less noticeable. / The temperatures weren't measured at the start of the experiment, they could have started off at different temperatures. / The drinks may not have been the same. They could have had different specific heat capacities, meaning that they store (and so release) different amounts of energy.

7. Efficiency
Page 42 — Fact Recall Questions
Q1 $\text{efficiency} = \dfrac{\text{useful output energy transfer}}{\text{total input energy transfer}}$

$\text{efficiency} = \dfrac{\text{useful power output}}{\text{total power input}}$

Q2 An electric heater.
Q3 It is transferred to thermal energy stores.

Page 42 — Application Questions
Q1 a) $\text{efficiency} = \dfrac{\text{useful power output}}{\text{total power input}}$
$= \dfrac{54}{90}$
$= \mathbf{0.6}$
   b) Useful energy out = 800 − 280 = 520 J
$\text{efficiency} = \dfrac{\text{useful output energy transfer}}{\text{total input energy transfer}}$
$= \dfrac{520}{800}$
$= \mathbf{0.65}$

Q2 a) $\text{efficiency} = \dfrac{\text{useful power output}}{\text{total power input}}$
$= \dfrac{12.6}{36}$
$= 0.35$
percentage efficiency = 0.35 × 100 = **35%**
   b) $\text{efficiency} = \dfrac{\text{useful output energy transfer}}{\text{total input energy transfer}}$
$= \dfrac{4.5}{7.5}$
$= 0.6$
percentage efficiency = 0.6 × 100 = **60%**
The energy transferred by heating to thermal energy stores and emitted as sound isn't useful — only the light is.
Q3 a) useful output energy transfer = 298 + 197 = **495 kJ**
The energy emitted as light and as sound is useful.
   b) wasted energy = 660 − 495 = **165 kJ**
   c) $\text{efficiency} = \dfrac{\text{useful output energy transfer}}{\text{total input energy transfer}}$
$= \dfrac{495}{660}$
$= \mathbf{0.75}$
Q4 convert percentage efficiency into decimal efficiency
efficiency = 68 % = 0.68
Rearrange $\text{efficiency} = \dfrac{\text{useful output energy transfer}}{\text{total input energy transfer}}$

$\text{total input energy transfer} = \dfrac{\text{useful output energy transfer}}{\text{efficiency}}$

$= \dfrac{816}{0.68}$

$= \mathbf{1200\ J}$

Q5 E.g. lubricate the workings to reduce friction in the motor, and therefore energy transferred to 'wasted' thermal stores. Reduce the amount of sound the vacuum makes, therefore reducing energy transferred to 'wasted' stores in the surroundings.

Pages 45-46 — Energy Transfers
Exam-style Questions

1.1 Eg. From the chemical energy store of the gas to the thermal energy store of the pot (and surroundings). From the thermal energy store of the pot to the thermal energy store of the water.
From the thermal energy store of the water (or pot) to the thermal energy stores of the surroundings.
(3 marks — 1 mark for each correct answer)

1.2 The second pan's material could have a higher thermal conductivity **(1 mark)**.

1.3 Air is an insulator / is a poor conductor / has a low thermal conductivity **(1 mark)** so energy is transferred across the gap slowly by conduction **(1 mark)**.

2.1 $E_p = mgh$
$E_p = (85.0 + 10.0) \times 9.8 \times 10.0$ **(1 mark)**
$= $ **9310 J (1 mark)**

2.2 $P = E \div t$
$P = 9310 \div 7.0$ **(1 mark)** $= $ **1330 W (1 mark)**
(Allow follow-through from part 2.1)
If you got 2.1 wrong, you still get all the marks for 2.2 if you've done everything else right.

2.3 $E_k = \frac{1}{2}mv^2$, so,
$v = \sqrt{\dfrac{2E_k}{m}}$ **(1 mark)** $= \sqrt{\dfrac{2 \times 153.9}{85.0 + 10.0}}$ **(1 mark)**
$= $ **1.8 m/s (1 mark)**

2.4 Assume there is no air resistance so all energy in g.p.e. store is transferred to the kinetic energy store, **(1 mark)**,
so by conservation of energy, $mgh = \frac{1}{2}mv^2$, so
$v = \sqrt{2gh}$ **(1 mark)** $= \sqrt{2 \times 9.8 \times 1.25}$ **(1 mark)**
$= 4.949... = $ **4.9 m/s (to 2 s.f.) (1 mark)**

3.1 $P = W \div t$ **(1 mark)**

3.2 $P = 4 \div 0.5$ **(1 mark)** $= $ **8 W (1 mark)**

3.3 Work done $= 4$ J $= $ energy transferred to elastic potential energy store **(1 mark)**,
$E_e = \frac{1}{2}ke^2$, so $k = 2E_e \div e^2 = (2 \times 4) \div 0.4^2$ **(1 mark)**
$= $ **50 N/m (1 mark)**

3.4 First, convert percentage efficiency to a decimal, $62.5\% = 0.625$
$\text{efficiency} = \dfrac{\text{useful power output}}{\text{total power input}}$ so,
total power input $=$ useful power output $\div$ efficiency
$= 25 \div 0.625$ **(1 mark)**
$= $ **40 W (1 mark)**

3.5 E.g. Friction between the wheel axle and its supports causes energy from the kinetic energy store of the car to be dissipated into thermal energy stores **(1 mark)**

3.6 E.g. The wheel axle of the car could be lubricated to decrease the friction **(1 mark)**.

4.1 How to grade your answer:
Level 0: There is no relevant information.
(No marks)
Level 1: There is a brief description of an experiment that could be used to calculate the specific heat capacity of water. **(1 to 2 marks)**
Level 2: There is a clear description of an experiment that could be used to calculate the specific heat capacity of water. A method describing how to process the results of this experiment to calculate the specific heat capacity is described briefly. **(3 to 4 marks)**
Level 3: There is a clear and detailed description of an experiment that could be used to calculate the specific heat capacity of water. A method describing how to process the results of this experiment to calculate the specific heat capacity is clearly described. **(5 to 6 marks)**

Here are some points your answer may include:
Put the beaker on a mass balance and zero the balance.
Fill the beaker with water and record the mass of the water.
Place the thermometer into the beaker and measure the temperature of the water.
Place the heating element into the water.
Turn on the heating element to heat the water.
As the water is heated, record the temperature of the water and current through the heating element periodically.
The current through the circuit shouldn't change.
Using the potential difference of the power supply, calculate the energy transferred by the heater at the time of each temperature reading using $P = VI$ and $E = Pt$.
Plot a graph of change in temperature against energy transferred.
Calculate the gradient (of the straight line part) of the graph, and calculate the specific heat capacity using:
specific heat capacity $= 1 \div$ (gradient $\times$ mass of water)

4.2 E.g. Care must be taken when handling the heating element and hot water to avoid being burnt **(1 mark)**.

4.3 $\Delta\theta = 100.0\ °C - 10.0\ °C = 90.0\ °C$
$\Delta E = mc\Delta\theta$, so
$c = \Delta E \div (m\Delta\theta)$ **(1 mark)**
$= 189\ 000 \div (0.50 \times 90.0)$ **(1 mark)**
$= $ **4200 J/kg°C (1 mark)**

Topic 1b — Energy Resources

1. Energy Resources and Their Uses
Page 48 — Fact Recall Questions

Q1 Coal, oil and (natural) gas.
Q2 a) Any four from: wind, the Sun (solar), water waves, geothermal, tides, hydroelectricity, bio-fuels.
b) E.g. they often produce much less energy than non-renewable resources / a lot of them are less reliable because they depend on the weather.

Q3 E.g. renewable: bio-fuels
        non-renewable: petrol/diesel/coal
Q4 Electromagnetic radiation from the Sun is used to
      heat water which is then pumped into radiators in the
      building.

2. Wind, Solar and Geothermal
Page 51 — Fact Recall Questions
Q1 The wind turns the blades of the turbine, which
      are connected to an electrical generator inside the
      turbine. This generates electricity as it turns.
Q2 E.g. they generate no electricity when there's no wind
      / they generate no electricity when the wind is too
      strong.
Q3 Supplying a device in a remote location with
      another source of energy could be very difficult and
      expensive.
Q4 Hot underground rocks (thermal energy stores).

Page 51 — Application Question
Q1 a) E.g. Solar cells, because the street light is remote,
          will get lots of sunshine during the day and
          requires a fairly small amount of electricity.
     b) E.g. Wind power, because the sign will be exposed
          to wind and requires a fairly small amount of
          electricity.
     c) E.g. Geothermal power, because the area is
          volcanic, so there will be hot rocks close to the
          surface that can be used to heat homes directly.

3. Hydroelectricity, Waves and Tides
Page 54 — Fact Recall Questions
Q1 Water is held behind a dam and allowed out through
      turbines. This turns the turbines, that are connected
      to generators which generate electricity.
Q2 E.g. flooding of a valley can result in rotting
      vegetation which releases methane and carbon
      dioxide.
      Hydroelectric power stations can have a large impact
      on local habitats (and wildlife).
Q3 Remote areas can be difficult to supply with fuel or
      connect to the national grid. Having a hydroelectric
      power station in a remote location to provide
      electricity avoids these problems, as it can be more
      easily connected to homes and needs no fuel to run.
Q4 Using wave-powered turbines on the coastline.
      When a wave reaches the coastline, the motion of the
      wave forces air up through a turbine which drives a
      generator and generates electricity. (When the wave
      retreats, the air is forced back out through the same
      turbine, generating more electricity.)
Q5 E.g. the electricity supply is not very reliable, as no
      electricity can be generated when the sea is calm.
Q6 Tidal barrages are dams with turbines in them. They
      stop the tide flowing into and out of a river estuary, so
      that a height difference of water builds up between
      the sides. The tide water is then allowed to flow
      through turbines from the higher side to the lower
      side. The motion of the turbines turns a generator,
      which produces electricity.

4. Bio-fuels and Non-renewables
Page 57 — Fact Recall Questions
Q1 a) If a process is carbon neutral, it removes as much
          CO_2 from the atmosphere as it releases.
     b) E.g. the plants used to make bio-fuels absorb CO_2
          from the atmosphere as they grow. If they absorb
          the same amount as is released when they burn
          they are said to be carbon neutral. The process
          may not be carbon neutral if the bio-fuel plants
          aren't being grown at the same rate as they are
          being used.
Q2 E.g. carbon dioxide and sulfur dioxide.
Q3 a) nuclear fuel
     b) E.g. highly radioactive waste is produced, which is
          difficult to dispose of.
          Possibility of nuclear disasters that can
          dramatically harm people/the environment.

5. Trends in Energy Resource Use
Page 59 — Fact Recall Questions
Q1 Appliances are being made more efficient, and
      people are more concerned about the amount of
      electricity they use.
Q2 Any two from: e.g. they're not as reliable as non-
      renewables / they can have large initial setup costs
      / a number of power stations that use renewable
      resources can only be built in certain locations /
      a lot of renewable resources cannot easily have
      their power output increased to meet demand /
      some renewables don't produce as much energy as
      non-renewables.

Page 61 — Energy Resources
Exam-style Questions
1.1 Non-renewable energy sources:
      Coal (49.9%), Oil (2.4%), Gas (20.3%) and Nuclear
      (19.6%).
      Total percentage = 49.9 + 2.4 + 20.3 + 19.6
                              = **92.2%**
      *(2 marks for correct answer, otherwise 1 mark for
      correctly identifying all of the non-renewable energy
      sources in the table)*
1.2 E.g. coal, oil and gas are the energy sources listed that
      emit the most harmful gases (e.g. CO_2, sulfur dioxide)
      when burned to generate electricity *(1 mark)*. A far
      greater percentage of Country 1's electricity comes
      from using these sources, so this country will produce
      more pollution *(1 mark)*.
      The reservoirs in hydroelectric power stations release a
      small amount of methane. This is nothing in comparison
      to the amount of harmful gases released by burning fossil
      fuels.
1.3 Any one from: e.g. burning coal releases CO_2 into
      the atmosphere (which contributes to the greenhouse
      effect) / burning coal releases sulfur dioxide (which
      causes acid rain) / coal mining destroys the landscape
      (1 mark).
1.4 E.g. crops used to make bio-fuels have to be grown,
      so they can't respond to immediate energy demands
      (1 mark).

2.1 A non-renewable energy resource is one that is not being replenished at the same rate as it is being used, so it will eventually run out *(1 mark)*.

2.2 E.g. coal is burnt in fireplaces to heat homes *(1 mark)*.

2.3 Any one from: e.g. burning coal produces greenhouse gases which contribute to global warming / burning coal produces sulfur dioxide which causes acid rain / coal mining makes a mess of the landscape *(1 mark)*.

2.4 E.g. Wind power could be an alternative *(1 mark)*. The winds and exposed hills would allow wind turbines to generate electricity *(1 mark)*. Wave power could be an alternative *(1 mark)*. It is an island, so it has a coastline, and the regular strong winds will produce powerful waves, allowing generation of electricity from wave power *(1 mark)*.

2.5 E.g. the renewable alternatives are not reliable, as the wind doesn't always blow, so there may be times when they generate no electricity *(1 mark)*.

Topic 2 — Electricity

Topic 2a — Circuits

1. Circuits, Current and Potential Difference

Page 64 — Fact Recall Questions

Q1 a)
   b) ──▭──
   c) ──▭──
   d) ──⊗──
   e) ──⊕──

Q2 A circuit is incomplete if you can't follow a wire from one end of the battery (or other power supply), through any components to the other end of the battery.

Q3 Current flows from positive to negative.

Q4 $Q = It$, where Q = charge in coulombs (C), I = current in amps (A) and t = time in seconds (s).

Q5 A source of potential difference.

Page 64 — Application Questions

Q1 E.g.

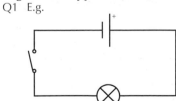

This is just one example of a correct circuit diagram — you could have the components in a different order in the circuit.

Q2 E.g.

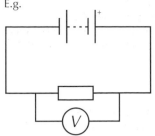

Remember that voltmeters are always connected across a component.

Q3 E.g.

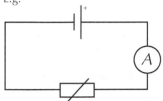

Remember that ammeters are always connected in series with a component.

Q4 Circuit B.
In the rest of the circuits, the lamp is in an incomplete part of the circuit. Don't worry about the extra bits coming off the circuit — as long as there's a complete cycle containing the lamp, current will flow through it and the lamp will light.

Q5 $I = 0.2$ A, $Q = 50$ C
$Q = It$, so
$t = Q \div I = 50 \div 0.2 = \textbf{250 s}$

Q6 $t = 1$ minute $= 60$ seconds
$Q = It$, so
$I = Q \div t = 102 \div 60 = \textbf{1.7 A}$

2. Resistance and *I-V* Characteristics

Page 69 — Fact Recall Questions

Q1 Resistance is anything in a circuit that opposes the flow of current. It is measured in ohms, Ω.

Q2 $V = IR$ (potential difference = current × resistance)

Q3 E.g.

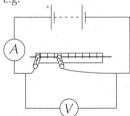

Q4 Wire length and resistance are directly proportional — as wire length increases, so does resistance.

Q5 E.g.

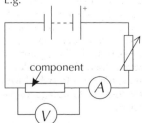

component

Q6 Read off the values of current and potential difference at a single point on the graph, and use $R = V \div I$ to calculate the resistance.

Q7 a) E.g.

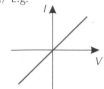

b) E.g.

Q8 a) The resistance is initially high but drops as V increases, allowing current to flow.
b) The resistance is very high, so current can't flow through a diode in this direction.

Page 69 — Application Questions
Q1 A diode.
Q2 $V = IR = 0.015 \times 2.0 = \textbf{0.03 V}$
Q3 $R = V \div I = 14.4 \div 0.60 = \textbf{24 } \Omega$
Q4 $R = V \div I$. The graph is a straight line, so the resistance is constant, so just pick a point on the line and use the values of V and I to work out the resistance:
E.g. $R = V \div I = 2 \div 0.25 = \textbf{8 } \Omega$

3. Series Circuits
Page 73 — Fact Recall Questions
Q1 Components are connected all in a line, end to end.
Q2 Ammeter
Q3 In series and in the same direction as each other (and any other cells in the circuit).
Q4 False, it is shared between all components.
Q5 It is always the same.
Q6 When a resistor is added in series, it has to take a share of the pd. This decreases the pd through each component, and hence decreases the current ($V = IR$). In a series circuit, the current is constant through each component, therefore the total current decreases, so the total resistance increases.

Page 73 — Application Question
Q1 a) 1.5 A
b) $V_1 = IR = 1.5 \times 7.0 = \textbf{10.5 V}$
c) $V = V_1 + V_2$ so $V_2 = V - V_1 = 12 - 10.5 = \textbf{1.5 V}$
d) $V = IR$ so $R = V \div I = 1.5 \div 1.5 = \textbf{1 } \Omega$

4. Parallel Circuits
Page 78 — Fact Recall Questions
Q1 Components are connected to the battery separately to the other components (on their own branches).

Q2 Each branch/component can be turned on and off separately, whereas in series circuits, turning off one component will turn them all off.
Q3 It is the same across each component and across the power supply.
Q4 By adding up the current in every branch.

Page 78 — Application Questions
Q1 a) Find the pd across the branch with the resistor in, and it will be the same as the pd shown by the voltmeter. The current through the branch is 0.75 A and the resistance is 8.0 Ω.
$V = IR = 0.75 \times 8.0 = \textbf{6.0 V}$
Remember, the resistance of an ammeter is so small that you don't need to consider it — you can pretend it's not there.
b) The total current in the circuit is 2.0 A, and the current through the branch with the resistor is 0.75 A. So the current through the bulb is 2.0 A – 0.75 A = 1.25 A. The voltage across the bulb is 6.0 V so the resistance can be found by rearranging $V = IR$.
$R = V \div I = 6.0 \div 1.25 = \textbf{4.8 } \Omega$
c) The total resistance is less than the answer to b).
Remember, the total resistance of a parallel circuit is always less than the resistance of any of the individual branches.
Q2 a) With the switch open, the circuit is just a simple series circuit with two resistors and an ammeter. The current in the circuit is 0.70 A and the total resistance is 12 + 8.0 = 20 Ω.
$V = IR = 0.70 \times 20 = \textbf{14 V}$
b) i) The potential difference is the same on each branch of the parallel part of the circuit. So you can just find the voltage on the lower branch by using the resistance of, and current through resistor R_1.
$V = IR_1 = 0.50 \times 8.0 = \textbf{4.0 V}$
ii) The total potential difference of the circuit is shared between resistor R_2 and the parallel loop with the other components on it. The potential difference across the parallel loop is 4.0 V and the total potential difference is 14 V, so the potential difference across resistor R_2 is $14 - 4.0 = \textbf{10 V}$.

5. Investigating Resistance
Page 81 — Application Question
Q1 a) E.g.

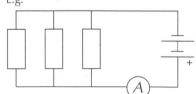

b)

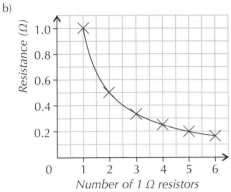

c) The total resistance of the circuit decreases as the number of resistors connected in parallel increases. When a resistor is added in parallel, it has the same potential difference across it as the source. But by adding another branch, the current has more than one direction to go in. This increases the amount of current flowing. Using $V = IR$, an increase in current at a constant pd means a decrease in the total resistance.

6. LDRs and Thermistors
Page 84 — Fact Recall Questions
Q1 Increase the light intensity.
An LDR is a light dependent resistor — its resistance decreases as the light hitting it gets brighter.

Q2 E.g.

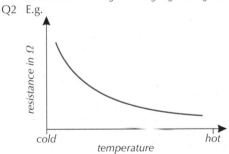

Q3 a) E.g. automatic night lights / outside lighting / burglar detectors.
  b) E.g. thermostats / temperature sensors.

Page 84 — Application Question
Q1 a) A light-dependant resistor (LDR).
  b) The cat blocks the light incident on the LDR, so the light intensity drops and the resistance increases, causing the alarm to sound.

Pages 87-88 — Circuits
Exam-style Questions
1.1 The ammeter should be connected in series with the filament lamp *(1 mark)* because the current is constant in a series loop, so it will measure the same current as is passing through the lamp *(1 mark)*.

1.2 E.g.

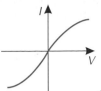

(1 mark)

1.3 No, the filament lamp is not an ohmic conductor as the current is not directly proportional to the voltage / the relationship between current and voltage is not linear *(1 mark)*.

1.4 E.g.

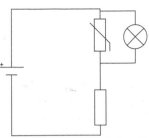

(1 mark for correct symbols for all components, 1 mark for cell, thermistor and resistor connected in series, 1 mark for filament lamp connected in parallel with the thermistor only.)
When the temperature drops, the thermistor's resistance will increase. So it'll take a higher share of the source pd. This means that putting the lamp across the thermistor will result in it getting brighter when the temperature decreases.

1.5 The student should move the filament lamp to be connected in parallel with the resistor *(1 mark)*.

2.1 The rate of flow of charge *(1 mark)*.

2.2 The potential difference across the battery is the same as the potential difference across the motor.
$V = 14$ V *(1 mark)*

2.3 $Q = It$, so the current passing through the motor is found by:
$I_{motor} = Q \div t$ *(1 mark)* $= 15 \div 30$ *(1 mark)*
$= 0.5$ A *(1 mark)*
The components are in parallel, so the sum of their currents is equal to the total current:
$I_{total} = I_{motor} + I_{lamp}$
$1.2 = 0.5 + I_{lamp}$ *(1 mark)*
$I_{lamp} = 1.2 - 0.5 = \textbf{0.7 A}$ *(1 mark)*

2.4 The potential difference through the filament bulb is the same as the power supply, $V = 14$ V. *(1 mark)*
$V = IR$
$R = V \div I$ *(1 mark)*
$= 14 \div 0.7$ *(1 mark)* $= \textbf{20 } \Omega$ *(1 mark)*

3.1 The resistance of the variable resistor can be increased or decreased, which will decrease or increase the current in the circuit *(1 mark)*.

3.2 A resistor OR a wire *(1 mark, accept 'an ohmic conductor')*.

3.3 Choose a point on the graph and read off the values — e.g. $I = 0.2$ A when $V = 1$ V.
$V = IR$
So $R = V \div I$ *(1 mark)*
$= 1 \div 0.2$ *(1 mark)* $= \textbf{5 } \Omega$ *(1 mark)*

3.4 Graph B *(1 mark)*. In graph B, Y has a lower value of *I* for each value of *V*, which means the resistance is higher *(1 mark)*.

Remember — a higher resistance means less current will flow for a given pd.

Topic 2b — Domestic Electricity

1. Electricity in the Home
Page 90 — Fact Recall Questions
Q1 Direct current is current that only flows in one direction.
Q2 Alternating current. The direction of ac is constantly changing whereas dc always flows in one direction.
Q3 The live wire is brown, the neutral wire is blue and the earth wire is green and yellow.
Q4 a) The live wire has a high potential difference (around 230 V). The blue neutral wire is around 0 V.
   b) The earth wire protects the wiring of an appliance and has a safety role — it stops the appliance from becoming live.

2. Power and Energy Transfer
Page 94 — Fact Recall Questions
Q1 a) E.g. any two of: a motor, a kettle, a speaker.
   b) E.g. any two of:
   motor — kinetic energy store
   kettle — thermal energy store
   speaker — kinetic energy store
   There are loads of options you could give here — these are just a few common examples.
Q2 It is transferred to the thermal energy store of the components and the surroundings.
Q3 $E = Pt$. *E* in energy transferred in joules, *P* is power in watts and *t* is time in seconds.
Q4 $P = VI$. Power (*P*) is measured in watts, potential difference (*V*) is in volts and current (*I*) is in amps.

Page 94 — Application Questions
Q1 Cooker B will use the most energy in 20 minutes because it has the highest power.
Q2 $P = VI = 230 \times 3.0 = $ **690 W**
Q3 $P = 2.0$ kW $= 2000$ W, $t = 30$ minutes $= 1800$ s
   $E = Pt = 2000 \times 1800 = $ **3 600 000 J**
Q4 $E = QV$, so $V = E \div Q = 22 \div 4.0 = $ **5.5 V**
Q5 $P = I^2R$
   $R = P \div I^2 = 40 \div (0.2)^2 = $ **1000 Ω**
Q6 Microwave A:
   $E = Pt = 900 \times (4 \times 60) = 216\,000$ J
   Microwave B:
   $E = Pt = 650 \times (6 \times 60) = 234\,000$ J
   So **microwave B** transfers the most energy.
   Remember to convert the time into seconds before carrying out any calculations.

3. The National Grid
Page 96 — Fact Recall Questions
Q1 a) A transformer changes the potential difference of an electrical supply.
   b) When electricity is transmitted, energy is lost through heating in the cables. This energy loss is greater for a higher current. Transmitting electricity at a high voltage reduces the current for the same amount of power, so transformers are used to increase the voltage, decrease the current and reduce the energy loss. Transformers are also used to decrease the voltage to safe levels again before the electricity reaches users.
Q2 *A* — power station
   B — step-up transformer
   C — pylons / electricity cables
   D — step-down transformer
   E — consumers
Q3 a) E.g. any two from: when people wake up in the morning / when it starts to get dark / when people get home from school or work / when it starts to get cold.
   b) E.g. Powers stations usually run below their maximum power output, so there's capacity to cope with higher demand. Smaller power stations which can be started quickly are kept on standby.

Page 98 — Domestic Electricity
Exam-style Questions
1.1 brown *(1 mark)*.
1.2 Power = potential difference × current / $P = VI$ *(1 mark)*
1.3 $I = P \div V$ *(1 mark)*
   $= 575 \div 230$ *(1 mark)* $= $ **2.5 A** *(1 mark)*
1.4 The exposed wire could provide a link to earth *(1 mark)* causing a huge current to flow *(1 mark)* which could heat the wire enough to start a fire *(1 mark)*.
2.1 Pd = 230 V *(1 mark)*, frequency = 50 Hz *(1 mark)*.
2.2 To transmit a high power, either a high current or high pd must be used *(1 mark)*. A high current would result in lots of energy being lost to thermal energy stores *(1 mark)*. Transmitting at a high voltage means a lower current can be used, which reduces energy loss and makes transmission more efficient *(1 mark)*.
2.3 Energy transferred = power × time / $E = Pt$ *(1 mark)*
2.4 Power = 2.55 kW = 2550 W
   $t = E \div P$ *(1 mark)*
   $= 7\,038\,000 \div 2550$ *(1 mark)*
   $= $ **2760 s** *(1 mark)*
2.5 Energy transferred = charge × potential difference / $E = QV$ *(1 mark)*
2.6 $Q = E \div V$ *(1 mark)*
   $= 7\,038\,000 \div 230$ *(1 mark)*
   $= $ **30 600 C** *(1 mark)*
2.7 $P = 1150$ W, $V = 230$ V
   $P = VI$
   $I = P \div V$ *(1 mark)*
   $= 1150 \div 230$ *(1 mark)*
   $= $ **5.00 A** *(1 mark)*

Topic 2c — Static Electricity and Electric Fields

1. Static Electricity
Page 102 — Fact Recall Questions
Q1 Negatively charged electrons may be rubbed off the first insulator and onto the other insulator, leaving the first insulator with a positive charge.
Q2 A large charge causes a large potential difference between the object and earth. The pd can get large enough for a spark to jump between the object and an earthed conductor.
Q3 a) Repel
   b) Repel
   c) Attract

Page 102 — Application Questions
Q1 a) All her hairs have the same charge (the opposite charge to the hair brush), so they will all repel each other, causing them to stand on end.
   b) + 0.5 nC
Q2 As the student walks up the staircase, the friction between their feet and the carpet causes electrons to be rubbed off one and deposited on the other. This causes the student to become charged. There is then a large pd between the student and earth. The metal hand-rail is an earthed conductor, so a spark jumps between it and the student as the student goes to touch it. This causes a small shock.

2. Electric Fields
Page 104 — Fact Recall Questions
Q1 Away from positive charge and towards negative charge.
Q2

Q3 An object with a large static charge has a large potential difference between it and the earth. A large potential difference means a strong electric field. The strong electric field ionises particles in the air. This makes the air conductive, and allows charge to flow in the form of a spark.

Page 104 — Application Question
Q1 a) A and B will move away from each other as they have like charge and so will be repelled by one another.
   b) As the particles move further apart, the repulsive force between them will get weaker, as the electric fields gets weaker with increasing distance.

Page 105 — Static Electricity and Electric Fields Exam-style Questions
1.1 The paint droplets are positively charged because they have **lost** electrons. The force between one paint droplet and another is **repulsive**. This means the droplets will **spread out**, creating an even layer of paint on the car. *(1 mark for each correct word)*
1.2 The force felt between the positively charged paint droplets and the negatively charged car body is attractive, so the droplets are attracted to the car body *(1 mark)*. If the paint droplets and car body were uncharged, there would be no electrostatic force between the droplets and the car, so less paint would reach the car body resulting in more waste *(1 mark)*.
1.3 The car has a large static charge and so there is a large potential difference between it and the earth *(1 mark)*. This causes a strong electric field *(1 mark)*. The electric field ionises air particles near the car, so that the air can conduct electricity *(1 mark)*. The engineer is an earthed conductor *(1 mark)*, so a spark jumps across the gap from the car to the engineer *(1 mark)*.
2.1 When the scarf and balloon are rubbed against each other, electrons are removed from the balloon and left on the scarf, giving the scarf a negative charge *(1 mark)*.
2.2 The balloon will move away from the rod *(1 mark)* because the balloon and rod are both positively charged and repel each other *(1 mark)*.
2.3

(1 mark for all field lines pointing outwards from the sphere)

Topic 3 — Particle Model of Matter

1. Density and States of Matter
Page 109 — Fact Recall Questions
Q1 density = mass ÷ volume ($\rho = m \div v$)
Q2 Solid, liquid and gas.
Q3 The particles in a solid are held close together by strong forces in a fixed, regular arrangement. The particles don't have much energy and so can't move around — they can only vibrate about fixed positions.
Q4 The particles in a liquid have more energy than in a solid. They are still close to each other, but unlike in a solid, they're able to move past each other and form irregular arrangements.
Q5 Gas
Q6 Eureka can, measuring cylinder and mass balance.

Page 109 — Application Questions

Q1 The volume of the block is:
$0.030 \times 0.045 \times 0.060 - 0.000001 \text{ m}^3$
$\rho = m \div v = 0.324 \div 0.000081 = \textbf{4000 kg/m}^3$
Don't forget to convert the lengths into metres, or convert the mass to g if you want to find the answer in g/cm³.

Q2 a) E.g. The balls represent the particles of a solid — they don't swap positions, they're close together and are arranged in a (fairly) regular pattern.
b) A gas.
c) The particles in a solid are packed closely together without much of a gap between them, like the balls in the box when the fan is switched off. So solids are generally quite dense (relative to the other states of matter). In a gas, the particles are spread much further apart, with large gaps between them, like the balls in the box when the fan is switched on. This means that gases are much less dense than solids.

2. Internal Energy and Changes of State
Page 112 — Fact Recall Questions

Q1 The internal energy of a system is the total energy that its particles have in their kinetic and potential energy stores.

Q2 By heating the system.

Q3 E.g. freezing, melting, boiling/evaporating, condensing and sublimating.

Q4 A physical change means that you don't end up with a new substance — it's the same substance as you started with, just in a different form.

Q5 Yes, a change of state does conserve mass. The number of particles in a substance doesn't change when the substance changes state. Only the arrangement and energy of the particles changes.

Q6

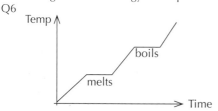

Page 112 — Application Questions

Q1 As the kettle heats the liquid, energy is transferred into the kinetic energy stores of the liquid's particles. The particles start to move around more. Eventually the particles have enough energy in their kinetic energy stores to be able to break the bonds between them. At this point the liquid starts to boil and become a gas.

Q2 When the steam reaches the cold window, it cools down. Energy is transferred away from the kinetic energy stores of the steam's particles. The particles move around less until bonds can form between them. At this point the steam condenses and becomes a liquid (water).

3. Specific Latent Heat
Page 113 — Fact Recall Questions

Q1 E.g. the specific latent heat of fusion is the amount of energy needed to change 1 kg of a solid to a liquid without changing its temperature.
The specific latent heat of vaporisation is the amount of energy needed to change 1 kg of a liquid to a gas without changing its temperature.

Q2 J/kg

Page 113 — Application Questions

Q1 $E = mL$ and $25.0 \text{ g} = 0.0250 \text{ kg}$
$E = 0.0250 \times 334\,000 = \textbf{8350 J}$

Q2 $E = mL$ so $m = E \div L = 4\,960\,000 \div 1\,550\,000$
$= \textbf{3.20 kg}$

4. Particle Motion in Gases
Page 116 — Fact Recall Questions

Q1 Random motion / random directions and speeds.

Q2 The particles move about at high speeds and collide with the walls of the container, exerting a force on them. Pressure is force per unit area, so the particles exert a pressure.

Q3 The higher the temperature of a gas, the higher the average energy in the kinetic stores of its particles (heating the gas will transfer energy into the kinetic stores of its particles). This means the particles will collide with the walls of a container more often and with greater speed, so they will exert a higher force. This means the pressure will be greater.

Q4 Inversely proportional

Q5 The pressure will decrease. This is because the particles will be more spread out, so there will be less frequent collisions with the walls of the container. This means the overall force exerted due to the collisions will decrease, and so will the pressure.

Q6 The internal energy increases.

Page 116 — Application Questions

Q1 Initially, $pV = p \times 75$
After compression, $pV = 110 \times 30 = 3300$
pV = constant, so:
$p \times 75 = 3300$
$p = 3300 \div 75 = \textbf{44 Pa}$

Q2 The volume of the balloon will decrease. When it is inside the refrigerator, the temperature of the helium inside the balloon will drop. Energy will be transferred away from the kinetic stores of the helium particles, and they will move more slowly. Less frequent collisions with the walls of the balloon means the gas pressure inside the balloon will decrease. When this pressure becomes less than the air pressure outside the balloon, the balloon will start to shrink until the pressures are balanced.

Pages 118-119 — Particle Model of Matter
Exam-style Questions

1.1 Density = mass ÷ volume / $\rho = m \div V$ *(1 mark)*

1.2 $V = m \div \rho$ *(1 mark)*
$= 450 \div 9$ *(1 mark)* $= $ **50 cm³** *(1 mark)*

1.3 It is a liquid *(1 mark)*. It started off as a solid, but the flat spot on the graph shows the point at which it changes state and melts into a liquid *(1 mark)*.
If there was a second flat spot on the graph, this would show another change of state, so the substance would've changed from liquid into gas.

1.4 The mass is 450 g *(1 mark)*. This is because a change of state conserves mass (the number of particles doesn't change, they're just arranged differently) *(1 mark)*.

2.1 $\rho = m \div V$
So $m = \rho V$ *(1 mark)*
$= 0.025 \times 0.08$ *(1 mark)*
$= $ **0.002 kg** *(1 mark)*

2.2 $E = mL$
So $L = E \div m$ *(1 mark)*
$= 5000 \div 0.002$ *(1 mark)*
$= $ **2 500 000** *(1 mark)* **J/kg** *(1 mark)*
If you got the mass wrong in part 2.1, you'll still get the marks in part 2.2 as long as you've done everything else correctly.

2.3 The density will be greater than 0.025 kg/m³ *(1 mark)*. Liquids are denser than gases, as their particles are more tightly packed together *(1 mark)*.

3.1 E.g. Place a measuring cylinder on a balance and zero the balance *(1 mark)*. Pour some liquid into the measuring cylinder and record the liquid's volume using the scale on the measuring cylinder *(1 mark)*. Record the mass of the liquid as shown on the balance *(1 mark)*. Use the formula $\rho = m \div v$ to find the density *(1 mark)*.

3.2 85 °C is the boiling point of the liquid *(1 mark)*. At this point, any energy transferred will go into breaking the bonds between particles, rather than raising temperature *(1 mark)*. The liquid will then change state, as it boils into a gas *(1 mark)*.

4.1 $E = mL$
So $m = E \div L$ *(1 mark)*
$= 67\ 500 \div 450\ 000$ *(1 mark)*
$= $ **0.15 kg** *(1 mark)*

4.2 If the solid was not already at its melting temperature, then some of the 67 500 J transferred would've gone into increasing the solid's temperature *(1 mark)*, rather than all of it being used to change the state *(1 mark)*.

5.1 Initially: $pV = 150 \times V$
After compression: $pV = 220 \times 30 = 6600$ *(1 mark)*
$pV = $ constant, so:
$150 \times V = 6600$ *(1 mark)*
$V = 6600 \div 150 = $ **44 cm³** *(1 mark)*

5.2 When the volume decreases, the particles have less distance to travel between the walls, so they collide with the walls of the container more often *(1 mark)*. This means the total force exerted due to collisions increases, and so the pressure increases *(1 mark)*.

5.3 Density before $= m \div V$
$= 0.132 \div 44$ *(1 mark)*
$= $ 0.003 g/cm³ *(1 mark)*
You'll still get the marks for using an incorrect value of V from part 5.1, as long as you do everything else right.
Density after $= m \div V$
$= 0.132 \div 30 = 0.0044$ g/cm³ *(1 mark)*
So the difference in density
$= 0.0044 - 0.003 = $ **0.0014 g/cm³** *(1 mark)*

6.1 Pumping air into the mattress means that the number of air particles inside the mattress will increase *(1 mark)*. This means there will be more frequent collisions between the air particles and the inside walls of the mattress — i.e. there will be an increase in gas pressure *(1 mark)*. As the gas pressure becomes greater than atmospheric pressure, the mattress/volume will expand, so that the pressure inside it remains at the same level as atmospheric pressure *(1 mark)*.

6.2 Pressure is applied to the plunger of the foot pump by the gas inside the mattress, and so a force is exerted on it *(1 mark)*. Work is done against this force *(1 mark)*, which transfers energy to the kinetic energy stores of the air particles, increasing the temperature *(1 mark)*.

Topic 4 — Atomic Structure

1. The History of the Atom
Page 121 — Fact Recall Questions

Q1 According to the model, an atom is made of a positively-charged sphere with tiny negative electrons stuck in it (like plums in a plum pudding).

Q2 It was expected that alpha particles fired at thin gold foil would be deflected at most by a very small amount by the electrons in the atoms.

Q3 Most alpha particles passed straight through and the odd one bounced straight back. The fact that most alpha particles passed straight through the foil showed that most of the atom is empty space. Some positively-charged alpha particles were deflected by the nucleus by a large angle, showing that the nucleus had a large positive charge. Very few alpha particles bounced back, showing that the nucleus is very small.

2. The Structure of the Atom
Page 125 — Fact Recall Questions

Q1 Protons, neutrons and electrons.

Q2 The electron.

Q3 The relative mass is 1 and the relative charge is 0.

Q4 They are equal. The atom has no overall charge (it's neutral), so the charges must be equal (and opposite) to cancel each other out.

Q5 Atomic number is the number of protons in the nucleus of an atom. Mass number is the number of protons and neutrons in the nucleus of an atom.

Q6 The atomic number is always the same (they have the same number of protons).

Q7 Another form of the same element which has atoms with the same number of protons (atomic number) but a different number of neutrons (mass number).

Page 125 — Application Question
Q1 a) $17 - 16 = \textbf{+1}$
Remember, electrons have a negative charge of −1 and protons have a positive charge of +1.
b) The particle has an overall charge/the particle has more protons than electrons.
c) They have the same number of protons but a different number of neutrons.

3. Radioactivity
Page 129 — Fact Recall Questions
Q1 It gives out radiation from the nuclei of its atoms no matter what is done to it.
Q2 You can't — it is random.
Sorry — that was sort of a trick question...
Q3 From strongest (most ionising) to weakest (least ionising): Alpha, beta, gamma.
Q4 Alpha radiation is made up of alpha particles, which are made up of two protons and two neutrons (a helium nucleus).
Beta radiation is made up of beta particles, which are electrons.
Gamma radiation is made up of gamma rays, which are electromagnetic waves with a very short wavelength.
Q5 The atomic number and mass number.

Page 129 — Application Questions
Q1 a) Radiation A, as it passes through the hand and radiation B doesn't.
b) Radiation B, as it doesn't even penetrate through the hand.
c) Gamma radiation, as it penetrates through all of the materials.
Q2 Alpha decay.
Q3 a) Beta decay
b) $^{228}_{88}\text{Ra} \rightarrow\ ^{228}_{89}\text{Ac} +\ ^{0}_{-1}\text{e}$
Remember you need to balance the atomic and mass numbers on each side of the equation.
Q4 a) Alpha particles would be stopped by the paper.
b) Gamma rays would pass straight through the paper, whatever its thickness.

4. Activity and Half-life
Page 132 — Fact Recall Questions
Q1 Because the activity never drops to zero.
Q2 Half-life is the time it takes for the number of nuclei of a radioactive isotope in a sample to halve.
Half-life is the time it takes for the count rate from a sample containing the isotope to fall to half its initial level.

Page 133 — Application Questions
Q1 After 1 hour (60 minutes), it will have had 4 half-lives because $60 \div 15 = 4$.
So it will have halved four times.
After one half-life: $240 \div 2 = 120$
After two half-lives: $120 \div 2 = 60$
After three half-lives: $60 \div 2 = 30$
After four half-lives: $30 \div 2 = 15$
So the count rate will be **15 cpm**.
Q2 Initial count rate = 16
After one half-life: $16 \div 2 = 8$
After two half-lives: $8 \div 2 = 4$
So 2 hours is 2 half-lives. So one half-life is **1 hour**.
Q3 Initial activity = 32 Bq
After one half-life: $32 \div 2 = 16$
After two half-lives: $16 \div 2 = 8$
After three half-lives: $8 \div 2 = 4$
Therefore, it will have dropped to 4 Bq after **3 half-lives**.
Q4 The initial count rate is 120, so after one half-life it will have decreased to 60. Find 60 on the activity axis and follow across to the curve and then down to the time axis.

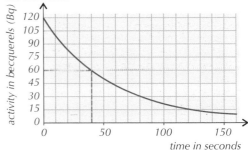

The half-life is **40 seconds**.
Q5 Initial activity = 9600 Bq
After one half-life, activity = $9600 \div 2 = 4800$
After two half-lives, activity = $4800 \div 2 = 2400$
After three half-lives, activity = $2400 \div 2 = 1200$
So the percentage = $(1200 \div 9600) \times 100 = \textbf{12.5\%}$

5. Irradiation and Contamination
Page 134 — Fact Recall Questions
Q1 Irradiation is exposure to radiation from a radioactive source.
Contamination is where unwanted atoms of a radioactive substance are present in or on another material.
Q2 Beta and gamma radiation are the most dangerous outside the body because they can penetrate into your body.
Q3 Alpha radiation is the most dangerous inside the body because it does damage in a localised area and is highly ionising.

6. Background Radiation
Page 136 — Fact Recall Questions
Q1 Naturally occurring unstable isotopes (e.g. in the air, food, building materials and rocks), radiation from space (cosmic rays), and radiation from man-made sources (e.g. nuclear weapons, waste and disasters).

Q2 Certain rocks can cause high levels of radioactive gases at the surface which can get trapped in homes.

Page 136 — Application Question
Q1 a) Cosmic rays.
b) The astronauts are exposed to a higher dose of radiation on board the ISS than in other occupations. Radiation can be damaging to health, so regularly changing the crew minimises their radiation dose and so minimises the possible health risks.

7. Risks and Uses of Radiation
Page 139 — Fact Recall Questions
Q1 Lower doses of radiation can damage cells by ionisation (without killing them), causing them to mutate so that they divide uncontrollably. This is cancer.
Q2 High doses of radiation can kill cells completely, by damaging them so much that they can't repair themselves.
Q3 A radioactive isotope is injected or swallowed and its progress around the body can be followed using an external radiation detector. A computer can produce an image from the detected radiation showing where the radiation is most concentrated. Gamma or beta emitters can be used as they can be detected outside the body.
Q4 Radiotherapy is the killing of cancer cells using high doses of radiation.

8. Nuclear Fission and Fusion
Page 141 — Fact Recall Questions
Q1 Nuclear fission is when an atomic nucleus splits up to form two smaller nuclei.
Q2 It must absorb a (slow-moving) neutron.
If absorbing a neutron makes the nucleus unstable enough, the nucleus will split.
Q3 When a nucleus splits, two or three neutrons are released. These neutrons may be absorbed by other nuclei, causing them to split, which will release more neutrons that can go on to cause further nuclei to split... and so the process goes on. This is a chain reaction.
Q4 Nuclear fusion is when two nuclei (e.g. hydrogen nuclei) join to create a larger nucleus.

Pages 144-146 — Atomic Structure Exam-style Questions
1.1 The nucleus contains <u>protons</u> and <u>neutrons</u>.
OR
The nucleus contains <u>neutrons</u> and <u>protons</u>.
The numbers of protons and <u>electrons</u> in a neutral atom are equal.
(2 marks for all correct otherwise 1 mark for one correct answer.)
1.2 The atoms have the same number of protons (atomic number) but a different number of neutrons (mass numbers) **(1 mark)**.

2.1

Radiation type:	Made up of:	Stopped by:
Alpha particles	2 protons and 2 neutrons **(1 mark)** OR helium nuclei **(1 mark)**	Thin paper
Beta particles **(1 mark)**	Electrons	Thin aluminium
Gamma rays	Short-wavelength EM waves	Thick lead

2.2 Ionising means it is capable of removing an electron from an atom and turning it into a positive ion. **(1 mark)**
2.3 Alpha radiation is the most highly ionising radiation and so if emitted inside the body it can badly damage cells in a localised area **(1 mark)**.
2.4 Its atomic number decreases by 2 **(1 mark)** and its mass number decreases by 4 **(1 mark)**.
It loses 2 protons and 2 neutrons.
2.5 A neutron changes into a proton **(1 mark)**.
3.1 The average time it takes for the number of nuclei in a caesium-137 sample to halve is 30 years **(1 mark)**.
OR
The time it takes for the count rate from a sample of caesium-137 to fall to half its initial level is 30 years **(1 mark)**.
3.2 90 years = 3 half-lives **(1 mark)**
Activity after 30 years = 24 ÷ 2 = 12
Activity after 60 years = 12 ÷ 2 = 6
Activity after 90 years = 6 ÷ 2 **(1 mark)** = **3 Bq** **(1 mark)**
3.3 E.g. any two of: cosmic rays (accept radiation from space) / naturally occurring isotopes in food / naturally occurring isotopes in building materials / naturally occurring isotopes in the air / radon gas (accept gases from rocks) / radiation used in medicine.
(Maximum 2 marks — 1 mark for each correct source.)
3.4 Some isotopes have very long half-lives (e.g. for caesium-137 it's 30 years) so it will take a long time for the radiation levels in affected areas to decrease to a safe level **(1 mark)**.
3.5 Beta **(1 mark)**.
3.6 $^{137}_{55}\text{Cs} \longrightarrow {}^{137}_{56}\text{Ba} + {}^{0}_{-1}\text{e}$ **(1 mark)**
4 How to grade your answer:
Level 0: There is no relevant information. **(0 marks)**
Level 1: There is a brief explanation of how the method would detect cracks with no explanation of the type of radiation used. **(1 to 2 marks)**
Level 2: There is an explanation of how the method would detect cracks and some explanation of the type of radiation or of an appropriate half-life. **(3 to 4 marks)**
Level 3: There is a clear and detailed explanation of how the method would detect cracks and the radiation type and half-life are fully explained. **(5 to 6 marks)**

Here are some points your answer may include:
The radioactive isotope will give out radiation.
The detector will detect how much radiation is
reaching it as it moves along above the pipe.
If there is a crack, the substance will leak out and
collect outside the pipe.
This means there will be a higher concentration of the
radioactive isotope around a crack.
This will be detected as a higher count rate by the
detector.
So the engineer will know where the crack is located
— it'll be directly below the detector when the
reading increases.
The source used should be a gamma source.
Gamma radiation penetrates far into materials without
being stopped and so will pass through the ground
and reach the detector.
Alpha and beta radiation would be stopped by the
ground as they have lower penetration levels, so they
wouldn't reach the detector.
The source should have a short half-life.
This will ensure it does not continue emitting lots of
radiation for a long time, which would possibly harm
people that come close to the pipes or the substance.

5.1 Ionising radiation enters living cells and collides with
molecules. This can knock electrons off the atoms
and molecules in the cell, damaging them *(1 mark)*.
A high enough dose of radiation will cause the cell
to be destroyed completely *(1 mark)*. So if a high
enough dose of radiation is directed at some cancer
cell, these cancer cells will be killed off *(1 mark)*.

5.2 The radiation will also kill some healthy cells
(1 mark).

5.3 The radioactive isotope could escape the casing and
travel to other parts of the body (or out of the body)
(1 mark).

5.4 Alpha radiation could be absorbed by the implant's
casing so no radiation would reach the tumour to
provide treatment *(1 mark)*.

6.1 To make sure that the radiation dose they are getting
at work is not too high as high doses are more likely
to damage or kill body cells *(1 mark)*.

6.2 E.g. any two of: limit exposure time / wear protective
clothing / wear breathing apparatus.
(2 marks — 1 mark for each correct answer)

7.1 (nuclear) fission *(1 mark)*

7.2 a (slow-moving) neutron *(1 mark)*

7.3 E.g.

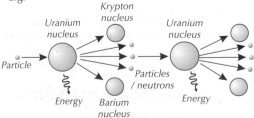

*(1 mark for showing small particles/neutrons being
produced when the uranium atom splits, 1 mark for
showing these particles going on to cause at least
one more fission.)*

Topic 5 — Forces

Topic 5a — Force Basics

1. Contact and Non-Contact Forces

Page 148 — Fact Recall Questions

Q1 A scalar quantity only has magnitude, but a vector
quantity has magnitude and direction.

Q2 a) Vector
b) Vector
c) Scalar
d) Vector
e) Scalar

Q3 A contact force is a force between two objects due to
the objects touching.

Q4 A non-contact force is a force between two objects
which exists when the objects are not touching.

Q5 An interaction pair is a pair of forces that act on
two interacting objects. The forces are equal and
opposite.

Page 148 — Application Question

Q1

The arrow should be longer than the original arrow and be
pointing in the opposite direction.

2. Weight, Mass and Gravity

Page 150 — Fact Recall Questions

Q1 Weight is the force acting on an object due to the
gravitational field strength at the object's location.

Q2 The point at which the weight of the object appears to
act.

Q3 A calibrated spring balance (or newton meter).

Page 150 — Application Question

Q1 a) $W = mg = 15 \times 9.8 =$ **147 N**
b) Its mass would stay the same.

3. Resultant Forces

Page 154 — Fact Recall Questions

Q1 When an object has more than one force acting on it,
the resultant force is a single force that has the same
effect as the original forces acting altogether.

Q2 A free body diagram shows all the forces acting on an
object, including the magnitude and direction of each
force.

Q3 An object is in equilibrium if drawing all the forces
tip-to-tail gives a complete loop.

Q4 Resolving a force means splitting the force into two
component forces that are at right angles to each
other. The combination of the two forces has the
same effect as the original single force.

Page 154 — Application Questions

Q1 If the forward direction is the positive direction, then the resultant force = 87 − 24 = 63 N
So the resultant force is **63 N forwards**.

Q2 The boat is not in equilibrium as drawing all the forces to scale and tip-to-tail doesn't create a complete loop:

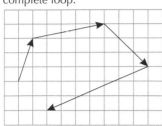

Q3 E.g.

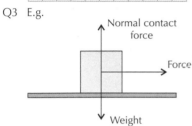

The weight and normal contact arrows should be vertical and pointing in opposite directions to each other. They should also be the same length. The force arrow should be horizontal. It could point in either direction and be any length, as you're not given its exact direction or magnitude.

Q4 First, resolve the 20 N force:

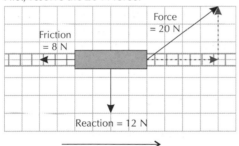

The frictional force that has a magnitude of 8 N is 2 squares long, so 1 square = 4 N.
The component of the pulling force in the direction of motion is 4 squares long, so has a magnitude of 4 × 4 = 16 N.
Resultant force in direction of motion = 16 − 8 = **8 N**.
You don't need to worry about the 12 N reaction force as it's at a right angle to the direction of the motion and doesn't contribute to the resultant force you're looking for.

4. Work Done

Page 156 — Fact Recall Questions

Q1 Work done = force × distance ($W = Fs$)
Work done is the energy transferred when a force acts on an object, causing it to move a certain distance.

Q2 1 joule is equal to 1 newton metre.

Q3 Doing work against friction cause. transferred to thermal energy stores. rise in temperature.

Page 156 — Application Questions

Q1 $W = Fs = 24 × 14 =$ **336 J**
Q2 a) $W = Fs = 250 × 20 =$ **5000 J**
   b) Rearrange equation:
   $s = W ÷ F = 750 ÷ 250 =$ **3 m**

Page 158 — Force Basics

Exam-style Questions

1.1 $W = mg$
   $= 1.8 × 9.8$ *(1 mark)*
   $= 17.64$
   $=$ **18 N (to 2 s.f.)** *(1 mark)*

1.2 Magnitude = 5 − 1.4 = **3.6 N** *(1 mark)*

1.3 $W = Fs$
   So $s = W ÷ F$ *(1 mark)* $= 19.8 ÷ 3.6$ *(1 mark)*
   $=$ **5.5 m** *(1 mark)*
   You're given the work done by the resultant force, so make sure you use the value you worked out in 1.2, not the 5 N or 1.4 N force.

2.1 Contact force *(1 mark)*

2.2 Resolve the forces of both horses:

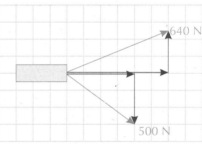

Find the scale of the diagram:
500 N = 2.5 cm, so 1 cm = 200 N
Measure the length of each component that is in the direction of the boat's motion (East):
Horse 1: East component = 3 cm
   3 × 200 N = 600 N
Horse 2: East component = 2 cm
   2 × 200 N = 400 N
So the total force provided by the horses in the direction of motion = 600 + 400 = **1000 N**
(4 marks for correct answer, otherwise 1 mark for resolving the forces, 1 mark for working out the scale and 1 mark for finding the magnitude of the force provided by each horse in the direction of the boat's motion)

2.3 Drag force = 1000 − 800 = 200 N *(1 mark)*

Topic 5b — Forces and their Effects

1. Forces and Elasticity

Page 162 — Fact Recall Questions

Q1 If only one force is applied to an object, then the object will just move in the direction of the force.

An elastic object will return to its original shape after all forces have been removed from it.

Q3 force (N) = spring constant (N/m) × extension (m)
(or $F = ke$)

Q4 The limit of proportionality is the point beyond which the extension of an object will no longer be proportional to the force applied to the object.

Q5 All the energy is released, because all the energy transferred ends up in the spring's elastic potential energy store, and all of the energy is transferred out of that store when the spring is released.

Page 162 — Application Questions

Q1 Extension = 0.75 − 0.50 = 0.25 m
$F = ke = 34 \times 0.25 = \textbf{8.5 N}$

Q2 Point B.
The limit of proportionality is always the point at which a force-extension graph starts to curve.

Q3 a) Compression = 16 − 12 = 4 cm
In metres, this is 4 ÷ 100 = 0.04 m
Rearrange $F = ke$ for k:
$k = F \div e = 0.80 \div 0.04 = \textbf{20 N/m}$
b) $E_e = \frac{1}{2}ke^2 = \frac{1}{2} \times 20 \times 0.04^2 = \textbf{0.016 J}$
c) Spring B's spring constant is higher than that of spring A.
You can work this out from $F = ke$. If e is smaller, then k must be larger to keep F the same.

2. Investigating Springs
Page 164 — Fact Recall Questions

Q1 A pilot experiment allows you to find the size of masses that should be used in order to record enough measurements to plot a straight line force-extension graph (the part of the graph up until the limit of proportionality is reached).

Q2 E.g. Safety goggles should be worn to protect the eyes if the spring snaps.

Q3 The gradient of the force-extension graph is equal to the spring constant of the spring.

3. Moments
Page 168 — Fact Recall Questions

Q1 The moment is the turning effect of the force.

Q2 moment of a force (Nm)
= force (N) × perpendicular distance (m)
(or $M = Fd$)

Q3 The sum of all clockwise moments acting on the object must be equal to the sum of all anticlockwise moments acting on the object.

Q4 When a small force is applied at a large distance from the pivot, there will be an equal moment created on the other side of the pivot. Because moment = force × distance, if the distance is small it will mean a large force.

Q5 A gear is a circular disc with teeth. In machinery, it is interlocked with other gears in order to transmit the rotational effects of a force.

Page 168 — Application Questions

Q1 B, D and E.

Q2 $M = Fd = 60.0 \times 2.70 = \textbf{162 Nm}$

Q3 Calculate the total moments clockwise:
$M = Fd = 420 \times 0.80 = 336$ Nm
Calculate the total moments anticlockwise:
$M = Fd = 350 \times 1.40 = 490$ Nm
The total moment anticlockwise is larger than the total moment clockwise, and the difference is:
490 − 336 = 154 Nm
So you need an extra 154 Nm clockwise moment provided by the third child. Rearrange $M = Fd$ for d, to find the distance from the pivot that the third child needs to sit:
$d = M \div F = 154 \div 308 = 0.5$ m
So the child should sit **0.5 m to the right of the pivot**.
Make sure you include which side of the pivot the child should sit in your answer. In this case, it's to the right of the pivot, because the moment needs to be clockwise for the seesaw to balance.

Q4 Any two from: the force could be moved further from the pivot / the force could be applied at a right angle to the lever / the pivot could be moved closer to the load.

4. Fluid Pressure
Page 171 — Fact Recall Questions

Q1 Pressure is force per unit area.

Q2 A fluid is a liquid or a gas.

Q3 The force due to pressure acts normal (at 90°) to the surface of the fluid.

Q4 pressure (Pa) = force (N) ÷ area (m²) (or $p = F \div A$)

Q5 The pressure of a liquid increases with increasing depth. This is because the weight of the liquid above adds to the pressure, and the greater the depth, the more liquid there is above.

Q6 The atmosphere is the (thin) layer of air that surrounds the Earth.

Q7 Atmospheric pressure decreases with increasing altitude. This is because the density of the air decreases, and the lower the density, the lower the pressure. There are also fewer molecules above a point at a higher altitude, so the weight of the air above that point is less, which in turn means the pressure is lower.

Page 171 — Application Questions

Q1 Area needs to be converted from cm² to m²:
320.0 ÷ 10 000 = 0.03200 m²
Rearrange $p = \frac{F}{A}$ for force:
$F = p \times A = 101\,000 \times 0.03200 = \textbf{3232 N}$

Q2 $p = h\rho g = 5.0 \times 1020 \times 9.8 = \textbf{49 980 Pa}$

5. Upthrust

Page 173 — Fact Recall Questions

Q1 When the object is partially (or completely) submerged in the liquid, there will be forces acting on the object from all sides due to pressure in the liquid. The pressure from the liquid on the bottom of the object is greater than the pressure at the top of the object. Upthrust is the resultant upwards force from this.

Q2 Float.

Q3 Sink.

Q4 The object will never be able to displace an amount of water that is equal to its weight, so the object's weight will always be greater than the upthrust, and so it sinks.

Page 173 — Application Question

Q1 a) The density of the object is higher than that of the water, so the object will sink.

b) The object has the same density as before, so will still sink.

The object has a uniform density, so changing its volume won't affect the density. The volume of the object is not important, it's just the density that affects whether it will sink or float.

Pages 176-177 — Forces and their Effects Exam-style Questions

1.1 When a force of 0.70 N was applied to the spring, the limit of proportionality of the spring may have been passed, and so the graph would have started to curve, and the data point would be in the right position *(1 mark)*.
E.g. The student could increase the force acting on the spring to see whether the data points continue to follow the straight line of best fit, or whether the line of best fit should start to curve *(1 mark)*.

1.2 If the spring has been stretched too far *(1 mark)*, it will no longer deform elastically, and so it could not be used again for the same experiment *(1 mark)*.

1.3 Spring constant = gradient of line of best fit *(1 mark)*

$$\text{gradient} = \frac{\text{change in } y}{\text{change in } x}$$

E.g. spring constant $= \frac{0.80 - 0}{0.300 - 0}$ *(1 mark)*

$= 2.66...$

$= \textbf{2.7 N/m (to 2 s.f.)}$ *(1 mark)*

Don't forget to convert the extension from cm to m. And remember to use as large a range of the line as possible when you're calculating the gradient.

2.1 The weight of the mass is the force that is acting on the spring, so use the equation $F = ke$.
Calculate the compression first: $e = 165 - 140$
$= 25$ mm *(1 mark)*
Convert the compression to m: $25 \div 1000 = 0.025$ m
Then substitute the values into the equation for force:
$F = ke = 160 \times 0.025$ *(1 mark)*
$= \textbf{4 N}$ *(1 mark)*

2.2 $E = \frac{1}{2}ke^2 = \frac{1}{2} \times 160 \times 0.025^2$ *(1 mark)*
$= \textbf{0.05 J}$ *(1 mark)*

3.1 $M = Fd = 32.0 \times 3.0$ *(1 mark)*
$= \textbf{96 Nm}$ *(1 mark)*

3.2 Calculate the moments of the other two forces:
$M = Fd = 18.0 \times (3.5 + 1.5)$ *(1 mark)*
$= 90$ Nm *(1 mark)*
$M = Fd = 4.0 \times 1.5$ *(1 mark)* $= 6$ Nm *(1 mark)*
So total moments anticlockwise $= 90 + 6 = 96$ Nm
Total moments clockwise (from 3.1) $= 96$ Nm
The total moment clockwise is equal to the total moment anticlockwise, so the plank of wood is balanced *(1 mark)*
The plank's centre of mass is directly above the pivot, so the weight of the plank doesn't have any turning effect and you don't need to worry about it in your calculations.

3.3 The total moments clockwise would decrease if the force labelled X was applied at any other angle *(1 mark)*, so the clockwise and anticlockwise moments would be unbalanced, and so the plank would turn anticlockwise *(1 mark)*.

4.1 The block in the water will float *(1 mark)*.
The block in the oil will sink *(1 mark)*.
The block will float in the water because its density is lower than the water, but it will sink in the oil because its density is higher than the oil.

4.2 The block will float, so the upthrust will be equal to the block's weight *(1 mark)*. So upthrust = 0.58 N *(1 mark)*
The block will float in the saltwater as the density of the saltwater is greater than the density of the block.

4.3 $p = \frac{F}{A}$
So $F = p \times A$ *(1 mark)*
$= 101\,000 \times 0.00800$ *(1 mark)*
$= \textbf{808 N}$ *(1 mark)*

Topic 5c — Forces and Motion

1. Distance, Displacement, Speed and Velocity

Page 180 — Fact Recall Questions

Q1 a) vector
b) scalar
c) vector
d) scalar

Q2 If an object is travelling at a constant speed but is changing direction, then its velocity is changing.
Remember — speed only has a magnitude, but velocity has magnitude and direction.

Q3 a) 6 m/s
b) 3 m/s
c) 1.5 m/s

Q4 distance travelled (m) = speed (m/s) × time (s)
(or $s = vt$)

Page 180 — Application Questions

Q1 a) Distance = 16 + 25 + 16 = **57 m**
b) The car has a displacement of 0 when it has travelled 16 m east and 16 m west, as it is travelling back on itself.
So displacement = **25 m south**.
Make sure you give both the magnitude and direction in your answer.

Q2 Typical walking speed = 1.5 m/s.
$s = vt = 1.5 \times 18 = \textbf{27 m}$

Q3 a) Use the train's speed between stations A and B and the time taken to travel between them.
$s = vt = 45 \times 120 = \textbf{5400 m}$

b) Use the speed the train is moving at between stations A and B, and the distance between them.
Change the distance into m:
$16.8 \times 1000 = 16\ 800$ m
Then rearrange the distance equation:
$s = vt$, so $t = s \div v = 16\ 800 \div 60 = \textbf{280 s}$

c) Total distance travelled = 5400 + 16 800 = 22 200 m
Total time taken to travel this distance = 120 + 280
= 400 s
$s = vt$, so $v = s \div t = 22\ 200 \div 400 = \textbf{55.5 m/s}$

2. Acceleration

Page 182 — Fact Recall Questions
Q1 Acceleration is a measure of the rate of change of velocity.
Q2 m/s^2

Page 182 — Application Questions
Q1 Δv = final velocity – initial velocity
= 10 – 25 = –15 m/s
$a = \dfrac{\Delta v}{t} = -15 \div 5 = \textbf{–3 m/s}^2$

Q2 $a = \dfrac{\Delta v}{t}$, so $\Delta v = at = 4 \times 5 = 20$ m/s
The cheetah is initially at rest, so its change in velocity is equal to its final velocity, $\Delta v = v$.
So $v = \textbf{20 m/s}$

Q3 Rearrange $v^2 - u^2 = 2as$ for acceleration:
$a = (v^2 - u^2) \div 2s = (7.1^2 - 5.6^2) \div (2 \times 38.1)$
$= \textbf{0.25 m/s}^2$

Q4 Rearrange $v^2 - u^2 = 2as$ for distance (height):
$s = (v^2 - u^2) \div 2a = (6.0^2 - 0^2) \div (2 \times 9.8)$
= 1.83...
= **1.8 m (to 2 s.f.)**
u = 0 m/s because the apple is at rest just before it falls.

3. Distance-Time Graphs

Page 185 — Fact Recall Questions
Q1 The speed of the object.
Q2 It tells you that the object is stationary.
Q3 It represents acceleration (or deceleration).
Q4 Draw a tangent to the curve at the given time, and then calculate the gradient of the tangent.

Page 185 — Application Questions
Q1 *A* — It's increasing (the object is speeding up).
B — It's not changing (the object is moving at a steady speed).
C — It's decreasing (the object is slowing down).
D — It's zero (the object is not moving).
Q2 For the first 10 seconds the graph will be curved with an increasing gradient. For the next 5 seconds it will be a sloped straight line.

Q3 a)
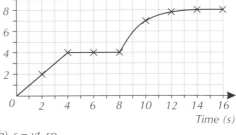

b) $s = vt$, so
$v = s \div t$
$= 4 \div 4 = \textbf{1 m/s}$

c) Speed at 10 s is equal to the gradient of the tangent to the curve at that point.
E.g.
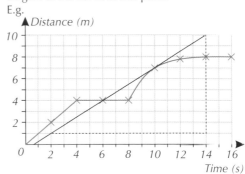
change in *y* = 10 – 1 = 9
change in *x* = 14 – 2 = 12
gradient = 9 ÷ 12 = 0.75
So, speed at 10 s = **0.75 m/s**
Tangents are difficult to draw accurately. Your answer may be slightly different to this one, but as long as it's close enough to this answer, you should be fine.

4. Velocity-Time Graphs

Page 188 — Fact Recall Questions
Q1 The acceleration (of the object).
Q2 The distance travelled (by the object).

Page 189 — Application Questions
Q1 *A* — It's not changing (the object has a steady acceleration).
B — It's zero (the object is moving at a steady speed).
C — It's increasing (the object is speeding up, but not at a steady rate).
D — It's not changing (the object has a steady deceleration).
Q2 a) 4 m/s
b) Between 100 and 120 seconds.
It's the part where the graph has a negative gradient (sloping downwards).
c) The acceleration is equal to the gradient, so:
$\text{gradient} = \dfrac{\text{change in the vertical}}{\text{change in the horizontal}} = \dfrac{8-4}{80-60}$
$= \textbf{0.2 m/s}^2$

d) Distance travelled between 0 and 20 s
   = ½ × base × height = ½ × 20 × 4 = 40 m
   Distance travelled between 20 and 60 s
   = base × height = (60 − 20) × 4 = 160 m
   So total distance travelled = 40 + 160 = **200 m**

Remember to break the area into separate shapes if you think it will make the calculation easier.

Q3
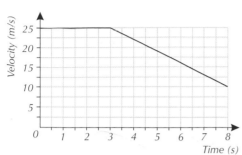

Acceleration = gradient = $\dfrac{\text{change in the vertical}}{\text{change in the horizontal}}$

$= \dfrac{10 - 25}{8 - 3} = $ **−3 m/s²**

The acceleration is negative as the car is decelerating.

Q4 Distance represented by one square
   = width of one square × height of one square
   = 2.5 s × 5 m/s = 12.5 m
   Number of squares under the graph between
   0 and 20 s ≈ 38
   Total distance travelled = 12.5 × 38 = **475 m**

5. Terminal Velocity

Page 193 — Fact Recall Questions

Q1 The resistive forces must be equal to the driving force.
Q2 Drag is the friction experienced by an object moving through a fluid.
Q3 The greater the object's speed, the greater the drag force.
Q4 As an object falls, it is acted on by a gravitational force which causes it to accelerate. It is also acted on by the resistive force of air resistance. As the object accelerates, the resistive force on it increases. Eventually, the resistive force acting upwards will equal the force due to gravity acting downwards. At this point the resultant force on the object becomes zero, so the object stops accelerating and reaches a constant velocity — the terminal velocity.

Q5
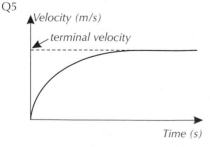

Page 193 — Application Questions

Q1 On Earth, air resistance causes objects to fall at different speeds depending on their shape and size — it has a greater effect on the feather than on the rock. On the Moon there's no air, so there's no air resistance and all objects accelerate due to gravity at the same rate.
Q2 A parachute increases the resistive forces acting by increasing the skydiver's area. Because the total resistive force is greater, drag will equal the skydiver's weight at a lower speed (i.e. the terminal velocity will be lower).

6. Newton's First Law

Page 195 — Fact Recall Question

Q1 a) Nothing will happen — it will remain stationary.
   b) The object will accelerate (start moving) in the direction of the resultant force.
   c) The object will keep moving at the same velocity.
   d) The object will accelerate in the direction of its motion.
   e) The object will decelerate.

Page 195 — Application Question

Q1 The resultant force is zero.

7. Newton's Second Law and Inertia

Page 198 — Fact Recall Questions

Q1 Newton's Second Law says that the resultant force acting on an object is directly proportional to the acceleration of the object, and that the acceleration of an object is inversely proportional to the mass of the object.
Q2 $F = ma$, where F = force in N, m = mass in kg and a = acceleration in m/s².
Q3 Inertia is the tendency of an object to remain at a constant velocity.
Q4 Inertial mass is the ratio of the resultant force on an object to its acceleration.

Page 198 — Application Questions

Q1 The car without the rock has a smaller mass, so it will win the race. Acceleration is inversely proportional to mass ($a = F \div m$), so a larger mass means a smaller acceleration for a given driving force.
Q2 $a = F \div m = 303 \div 1.5 = $ **202 m/s²**
Q3 No resistive forces, so driving force = resultant force.
   Resultant force acting on left-hand car:
   $F = ma = 1200 × 2.3 = 2760$ N
   Resultant force acting on right-hand car:
   $F = ma = 820 × 3.1 = 2542$ N
   The resultant force acting on the **left-hand car** is greater, so that car's engine is providing the greater driving force.
Q4 Typical speed of a car = 25 m/s
   Typical mass of a car = 1000 kg
   $\Delta v = 25 - 5 = 20$ m/s
   $a = \dfrac{\Delta v}{t} = 20 \div 20 = 1$ m/s²
   $F = ma = 1\,000 × 1 = $ **1000 N**

8. Investigating Motion

Page 200 — Fact Recall Question

Q1 a) It would decrease.
a = F ÷ m, so an increase in mass would give a decrease in acceleration.
b) It would decrease.
a = F ÷ m, so a decrease in force would give a decrease in acceleration.

9. Newton's Third Law

Page 202 — Fact Recall Question

Q1 When two objects interact, the forces they exert on each other are equal and opposite.

Page 202 — Application Question

Q1 No, as the two forces are not of the same type and they are both acting on the ball.

Pages 205-207 — Forces and Motion Exam-style Questions

1.1 Between 70 and 120 seconds *(1 mark)*
1.2 Acceleration = gradient *(1 mark)*
$$= \frac{\text{change in the vertical}}{\text{change in the horizontal}} = \frac{10-6}{40-30}$$ *(1 mark)*
$$= \textbf{0.4 m/s}^2 \textbf{ (1 mark)}$$
You could have also used the formula a = Δv ÷ t, using the graph to find Δv and t.
1.3 $F = ma$
$= 980 \times 0.4$ *(1 mark)*
$= \textbf{392 N}$ *(1 mark)*
1.4 The distance travelled between 30 and 40 s is equal to the area under the graph *(1 mark)*, which can be found by splitting the area up into a triangle and a rectangle:
Area = $(\frac{1}{2} \times b \times h) + (b \times h)$
$= (\frac{1}{2} \times 10 \times 4) + (10 \times 6)$ *(1 mark)*
$= \textbf{80 m}$ *(1 mark)*
1.5 The graph between 30 and 40 s should show an acceleration (i.e. a steepening upwards curve), but it is showing that the car is stationary *(1 mark)*. The graph between 40 and 70 s should have a steeper gradient than the graph between 10 and 30 s, as the velocity is greater for this range *(1 mark)*. The graph between 70 and 120 s should show a deceleration (i.e. a levelling off curve upwards), instead of the straight line which represents a constant speed *(1 mark)*.
The graph between 70 and 120 s is also going downhill, which isn't possible for a distance-time graph.
2.1 The acceleration of the trolley is due to the weight of the hook and the masses attached to it *(1 mark)*.
2.2 The force can be decreased by moving a mass from the hook to the trolley *(1 mark)*.
2.3 The gap makes sure that the card interrupts the light gate signal twice, which allows the light gate to measure the velocity of the trolley at two different times during its motion *(1 mark)*. This information can then be used to calculate the acceleration of the trolley *(1 mark)*.

2.4 Doubling the force will double the acceleration *(1 mark)*. This is because an object's acceleration is directly proportional to the resultant force acting on it ($F = ma$) *(1 mark)*.
3.1 0 N *(1 mark)*
If it's travelling at a steady speed in a straight line, the resultant force acting on it must be zero.
3.2 $s = vt$, so:
$t = s ÷ v$ *(1 mark)* $= 448 ÷ 28$ *(1 mark)*
$= \textbf{16 s}$ *(1 mark)*
3.3 Making the truck more streamlined reduces the air resistance acting on the truck *(1 mark)*. So the speed at which resistive forces match the driving force will be greater *(1 mark)*.
3.4 First calculate the acceleration of the truck.
$F = ma$, so:
$a = F ÷ m$ *(1 mark)* $= 35\ 000 ÷ 25\ 000$ *(1 mark)*
$= \textbf{1.4 m/s}^2$ *(1 mark)*
Then rearrange the equation $v^2 - u^2 = 2as$:
$s = (v^2 - u^2) ÷ 2a = (0^2 - 28^2) ÷ (2 \times -1.4)$ *(1 mark)*
$= \textbf{280 m}$ *(1 mark)*
It's a deceleration, so put a = −1.4 into the equation.
4.1 Before reaching terminal velocity, the force due to gravity is **greater than** the resistive force due to air resistance. *(1 mark)*
4.2 After reaching terminal velocity, the force due to gravity is **the same as** the resistive force due to air resistance. *(1 mark)*
4.3
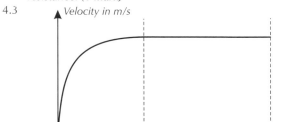
(3 marks available — 1 mark for a line with positive gradient between 0 and 15 seconds, 1 mark if the line is sloped with decreasing gradient and 1 mark for a horizontal line between 15 and 40 seconds.)
4.4 It will decrease *(1 mark)*
4.5 They will increase *(1 mark)*
5.1 The force is equal in size and in the opposite direction to F *(1 mark)*.
5.2 Change in velocity = 3 − 0 = 3 m/s
$a = \frac{\Delta v}{t} = 3 ÷ 1.2$ *(1 mark)*
$= \textbf{2.5 m/s}^2$ *(1 mark)*
5.3 $F = ma = 100 \times 2.5$ *(1 mark)*
$= \textbf{250 N}$ *(1 mark)*

Topic 5d — Car Safety and Momentum

1. Stopping Distances

Page 211 — Fact Recall Questions

Q1 The stopping distance is the distance covered by the vehicle in the time between the driver first spotting a hazard and the vehicle coming to a complete stop. It's the sum of the thinking distance and the braking distance.

The thinking distance is the distance the vehicle travels during the driver's reaction time.
The braking distance is the distance the vehicle travels after the brakes are applied until it comes to a complete stop.
Q2 Any three from: e.g. ice (or water, oil, leaves etc.) on the road, quality of the tyres, quality of the car's brakes, weather conditions.
Q3 a) Braking
b) Thinking
c) Thinking
d) Braking

Page 211 — Application Questions
Q1 Stopping distance = 15 + 38 = **53 m**
Q2 E.g. Her colleagues could talk to her and distract her, delaying when she spots the hazard. She could be tired from the day at work, increasing her reaction time and so her thinking distance. It may be icy on the roads, increasing her braking distance.
Q3 a) 60 mph is twice 30 mph.
Thinking distance is directly proportional to speed, so new thinking distance = 9×2 = **18 m**
b) Braking distance increases by the square of the scale factor of the speed increase, so new braking distance = 14×2^2 = **56 m**
c) Stopping distance at 60 mph = 18 + 56 = **74 m**

2. Graphs of Stopping Distance
Page 212 — Application Questions
Q1 Vehicle A stopping distance = 36 m
Vehicle B stopping distance = 26 m
36 – 26 = 10 m
The stopping distance of vehicle A is 10 m longer than that of vehicle B at 40 mph.

3. Reaction Times
Page 214 — Fact Recall Question
Q1 Sit with your arm resting on the edge of a table. Have someone hold a ruler end-down so that the 0 cm mark hangs between your thumb and forefinger. The ruler should be dropped without warning. Grab the ruler between your thumb and forefinger as quickly as possible. Measure the distance at which you have caught the ruler. Use $v^2 - u^2 = 2as$ and $a = \Delta v \div t$ to calculate the time taken for the ruler to fall that distance. This is your reaction time.

Page 214 — Application Question
Q1 a) Average distance = $(3.0 + 5.0 + 7.0) \div 3$
$= 15.0 \div 3$
$= \textbf{5.0 cm}$
b) $v^2 - u^2 = 2as$
$s = 5.0$ cm = 0.050 m
so $v = \sqrt{2as + u^2} = \sqrt{(2 \times 9.8 \times 0.050) + 0^2}$
$= 0.989...$ m/s
$a = \Delta v \div t$, so $t = \Delta v \div a = 0.989... \div 9.8$
$= 0.101...$
$= \textbf{0.10 s (to 2 s.f.)}$
Remember to convert the distance from cm to m.

c) E.g. They are tired from the school day, so their reaction time is slower.

4. Braking and Energy Transfer
Page 215 — Application Questions
Q1 When the brakes are applied, work is done. This transfers energy from the kinetic energy stores of the wheels to the thermal energy stores of the brakes. This means the temperature of the brakes increases. The higher the speed of the vehicle, the greater the work done in stopping the vehicle. So stopping suddenly from a high speed causes a large energy transfer, and the brakes to overheat.
Q2 $v^2 - u^2 = 2as$ so
$a = (v^2 - u^2) \div 2s = (0^2 - 30^2) \div (2 \times 45)$
$= -10$ m/s^2
Typical car mass = 1000 kg
$F = ma = 1000 \times 10 = \textbf{10 000 N}$

5. Momentum
Page 218 — Fact Recall Questions
Q1 The object's mass and velocity.
Q2 $p = mv$. p is momentum in kg m/s, m is mass in kg and v is velocity in m/s.
Q3 They are the same.

Page 218 — Application Questions
Q1 a) $p = mv = 0.1 \times 0.6 = \textbf{0.06 kg m/s north}$
Don't forget to give a direction.
b) $p = mv = 0.00080 \times 12 = \textbf{0.0096 kg m/s left}$
c) $p = mv = 5.2 \times 8.0 = \textbf{42 kg m/s down (2 s.f.)}$
Q2 The momentum after the collision will be the same as the momentum before the collision.
Say positive means to the right.
Total momentum before the collision
$= (m_A v_A) + (m_B v_B) = (98.0 \times 1.75) + (53.0 \times -2.31)$
$= \textbf{49.1 kg m/s to the right (3 s.f.)}$
Q3 The total momentum before and after the event is 0. Momentum depends on mass and velocity, so if the gas canister's velocity before the explosion is zero, the momentum is 0. Assuming it's a closed system, the total momentum after the explosion will be equal to the momentum before, so it must be 0 as well.
Q4 a) Rearranging $p = mv$,
$v = p \div m = 3.04 \div 0.95 = \textbf{3.2 m/s south}$
b) Rearranging $p = mv$,
$v = p \div m = 45\,000 \div 2000 = \textbf{22.5 m/s east}$
Q5 a) Rearranging $p = mv$,
$m = p \div v = 31.5 \div 0.75 = \textbf{42 kg}$
b) Rearranging $p = mv$,
$m = p \div v = 210 \div 7.5 = \textbf{28 kg}$
Q6 Total momentum before = total momentum after.
Total momentum before = 0
Total momentum after = $(m_{gun} v_{gun}) + (m_{bullet} v_{bullet})$
So $0 = (1 \times -2) + (m \times 200)$
$m = 2 \div 200 = \textbf{0.01 kg}$

Q7 Say positive means to the right.
Total momentum after the collision = $m_{C+D} \, v_{C+D}$
= $(56.0 + 70.0) \times -1.50 = -189$ kg m/s
Total momentum before the collision
= $(m_c \, v_C) + (m_D \, v_D)$
$-189 = (56.0 \times 1.30) + (70.0 \times v)$, so:
$v = (-189 - 72.8) \div 70.0 = -3.74$ m/s
= **3.74 m/s to the left**
*The minus sign means that the skier D moves to the left
(because we said positive was to the right).*

6. Changes in Momentum
Page 220 — Fact Recall Question
Q1 Air bags provide a cushion of compressible air which cushions the body on impact, extending the time taken for the body to come to rest. This reduces the rate of change of momentum, and so results in a smaller the force on the body and reduced risk of injury.

Page 220 — Application Questions
Q1 $F = (m\Delta v) \div \Delta t = 27 \div 3 = $ **9 N**
Q2 $F = (m\Delta v) \div \Delta t$
so, $\Delta t = (m\Delta v) \div F = 32 \div 20 = $ **1.6 s**
Q3 $F = (m\Delta v) \div \Delta t$
so $\Delta v = F\Delta t \div m = (100 \times 0.05) \div 0.05 = 100$ m/s
$\Delta v = v - u$
$100 = v - 0$, $v = $ **100 m/s**
*You know $u = 0$ m/s because the paintball started from
rest.*

Page 223 — Car Safety and Momentum
Exam-style Questions
1.1 The other driver may have a different reaction time, which would result in a different thinking distance, and so a different stopping distance *(1 mark)*.
1.2 $v^2 - u^2 = 2as$
so $a = (v^2 - u^2) \div 2s = (0^2 - 30^2) \div (2 \times 60)$ *(1 mark)*
= -7.5 m/s^2 *(1 mark)*
$F = ma = 2000 \times 7.5$ *(1 mark)*
= **15 000 N** *(1 mark)*
1.3 Thinking distance = stopping distance – braking distance = $84 - 60 = 24$ m *(1 mark)*
Thinking distance is directly proportional to speed. The speed is halved (multiplied by 0.5), so:
new thinking distance = $24 \times 0.5 = 12$ m *(1 mark)*
Braking distance increases by the square of the scale factor of the speed increase, so:
new braking distance = $60 \times 0.5^2 = 15$ m *(1 mark)*
New stopping distance = $12 + 15 = $ **27 m** *(1 mark)*
1.4 Work is done against friction between the brakes and the wheels *(1 mark)*. This transfers energy from the kinetic energy stores of the wheels to the thermal energy stores of the brakes, causing a decrease in speed *(1 mark)*.
1.5 E.g. poor visibility due to rainfall may increase the time it takes the driver to spot the obstacle / the stopping distance will be increased because wet roads increase braking distance *(2 marks, 1 mark for each correct answer)*.

2.1 In a closed system, the total momentum before an event is the same as after the event *(1 mark)*.
2.2 Say to the right is positive.
The blue ball is at rest, so it has zero momentum
Total momentum = momentum of white ball *(1 mark)*
$p = mv$, so total momentum = $(m_{white} \times v_{white})$
$p = [0.16 \times 0.5]$ *(1 mark)*
$p = $ **0.08 kg m/s to the right** *(1 mark)*
2.3 Total momentum before the collision = 0.08 kg m/s
Total momentum after the collision
= $(m_{white} \times v_{white}) + (m_{blue} \times v_{blue})$
= $(0.16 \times 0.1) + (0.16 \times v)$ *(1 mark)*
so $0.08 = (0.16 \times 0.1) + (0.16 \times v)$ *(1 mark)*
$v = (0.08 - 0.016) \div 0.16$
= **0.4 m/s to the right** *(1 mark)*

Topic 6 — Waves

Topic 6a — Properties of Waves

1. Wave Basics
Page 225 — Fact Recall Questions
Q1 a) A wave in which the oscillations are perpendicular to the direction in which the wave travels/transfers energy.
  b) E.g. any electromagnetic wave, ripples on water, waves on a string, a slinky wiggled up and down.
  c) A wave in which the oscillations are parallel to the direction in which the wave travels/transfers energy.
Q2 In areas of compression, the particles are bunched up close together. In areas of rarefaction, the particles are more spread out.
Q3 B
Q4 No. An object will bob up and down on the ripples rather than move across the water.

2. Features of Waves
Page 227 — Fact Recall Question
Q1 a) The amplitude of a wave is the maximum displacement of a point on the wave from its undisturbed (or rest) position.
  b) The wavelength is the distance between the same point on two adjacent waves.
  c) The frequency is the number of waves passing a certain point per second.
  d) The period is the time taken for one cycle of a wave to be completed.

Page 227 — Application Questions
Q1 The wavelength is the length of one full cycle. The distance shown on the diagram is only half a cycle, so the wavelength is twice the distance shown on the diagram, **4.0 m**.
Q2 a) Amplitude = **1 cm**, Period = **4 s**
  b) $f = 1 \div T = 1 \div 4 = $ **0.25 Hz**

3. Wave Speed
Page 231 — Fact Recall Questions
Q1 $v = f\lambda$, v = speed in m/s, f = frequency in Hz,
λ = wavelength in m

Q2 E.g. Use a signal generator connected to a speaker to make a sound of known frequency. Put two microphones next to the speaker. The microphones should be connected to an oscilloscope. Slowly move one microphone away from the speaker. Its wave will shift sideways on the oscilloscope. Keep moving it until the two waves on the oscilloscope are aligned once more. At this point the microphones will be exactly one wavelength apart, so measure the distance between them. You can then use the formula $v = f\lambda$ to find the speed (v) of the sound wave passing through the air, using the wavelength you measured and the frequency of the signal generator.

Q3 330 m/s

Q4 Adjust the frequency of the vibration transducer until a clear wave can be seen on the string. Record this frequency, f. Use a metre ruler to measure the length of multiple (at least 4) half-wavelengths. Divide this length by the number of half-wavelengths to get the average half-wavelength, and then double it to find the average wavelength, λ. Then use the formula $v = f\lambda$ to find the speed (v) of the waves.

Page 231 — Application Questions
Q1 $v = f\lambda = 15 \times 0.45 = \mathbf{6.75\ m/s}$

Q2 $f = 3$ kHz = 3000 Hz
$v = f\lambda$, so $\lambda = v \div f = 1500 \div 3000 = \mathbf{0.5\ m}$

Q3 $v = f\lambda$, so $f = v \div \lambda = (3.0 \times 10^8) \div (7.5 \times 10^{-7})$
$= \mathbf{4 \times 10^{14}\ Hz}$

Q4 $\lambda = 1 \div 10 = 0.1$ cm, which is 0.001 m
$v = f\lambda = 100 \times 0.001 = \mathbf{0.1\ m/s}$
Make sure the frequency is in Hz and the wavelength is in metres before you use the wave equation.

4. Refraction of Waves
Page 234 — Fact Recall Questions
Q1 When a wave changes direction as it passes across the boundary between two materials at an angle to the normal.

Q2 E.g.

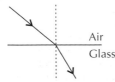

Air
Glass

The exact angles aren't important, but it must bend towards the normal.

Q3 Its path doesn't change direction.

Q4 Its speed and wavelength will increase. Its frequency won't change.

Q5 The wavefronts are closer together.

Q6 It bends away from the normal.

Page 234 — Application Questions
Q1

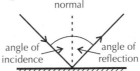

40°
60°

Q2 The wave in Q1 **speeds up** as it enters the new substance.

5. Reflection of Waves
Page 236 — Fact Recall Questions
Q1 E.g.

normal

angle of incidence   angle of reflection

Q2 Specular reflection is when parallel waves are reflected in a single direction by a smooth surface. Diffuse reflection is when parallel waves are reflected by a rough surface (e.g. a piece of paper) and the reflected rays are scattered in lots of different directions.

Page 236 — Application Questions
Q1 The glass is much smoother than the wood so that parallel light rays are all reflected in the same direction (i.e. the reflection is specular). The rough wood scatters light rays in all directions (i.e. the reflection is diffuse).

Q2

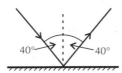

40°   40°

The angle of incidence is always equal to the angle of reflection.

6. Investigating Light
Page 238 — Fact Recall Questions
Q1 E.g. Place a transparent rectangular block on a piece of paper and trace around it. Use a ray box or a laser to shine a ray of light at the middle of one side of the block at an angle to it. Trace the incident ray and mark where the light ray emerges on the other size of the block. Remove the block and, with a straight line, join up the incident ray and the emerging point to show the path of the refracted ray through the block.
Draw the normal at the point where the light ray entered the block. Use a protractor to measure the angles of incidence and refraction. Repeat this experiment using rectangular blocks made from different materials, keeping the incident angle the same throughout. The angles of refraction will show you how much different materials refract light.

Q2 E.g. Take a piece of paper and draw a straight line across it. Place an object so one of its sides lines up with this line. Shine a ray of light at the object's surface at an angle. Make a note of the width and brightness of the reflected light ray. Repeat with other surfaces and compare the width and brightness of the reflected rays.

Page 238 — Application Question
Q1 The reflected beam from the polystyrene will be wide and dim (or not visible at all), whereas the reflected beam from the aluminium foil will be thin and bright.

Pages 240-241 — Properties of Waves
Exam-style Questions
1.1 $3.0 \div 2 = $ **1.5 m *(1 mark)***
1.2 frequency *(1 mark)*
1.3 $T = 1 \div f = 1 \div 2$ *(1 mark)* = **0.5 s *(1 mark)***
1.4 wave speed = frequency × wavelength / $v = f\lambda$ *(1 mark)*
1.5 $v = 2 \times 1.5$ *(1 mark)* = **3 m/s *(1 mark)***
1.6 E.g. they transfer energy / can be reflected / can be refracted *(1 mark)*.
1.7

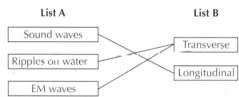

(2 marks for matching all three waves in List A correctly, otherwise 1 mark for matching two waves correctly.)

2.1

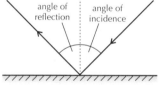

(1 mark for reflected ray drawn correctly, 1 mark for angles labelled correctly.)
Use a protractor to make sure the angle of incidence is equal to the angle of reflection.

2.2 The angles of reflection don't match the angles of incidence. / The normals aren't drawn perpendicular to points where the rays meet the rough surface. *(1 mark)*
2.3 Diffuse reflection *(1 mark)*
3.1 E.g. Laser / ray box *(1 mark)*. A thin beam is needed so it can be traced accurately *(1 mark)*.

3.2

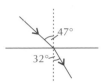

(1 mark for basic construction of ray diagram, with incident ray at an angle to the boundary and the refracted ray bent towards the normal, 1 mark for accurately drawn angles of incidence and refraction.)

3.3 How to grade your answer:
Level 0: There is no relevant information.
          (No marks)
Level 1: There is a brief description of how the wave changes. *(1 to 2 marks)*
Level 2: There is some explanation of how the diagram relates to the changes in the wave. *(3 to 4 marks)*
Level 3: There is a clear and detailed explanation of how the diagram relates to the changes in the wave. *(5 to 6 marks)*
Here are some points your answer may include:
The wavefronts are shown closer together in the glass block. This indicates a decrease in wavelength.
The frequency does not change.
Because frequency doesn't change, the speed of the wave must also decrease in the glass block.
This is because the wave obeys the wave equation, $v = f\lambda$, so a decrease in wavelength at a constant frequency must mean a decrease in velocity.
The direction of the wavefronts changes in the block.
The wave has bent towards the normal. This is because the part of the wave that reaches the glass block first starts slowing down before the rest of the wave.

Topic 6b — Electromagnetic Waves

1. What are Electromagnetic Waves?
Page 244 — Fact Recall Questions
Q1 Transverse
Q2 A continuous spectrum of all the possible wavelengths (or frequencies) of electromagnetic waves.
Q3 Gamma rays, X-rays, ultraviolet, visible light, infrared, microwaves, radio waves.
Q4 Gamma rays
Q5 The speed of any electromagnetic wave in a vacuum is the same (about 3×10^8 m/s).
Q6 E.g. A hot object (the source) emits infrared radiation as it cools down. This radiation is absorbed by the air around it (the absorber) and transfers energy to the thermal energy stores of the air.
Q7 Electrons falling to a lower energy level from a higher one and a nucleus releasing excess energy after a radioactive decay.
Q8 It can be absorbed, reflected or transmitted. No, the same thing won't always happen to any EM wave — different wavelengths may behave differently at the same boundary.

Page 244 — Application Question
Q1 a) The wavelength increases.
    b) Radio waves

2. Radio Waves
Page 246 — Fact Recall Questions
Q1 Alternating current.
Q2 TV and radio communications.
Q3 They can be reflected back and forth between the atmospheric layers and the Earth to travel long distances.

3. More EM Waves and their Uses
Page 250 — Fact Recall Questions
Q1 Microwaves and infrared radiation.
Q2 An electric heater contains a wire that gets hot when current flows through it. The wire then emits infrared radiation, which is absorbed by the air and objects in the room / is transferred to the thermal energy stores of the air and objects in the room. This results in an increase in temperature.
Q3 It travels along the fibre by reflecting off the fibre's walls.
Q4 E.g. in sun tanning lamps and in energy efficient lamps.
Q5 They pass easily through soft tissue like flesh, but less easily through denser material like bone, so they're useful for forming an image of the body (particularly the skeleton).

Page 250 — Application Question
Q1 Image B was taken at night. In image B, the background is darker. This means the surroundings must be cooler, so the photo was taken at night-time.

4. Dangers of EM Radiation
Page 252 — Fact Recall Questions
Q1 Any two from: sunburn, causing skin to age prematurely, damage to eyes, increased risk of skin cancer.
Q2 Radiation that has enough energy to knock electrons off atoms.
Q3 X-rays and gamma rays
Q4 They can kill cells or cause gene mutations which can lead to cancer.
Q5 The risk of harm from exposure to radiation.
Q6 How much radiation is absorbed, the type of radiation, the body tissue absorbing the radiation.

Page 252 — Application Question
Q1 a) 10 times larger
    b) E.g. the body tissues exposed to radiation in the X-ray of the pelvis are more sensitive to radiation damage than the tissues exposed in a dental X-ray. / In a pelvic X-ray, more X-rays must be used as the X-rays pass through a lot more body tissue, because the pelvis is a much larger part of the body than the mouth.

5. Visible Light and Colour
Page 255 — Fact Recall Questions
Q1 Its wavelength (or frequency).
Q2 Some (or all) wavelengths are absorbed and the rest (if any) are reflected.
Q3 a) none
    b) all
Q4 A translucent material transmits some light, whereas an opaque material doesn't.
Q5 It only transmits certain colours (wavelengths) and absorbs the others.

Page 255 — Application Questions
Q1 It most strongly reflects the wavelengths corresponding to the green part of the visible spectrum and absorbs the others.
Q2 The red light will appear red, but the green light will appear black.
Q3 The red bag only reflects red light which is absorbed by the blue filter, so the bag appears black. The blue buckle reflects blue light which is transmitted by the blue filter, so the buckle appears blue.

6. Infrared Radiation
Page 259 — Fact Recall Questions
Q1 The hotter the object, the more infrared radiation it emits in a given time.
Q2 False. All objects are continually absorbing IR radiation. Objects that are warmer than their surroundings just emit a lot of radiation too, so they cool down even though they are absorbing radiation.
Q3 They are equal.
Q4 E.g. a hollow, metal, water-tight cube. Its vertical faces have different surfaces. It is filled with hot water and readings of the infrared radiation emitted by each vertical face are taken, at a fixed distance. This is used to determine which colour and texture of face is the best radiation emitter.
Q5 At night, that side of the Earth is facing away from the Sun, so less IR radiation is being absorbed than emitted.

Page 259 — Application Questions
Q1 The person will be at a higher temperature than their surroundings, so will emit more IR radiation. This will be detected by the sensor, causing the light to come on.
Q2 The ice cream is colder than the surroundings, so it absorbs more IR radiation than it emits. This leads to a rise in its temperature. Eventually it will reach the same temperature as its surroundings (the room) and will be absorbing and emitting IR radiation at the same rate.

7. Black Bodies and Radiation
Page 261 — Fact Recall Questions
Q1 An object that absorbs all of the electromagnetic radiation that hits it.

Q2 All objects.

Q3 Intensity is power per unit area.

Q4 The intensity of every wavelength increases and the peak wavelength decreases.

Page 261 — Application Question

Q1 The Sun. Peak wavelength decreases as temperature increases, and the Sun has the lower peak wavelength of the two.

Pages 264-265 — Electromagnetic Waves
Exam-style Questions

1.1 Wavelength *(1 mark)*

1.2 Ultraviolet *(1 mark)*

1.3 Radio waves *(1 mark)* and microwaves *(1 mark)*

2.1 0.7 ÷ 0.004 *(1 mark)* = **175** times greater *(1 mark)*

2.2 X-rays are ionising — they can cause electrons to be removed from an atom *(1 mark)*. If this happens in a human cell, it can cause the cell to die, or mutate and cause cancer *(1 mark)*.

2.3 The radiation dose from a spinal X-ray is much higher than from one of the teeth *(1 mark)*. Having regular spinal X-rays could be very dangerous due to the high radiation dose, so the benefits would be outweighed by the risks *(1 mark)*.

3.1 Alternating current is made up of oscillating electrons — these produce radio waves of the same frequency as the current *(1 mark)*. These radio waves are emitted by a transmitter and absorbed by a receiver *(1 mark)*. The energy carried by the waves is transferred to the kinetic energy stores of electrons in the receiver *(1 mark)*. This causes the electrons to oscillate and, if the receiver is part of a complete circuit, generates an alternating current with the same frequency as the absorbed radio waves *(1 mark)*.

3.2 X-rays are absorbed by the Earth's atmosphere *(1 mark)*, so telescopes must be located outside the Earth's atmosphere to detect X-rays from space *(1 mark)*.

3.3 The microwaves used in a microwave oven have to be absorbed by water molecules to cook the food *(1 mark)*. Microwaves of these wavelengths could not be used to communicate with satellites because the Earth's atmosphere contains water, and so the microwaves would be absorbed by water molecules in the atmosphere *(1 mark)*.

4.1 The whole surface of the can is the same temperature, as it is all heated by the same water, so the thermometer readings are equal *(1 mark)*. The two different surface colours emit different levels of IR radiation — shiny silver is a worse emitter than matt black, so the IR sensor readings are different *(1 mark)*.

4.2 The amount of radiation detected would be lower than before *(1 mark)*. This is because the can would have cooled/decreased in temperature during the 15 minute period, and cooler objects emit less infrared radiation *(1 mark)*.

4.3 The amount of radiation detected would be less than from the matt black side *(1 mark)* but more than from the shiny silver side *(1 mark)*. This is because matt objects are better emitters than shiny objects, but black objects are better emitters than lighter-coloured ones *(1 mark)*.

5 How to grade your answer:

Level 0: There is no relevant information. *(No marks)*

Level 1: There is a brief description of how the ball will appear. *(1 to 2 marks)*

Level 2: There is some explanation of how the transmitted and reflected light relates to the appearance of the ball. *(3 to 4 marks)*

Level 3: There is a clear and detailed explanation of how the light transmitted and reflected relates to the appearance of the ball. *(5 to 6 marks)*

Here are some points your answer may include:

Light with wavelengths corresponding to the red part of the visible light spectrum will be strongly reflected by the red part of the ball.

All wavelengths of visible light will be reflected equally by the white areas of the ball.

So, when viewed from point A, the ball will appear white with a red stripe, as normal.

Only red light will be transmitted by the red filter, so the whole ball will appear red when viewed from point B. The other colours reflected from the white part of the ball will be absorbed by the red filter.

When the red light transmitted by the red filter reaches the green filter, it will be absorbed.

None of the light that has been reflected by the ball will be transmitted by the green filter, so the ball will appear black when viewed from point C.

Topic 6c — Lenses

1. Lenses and Images

Page 269 — Fact Recall Questions

Q1 The distance between the centre of the lens and the principal focus.

Q2 Concave lens

Q3 E.g.

convex lens   *concave lens*

The convex lens should bulge outwards and the concave lens should curve inwards.

Q4 The principal focus of a convex lens is the point where rays hitting the lens parallel to the axis all meet.
The principal focus of a concave lens is the point where rays hitting the lens parallel to the axis appear to have come from.

Q5 Real and virtual images.

Q6 Its size, which way up it is (upright or inverted) and whether it's real or virtual.

Page 269 — Application Question

Q1 a) A convex lens
   b) The image is smaller than the object, inverted and real.

2. Ray Diagrams
Page 274 — Fact Recall Questions

Q1

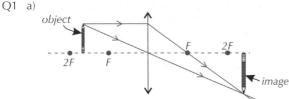

Q2 a) Real, inverted, smaller than the object and between F and 2F on the far side of the lens.
   b) Real, inverted and the same size as the object, positioned at 2F on the far side of the lens.
   c) Real, inverted and bigger than the object, and beyond 2F on the far side of the lens.
   d) Virtual, upright and bigger than the object and on the same side of the lens as the object.

Q3 Virtual, upright, smaller than the object and the same side of the lens as the object.

A concave lens always produces a virtual image that's the right way up, smaller than the object and on the same side of the lens as the object, no matter where the object is placed.

Page 274 — Application Questions

Q1 a)

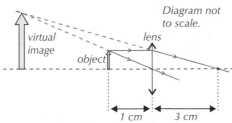

   b) Real, inverted and bigger than the object.
   c) A convex lens.

Q2 a)

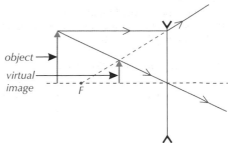

Diagram not to scale.

   b) Virtual, upright, larger than the object and on the same side of the lens as the object.

Q3 a)

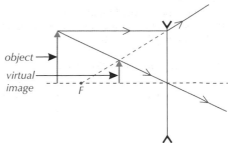

   b) Virtual, upright, smaller than the object and on the same side of the lens as the object.
   c) The image produced will still be virtual, upright, smaller than the object and on the same side of the lens as the object.

3. Magnification
Page 276 — Application Questions

Q1 a) A convex lens.
   b) magnification = $\dfrac{\text{image height}}{\text{object height}} = \dfrac{4.5}{3} = $ **1.5**

Q2 object height = $\dfrac{\text{image height}}{\text{magnification}} = \dfrac{1.5}{3} = $ **0.5 cm**

Just rearrange the equation — use a formula triangle if it helps.

Page 278 — Lenses
Exam-style Questions

1.1

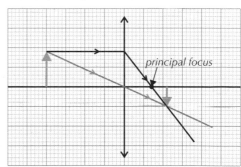

   (3 marks available — 1 mark for the second ray drawn correctly with arrows, 1 mark for arrows added to the end of the lens correctly, 1 mark for labelling the principal focus correctly.)

1.2 The image is real *(1 mark)*, inverted *(1 mark)* and smaller than the object *(1 mark)*.

1.3 magnification = $\dfrac{\text{image height}}{\text{object height}}$
   magnification = 5 ÷ 9 *(1 mark)*
   = **0.56** (to 2 significant figures) *(1 mark)*

   This answer used the number of little squares as its unit for height. Remember, you can use any unit for the height provided you use the same unit for both the image and the object. So you could have done 1 ÷ 1.8 if you'd measured them in centimetres.

1.4 The new image is virtual *(1 mark)*, upright *(1 mark)* and bigger than the object *(1 mark)*.

The object has been moved so that it is between the lens and F (i.e. closer to the lens than the principal focus). This always forms a virtual, upright image which is bigger than the object. These rules are very handy to remember — they can save you the time of having to draw out the whole diagram again. (Although don't skip the diagram if you're asked to draw it in one of these questions.)

2.1 The image is virtual *(1 mark)*, upright *(1 mark)* and smaller than the object *(1 mark)*.

2.2

(5 marks available — 1 mark for correct symbol for concave lens, 1 mark for object, lens, and principal focus in correct positions, 1 mark for correctly drawn rays parallel to the axis and through the centre, 1 mark for refracted ray traced back through principal focus, 1 mark for image drawn at intercept.)

2.3 1.7 cm *(1 mark for any value between 1.6 cm and 1.8 cm)*

You need to measure the height of the image you've drawn, and then divide it by any scale factor you used to draw your diagram.

Topic 6d — Other Waves

1. Sound Waves

Page 281 — Fact Recall Questions

Q1 When a sound wave enters a denser medium, the frequency stays the same, the speed increases, and the wavelength increases.

Q2 Sound waves hit the solid object and cause the closest particles in the solid to vibrate. These particles hit the particles next to them, causing them to vibrate. This series of vibrations is then passed through the object, transmitting the sound wave.

Q3 When sound waves reach the ear drum, they cause it to vibrate. These vibrations are passed on to other parts of the ear, which causes the sensation of sound.

Q4 Humans can hear sounds of frequency 20 Hz to 20 kHz.

Q5 They cannot hear all frequencies because the ear drum and other parts within the ear can only vibrate at certain frequencies due to their size, shape and structure.

Page 281 — Application Questions

Q1 Lead has a higher density than aluminium. So, when the sound wave moves from the lead to the aluminium, the speed of the wave will decrease. The frequency of the wave stays the same, so the wavelength must decrease.

Q2 Humans cannot hear dog whistles because the sound is outside the range of human hearing — the human ear cannot convert waves of such a high frequency into vibrations of parts of the ear, due to the size, shape and structure of these parts.

2. Ultrasound

Page 284 — Fact Recall Questions

Q1 Ultrasound is sound waves with a frequency higher than the upper limit of human hearing / with a frequency above 20 kHz.

Q2 Some of the sound waves are reflected at the boundary and some are transmitted (and possibly refracted).

Q3 E.g. prenatal scanning / forming images of internal organs.

Q4 Ultrasound pulses can be sent from a submarine to the ocean floor. The ultrasound waves will reflect off the ocean floor, and can be detected when they reach the submarine again. Using the speed of ultrasound in water, and the time between sending the ultrasound and detecting the reflected waves, the distance to the ocean floor can be calculated.

Page 284 — Application Questions

Q1 Ultrasound waves can be directed at the tins. The ultrasound will partially reflect at each boundary between materials. In a tin without a plastic shard, the ultrasound should only reflect at the boundaries between the tin and the syrup (at the near side, and the far side). If a plastic shard is present, ultrasound will reflect off the plastic shard, and a reflected ultrasound pulse will be detected between the two expected reflected pulses from the sides of the tin.

Q2 $s = vt = 1720 \times 15.0 \times 10^{-6} = 0.0258$ m
So the distance is $0.0258 \div 2 = 0.0129$ m = **12.9 mm**
Remember, the ultrasound pulse travels the distance between the transmitter and the boundary twice (there and back), so you need to divide the distance by two.

3. Seismic Waves

Page 286 — Fact Recall Questions

Q1 E.g. P-waves and S-waves.

Q2 S-waves.

Q3 P-waves.

Page 287 — Application Question

Q1 The seismometer must be placed at the bottom of the lake so that it can detect S-waves. S-waves cannot travel through liquids, so if it was placed above the bottom of the lake, S-waves could not reach the seismometer.

Page 288 — Other Waves
Exam-style Questions

1.1 Ultrasound is sound with a frequency **higher than** the range of human hearing. 20 – 20 000 **Hz** is the range of frequencies of human hearing **(2 marks)**

1.2 The distance travelled by the pulse can be found by using $s = vt$.
The speed, $v = 1500$ m/s, and the time between transmission and detection is $t = 0.000020$ s.
So $s = vt = 1500 \times 0.000020$ **(1 mark)**
$= 0.030$ m **(1 mark)**
So distance is $0.030 \div 2 = $ **0.015 m** **(1 mark)**
The distance travelled by the ultrasound pulse between transmission and detection is the total distance there and back. So you need to divide your calculated distance by two to find the answer you're looking for.

2 To reach point B from point A, the waves must travel through the liquid outer core of the Earth **(1 mark)**. P-waves can pass through both solids and liquids, so can reach B **(1 mark)** but S-waves can only pass through solids, so cannot reach B **(1 mark)**.

3.1 Find the frequency in concrete, as frequency stays the same:
$v = f\lambda$, so $f = v \div \lambda$ **(1 mark)**
$f = 3300 \div 3.3$ **(1 mark)** $= 1000$ Hz
For the steel beam:
$v = f\lambda = 1000 \times 6.0 = $ **6000 m/s** **(1 mark)**

3.2 The speed of the wave decreases **(1 mark)**, because sound travels slower in gases than solids / because it is moving into a less dense material **(1 mark)**.

Topic 7 — Magnetism and Electromagnetism

1. Magnetic Fields
Page 290 — Fact Recall Questions

Q1 It's a region where magnetic materials (like iron and steel), and also wires carrying currents, experience a force acting on them.

Q2

Q3 When not in the field of another magnet, the magnet in a compass points north. It is experiencing a magnetic force, so the Earth must have a magnetic field. For the Earth to have a magnetic field, its core must be magnetic.

Q4 An induced magnet is a magnetic material which becomes a magnet (with its own magnetic field) when it is placed inside another magnetic field, but which has no magnetic field otherwise.

2. Electromagnetism
Page 293 — Fact Recall Questions

Q1 In the right-hand thumb rule, your thumb represents the current. Your fingers represent the magnetic field.

Q2

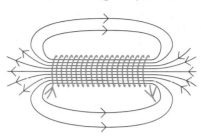

All the lines on the inside of the coil should be straight and parallel. Outside the coil, it looks just like a bar magnet.

Q3 You can increase the strength of a solenoid by placing an iron core inside the coil.
You could also increase the current in the coil.

Q4 An electromagnet is a magnet whose magnetic field can be turned on and off with an electric current. Their most common form is a coil of wire around an iron core. When a current flows through the coil of wire, a magnetic field is formed around the wire and the core is magnetised.

Q5 E.g. in scrap yard cranes used for picking up and moving scrap metal.

Page 294 — Application Questions

Q1 a) The current is flowing into the page.
Curl the fingers of your right hand in the direction of the magnetic field (i.e. clockwise). Your thumb should point into the page.
   b) The magnetic field lines would also change direction — i.e. they'd point anti-clockwise.

Q2 a) When current flows in the circuit, the electromagnet has a magnetic field. The iron bolt is magnetised and attracted to the electromagnet so it moves towards it, bolting the door.
   b) When the circuit is opened, current stops flowing through the electromagnet. The electromagnet demagnetises and the iron bolt is no longer attracted to it. The spring pulls the iron bolt back across the doorway and the door can be opened.

3. The Motor Effect
Page 298 — Fact Recall Questions

Q1 By increasing the strength of the magnetic field and increasing the current in the wire.

Q2 Zero
The wire won't feel any force if it's parallel to the magnetic field — it'll experience the greatest force when it's at 90° to the magnetic field.

Q3 $F = BIl$. F is in N (newtons), B is in T (tesla), I is in A (amperes) and l is in m (meters) This equation is only valid when the current is flowing at 90° to the direction of the magnetic field.

Q4 The first finger is the field, the second finger is the current and the thumb is the force (or motion).

Page 298 — Application Questions

Q1 a) Out of the page.
    b) Out of the page.
    c) Into the page.

Q2 Into the page.
Swapping both the current and the field will swap the direction of the force twice, with no overall effect.

Q3 $F = BIl$, so $l = F \div (B \times I)$
    $l = 0.25 \div (0.20 \times 5.0) = 0.25$ m = **25 cm**

4. Using the Motor Effect
Page 302 — Fact Recall Questions

Q1 It reverses the direction of the current every half turn so that the force on each arm of the wire loop changes direction every half turn. This way the force is always acting in a direction that causes the coil to keep rotating.

Q2 By swapping the polarity of the dc supply or swapping the magnetic poles over.
It'll only be reversed if you swap one of these things. If you swap both it'll just carry on in the same direction. If you don't believe me, check it using Fleming's left-hand rule.

Q3 The paper cone of a loudspeaker is a attached to a coil of wire that sits around one pole of a permanent magnet, and is surrounded by the other pole. When a current flows through the coil, it experiences a force due to the motor effect. With an alternating current, the force changes direction every time the current does, moving the coil, and so the cone, backwards and forwards. This makes the air around the cone vibrate, creating variations in pressure (sound waves).

Page 302 — Application Question

Q1 a) anticlockwise
    b) It'll increase.
    c) The direction of rotation will be reversed.

5. The Generator Effect
Page 307 — Fact Recall Questions

Q1 It's the creation of a potential difference across a conductor which is experiencing a change in an external magnetic field.

Q2 By moving a coil of wire in a magnetic field or by moving or changing an external magnetic field around a coil of wire.

Q3 E.g.

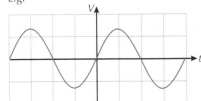

Q4 Sound waves hit a flexible diaphragm that is attached to a coil of wire, wrapped around one pole of a cylindrical magnet, and entirely surrounded by the other pole. This causes the coil of wire to move back and forward in the magnetic field, which generates an alternating current. The generated current depends on the properties of the sound wave, so variations in sound are converted into variations in current.

Page 307 — Application Question

Q1 a) The windmill rotating will cause the axle, and therefore the coil of wire, to rotate. As the coil of wire turns inside the magnetic field, it cuts the field lines and a potential difference is induced across the ends of the coil of wire by the generator effect. The ends of the wire are connected to the bulb circuit by the slip rings, so a current will flow in the bulb circuit.
    b) Alternating current.

6. Transformers
Page 313 — Fact Recall Questions

Q1 A transformer uses the generator effect to change the size of the potential difference of an ac supply.

Q2 A — primary coil
    B — iron core
    C — secondary coil

Q3 When an ac current flows through the primary coil it produces a magnetic field within the iron core. This field is constantly changing as the current changes direction, so it induces an alternating potential difference across the secondary coil by the generator effect. This pd will cause an alternating current to flow in the secondary coil if it is part of a complete circuit.

Q4 It's a transformer with more turns on the secondary coil than on the primary, so the pd across the secondary coil is greater than across the primary.

Q5 $\dfrac{V_p}{V_s} = \dfrac{n_p}{n_s}$ or $\dfrac{V_s}{V_p} = \dfrac{n_s}{n_p}$

Q6 The electrical power input is equal to the electrical power output.

Page 313 — Application Questions

Q1 a) A step-up transformer — it has more turns on the secondary coil than on the primary coil.
    b) greater

Q2 The output power will be equal to the electrical power input, so $P = V_s I_s = 12 \times 10 =$ **120 W.**

Q3 $\dfrac{V_s}{V_p} = \dfrac{n_s}{n_p}$, so $V_s = \dfrac{n_s}{n_p} \times V_p = \dfrac{12}{8} \times 4 =$ **6 V**

Q4 $V_p I_p = V_s I_s$
    So $I_p = \dfrac{V_s I_s}{V_p} = \dfrac{20 \times 100}{400} =$ **5 A**

Pages 316-317 — Magnetism and Electromagnetism
Exam-style Questions

1.1 The motor effect is when a current-carrying wire in a magnetic field experiences a force *(1 mark)*.

1.2 Reversing the direction of the magnetic field reverses the direction of the force caused by the motor effect *(1 mark)*, so the motor turns in the opposite direction (anticlockwise) *(1 mark)*.

1.3 $F = BIl$
   $F = 0.2 \times 3 \times 0.005$ *(1 mark)*
   $F = 0.003$ N
   There are 4 wires in total, and the above force on each wire, so the total force is
   $F = 0.003 \times 4 = $ **0.012 N** *(1 mark)*

1.4 When the dc supply is turned on a current flows through the coil of wire *(1 mark)*. When current flows through a wire, a magnetic field is produced around it, so the electromagnet becomes magnetic *(1 mark)*.

1.5 An alternating current supply would not be suitable because the direction of the current would be constantly changing *(1 mark)*. This would cause the direction of the magnetic field to constantly change *(1 mark)*, which would cause the direction of rotation of the coil (and hence the screwdriver) to constantly change, making it unable to function as intended *(1 mark)*.

2.1 a changing magnetic field *(1 mark)*

2.2 E.g. If direct current is supplied, a magnetic field will be induced in the core but it won't be constantly changing *(1 mark)*. If the magnetic field in the core isn't constantly changing, no potential difference will be induced in the secondary coil by the generator effect *(1 mark)*.

2.3 A step-down transformer reduces the size of the potential difference of a power supply *(1 mark)*. It must have more turns on the primary coil than on the secondary coil *(1 mark)*.

2.4 Rearranging $\frac{V_s}{V_p} = \frac{n_s}{n_p}$:
   $V_s = \frac{n_s}{n_p} \times V_p$ *(1 mark)*
   $V_s = (60 \div 50) \times 100$ *(1 mark)* = **120 V** *(1 mark)*

2.5 Rearranging $V_s I_s = V_p I_p$:
   $I_p = (V_s \times I_s) \div V_p$ *(1 mark)*
   $I_p = (120 \times 10) \div 100$ *(1 mark)*
   $= $ **12 A** *(1 mark)*

3.1 As the coil in the dynamo rotates, it induces a current which changes direction every half turn *(1 mark)*. The split-ring commutator swaps the electrical contacts every half-turn *(1 mark)*. This keeps the induced current flowing in the same direction *(1 mark)*.

3.2 E.g.

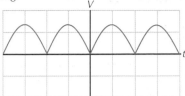

(1 mark for all points above or on the horizontal axis, 1 mark for a series of peaks which all begin and end at V = 0.)

3.3 The height of the peaks will increase *(1 mark)* and the horizontal distance between peaks will decrease (so the frequency of peaks will increase) *(1 mark)*.

Topic 8 — Space Physics

1. The Life Cycle of Stars
Page 319 — Fact Recall Questions

Q1 Gravity causes clouds of dust and gas (a nebula) to be pulled together and form a protostar. Within the protostar, gravity increases the density and causes particles to collide more frequently, increasing the temperature. When the temperature is high enough nuclear fusion begins, causing the star to give out heat and light. Smaller masses of dust and gas may also be pulled together by gravity to form planets.

Q2 Nuclear fusion

Q3 Main sequence stars are stable because the energy released by nuclear fusion provides an outward pressure which balances the force of gravity trying to pull everything inwards.

Q4 It will become a red giant. The red giant will eventually become unstable and explode, ejecting its outer layer of dust and gas. This will leave behind a white dwarf, which will eventually cool down to form a black dwarf.

Q5 Elements as heavy as iron are formed by nuclear fusion in stable stars, elements heavier than iron are formed in supernovae. Elements have been spread through the universe by supernovae.

Page 319 — Application Question
Q1 A black hole.

2. The Solar System and Orbits
Page 321 — Fact Recall Questions
Q1 A natural satellite is a naturally occurring object which orbits a planet, with smaller mass than the planet. E.g. the Moon.

Q2 Gravity.

Q3 The force acts towards the centre of the Sun / the centre of the orbit.

Page 321 — Application Question

Q1 The speed of orbit would increase, as a greater force is acting, so a greater speed is needed to keep the Earth in a stable orbit.

3. Red-shift and the Big Bang

Page 323 — Fact Recall Questions

Q1 a) The observed wavelength is greater than the wavelength of the light when it is emitted.

    b) Red-shift

Q2 a) E.g. The light observed from distant galaxies is red-shifted and it is more red-shifted the further the galaxy is away from us. This suggests the universe is expanding, not that is has remained at the same size forever.

    b) The universe began as a small region of space that was very hot and very dense. It exploded, and has been expanding ever since. This is supported by the red-shift observations described in a) which suggest the universe is expanding.

Q3 Supernova observations show that distant galaxies are receding faster and faster, hence the universe is expanding at an accelerated rate.

Page 326 — Space Physics

Exam-style Questions

1.1 The element that undergoes fusion is hydrogen *(1 mark)*. The element that is produced is helium *(1 mark)*.

1.2 The outwards pressure of the energy released by nuclear fusion *(1 mark)* balances the inward pull of gravity *(1 mark)*.

1.3 The main sequence star begins to run out of hydrogen for fusion *(1 mark)* and expands to a red super giant (fusing heavier nuclei) *(1 mark)*. The red super giant eventually becomes unstable and explodes in a supernova *(1 mark)*, leaving behind a black hole *(1 mark)*.

1.4 No, the Sun is not massive enough to become a (red super giant and hence a) black hole *(1 mark)*.

1.5 A supernova creates elements heavier than iron *(1 mark)* and ejects them into the universe *(1 mark)*.

2 How to grade your answer:
    Level 0: There is no relevant information.
               (No marks)
    Level 1: There is a brief description of why the planet orbits and how the orbital speed is related to the orbital radius. *(1 to 2 marks)*
    Level 2: There is some explanation of why the planet orbits and how the orbital speed is related to the orbital radius. *(3 to 4 marks)*
    Level 3: There is a clear and detailed explanation of why the planet orbits and how the orbital speed and orbital radius are related in a stable orbit. *(5 to 6 marks)*

Here are some points your answer may include:
There is a gravitational force of attraction between the planet and the star.
The gravitational force causes an acceleration of the planet towards the star.
The planet keeps accelerating towards the star, but the planet also has an instantaneous velocity (forward motion) that keeps it travelling in a circular path.
The closer the planet is to the star, the stronger the gravitational force between them.
A greater gravitational force requires a greater instantaneous velocity to keep the object in orbit.
So, to maintain a stable circular orbit, the closer the planet is to the star, the faster its orbital speed must be. The further away it is, the slower its orbit will be.

3.1 The Milky Way *(1 mark)*

3.2 The wavelength of observed light from a source is longer than that of the light emitted by the source *(1 mark)* when the source is moving away from the observer *(1 mark)*.

3.3 It light from most galaxies is red-shifted, then most galaxies are moving away from us *(1 mark)*. This suggests that the universe is expanding *(1 mark)*.

3.4 Observations of supernovae *(1 mark)* showed that the expansion is accelerating / distant galaxies are receding faster and faster *(1 mark)*.

3.5 No strong evidence found to support them *(1 mark)*.

Glossary

A

Absorption
When a wave transfers energy to the energy stores of a material.

Acceleration
A measure of how quickly velocity is changing.

Accurate result
A result that is very close to the true answer.

Activity (radioactive)
The number of nuclei of a sample that decay per second.

Air resistance
The frictional force caused by air on a moving object.

Alpha decay
A type of radioactive decay in which an alpha particle is given out from a decaying nucleus.

Alpha particle
A positively-charged particle made up of two protons and two neutrons (a helium nucleus).

Alpha particle scattering experiment
An experiment in which alpha particles were fired at gold foil to see if they were deflected. It led to the plum pudding model being abandoned in favour of the nuclear model of the atom.

Alternating current (ac)
Current that is constantly changing direction.

Alternator
A device which generates an ac supply using the generator effect.

Ammeter
A component used to measure the current through a component. It is always connected in series with the component.

Amplitude
The maximum displacement of a point on a wave from its rest position.

Angle of incidence
The angle the incident ray of a wave makes with the normal at a boundary.

Angle of reflection
The angle a reflected ray makes with the normal at a surface.

Angle of refraction
The angle a refracted ray makes with the normal when a wave refracts at a boundary.

Anomalous result
A result that doesn't seem to fit with the rest of the data.

Artificial satellite
A man-made satellite (normally orbiting the Earth).

Atmosphere
A relatively thin layer of air that surrounds the Earth.

Atmospheric pressure
The pressure felt by any surface within the atmosphere, due to air molecules colliding with the surface.

Atom
Particles that make up matter.

Atomic number
The number of protons in the nucleus of an atom.

Axis (of a lens)
A line passing through the middle of a lens, perpendicular to the lens.

B

Background radiation
The radiation which surrounds us at all times, arising from both natural and man-made sources.

Beta decay
A type of radioactive decay in which a beta particle is given out from a decaying nucleus.

Beta particle
A high-speed electron emitted in beta decay.

Bias
Unfairness in the way data is presented, possibly because the presenter is trying to make a particular point (sometimes without knowing they're doing it).

Big Bang theory
The idea that the universe began from a small, very hot and dense region of space, which exploded and has been expanding ever since.

Bio-fuel
A renewable energy resource made from plant products or animal dung.

Black body
An object that absorbs all the electromagnetic radiation that hits it. A black body is also the best possible emitter of radiation.

Black dwarf
The remains of a star that are left behind when a white dwarf cools.

Black hole
A super dense point in space that light cannot escape from.

Braking distance
The braking distance is the distance a vehicle travels after the brakes are applied until it comes to a complete stop, as a result of the braking force.

C

Calibrate
Measure something with a known quantity to see if the instrument being used to measure that quantity gives the correct value.

Carbon neutral fuel
A fuel is carbon neutral if it absorbs as much CO_2 from the atmosphere (when it's produced/grown) as it releases when it's burned.

Categoric data
Data that comes in distinct categories, e.g. blood type (A+, B–, etc.), metals (copper, zinc, etc.).

Chain reaction
A reaction which keeps going (without any outside input) because the products of the reaction cause further reactions (e.g. nuclear fission).

Closed system
A system where neither matter nor energy can enter or leave. The net change in total energy in a closed system is always zero.

Concave lens
A lens that curves inwards and causes rays of light parallel to the axis to diverge (spread out) so they appear to have come from the principal focus.

Conduction
A method of energy transfer by heating where vibrating particles transfer energy through a material by colliding with neighbouring particles and transferring energy between their kinetic energy stores.

Conductor (electrical)
A material in which electrical charges can easily move.

Conservation of energy principle
Energy can be transferred usefully from one energy store to another, stored or dissipated — but it can never be created or destroyed.

Conservation of momentum
In a closed system, the total momentum before an event is the same as the total momentum after the event.

Contamination (radioactive)
The presence of unwanted radioactive atoms on or inside an object.

Continuous data
Numerical data that can have any value within a range (e.g. length, volume or temperature).

Control experiment
An experiment that's kept under the same conditions as the rest of the investigation, but where the independent variable isn't altered.

Control group
A group that matches the one being studied, but where the independent variable isn't altered. The group is kept under the same conditions as the group in the experiment.

Control variable
A variable in an experiment that is kept the same.

Convection
A method of energy transfer by heating in liquids and gases in which energetic particles move away from hotter regions to cooler regions.

Conversion factor
A number which you must multiply or divide a unit by to convert it to a different unit.

Convex lens
A lens that bulges outwards and causes rays of light parallel to the axis to converge (come together) at the principal focus.

Correlation
A relationship between two variables.

Cosmic ray
Radiation from space.

Current
The flow of electric charge. The size of the current is the rate of flow of charge. Measured in amperes (A).

D

Decommissioning
The process of shutting down a power station so that it's completely safe and poses no risk to people or the environment.

Density
A substance's mass per unit volume.

Dependent variable
The variable in an experiment that is measured.

Diffuse reflection
When parallel waves are reflected by a rough surface (e.g. a piece of paper) and the reflected rays are scattered in lots of different directions.

Diode
A circuit component that only allows current to flow through it in one direction. It has a very high resistance in the other direction.

Direct current (dc)
Current that always flows in the same direction.

Discrete data
Numerical data that can only take a certain value, with no in-between value (e.g. number of people).

Displacement
The straight-line distance and direction from an object's starting position to its finishing position.

Distance-time graph
A graph showing how the distance travelled by an object changes over a period of time.

Drag
The frictional force caused by any fluid (a liquid or gas) on a moving object.

Dwarf planet
A planet-like object in space that orbits a star but which doesn't match all of the rules for being a planet.

Dynamo
A device which generates a dc supply using the generator effect.

E

Ear drum
The part of the ear which vibrates when sound waves enter the ear. It passes on these vibrations to other parts of the ear, which convert them to electrical signals that cause the sensation of hearing.

Earth wire
The green and yellow wire in an electrical cable that only carries current when there's a fault. It stops exposed metal parts of an appliance from becoming live.

Efficiency
The proportion of input energy transfer which is usefully transferred. Also the proportion of input power which is usefully output.

Elastic deformation
An object undergoing elastic deformation will return to its original shape once any forces being applied to it are removed.

Elastic object
An object which can be elastically deformed.

Elastic potential energy store
Anything that has been stretched or compressed, e.g. a spring, has energy in its elastic potential energy store.

Electric field
A region in which an electrically charged object experiences an electrostatic force.

Electromagnet
A solenoid with an iron core.

Electromagnetic (EM) spectrum
A continuous spectrum of all the possible wavelengths of electromagnetic waves.

Electron
A subatomic particle with a relative charge of −1 and a relative mass of 1/2000.

Electrostatic attraction/repulsion
The non-contact force which acts to bring together opposite charges (attraction) / push apart like charges (repulsion).

Energy store
A means by which an object stores energy. There are 8 different types of energy store: thermal, kinetic, gravitational potential, elastic potential, chemical, magnetic, electrostatic and nuclear.

Equilibrium
A state in which all the forces acting on an object are balanced, so the resultant force is zero.

Fair test
A controlled experiment where the only thing being changed is the independent variable.

Fleming's left-hand rule
The rule used to work out the direction of the force produced by the motor effect. Your first finger points in the direction of the magnetic field, your second finger points in the direction of the current and your thumb points in the direction of the force (or motion).

Fluid
A substance that can flow — either a liquid or a gas.

Focal length (of a lens)
The distance from the centre of a lens to the principal focus.

Force
A push or a pull on an object caused by it interacting with something.

Fossil fuel
The fossil fuels are coal, oil and natural gas. They're non-renewable energy resources that we burn to generate electricity.

Free body diagram
A diagram that shows all the forces acting on an isolated object, the direction in which the forces are acting and their (relative) magnitudes.

Frequency
The number of complete waves passing a certain point per second, or the number of waves produced by a source per second. Measured in hertz, Hz.

Frequency density
The height of a bar on a histogram. It is found by the frequency divided by the class width.

Friction
A force that opposes an object's motion. It acts in the opposite direction to motion.

Gamma decay
A type of radioactive decay in which a gamma ray is given out from a decaying nucleus.

Gamma ray
A high-frequency, short-wavelength electromagnetic wave.

Gear
A circular disc with teeth round its edge. It can be used to transmit the rotational effect of a force.

Geiger-Müller tube
A particle detector that is used with a counter to measure count rate.

Generator effect
The induction of a potential difference across a conductor which is experiencing a change in external magnetic field. If the conductor is part of a complete circuit, this will cause a current to flow.

Geothermal power
A renewable energy resource where energy is transferred from the thermal energy stores of hot rocks underground and is used to generate electricity or to heat buildings.

Gradient
The slope of a line graph. It shows how quickly the variable on the y-axis changes with the variable on the x-axis.

Gravitational potential energy (g.p.e) store
Anything that has mass and is in a gravitational field has energy in its gravitational potential energy store.

Greenhouse effect
The process by which gases in the Earth's atmosphere block radiation from the Sun from leaving the atmosphere. This causes the overall temperature of the atmosphere to rise (i.e. global warming).

H

Half-life
The time it takes for the number of nuclei of a radioactive isotope in a sample to halve.
OR
The time it takes for the count rate (or activity) of a radioactive sample to fall to half its initial level.

Hazard
Something that has the potential to cause harm (e.g. fire, electricity, etc.).

Hydroelectric power station
A power station in which a dam is built across a valley or river which holds back water, forming a reservoir. It allows water to flow out of the reservoir through turbines at a controlled rate. This turns the turbines, which are attached to generators and can generate electricity.

Hypothesis
A possible explanation for a scientific observation.

I

Independent variable
The variable in an experiment that is changed.

Induced magnet
A magnetic material that only has its own magnetic field, and behaves as a magnet, while it is inside another magnetic field.

Inelastic deformation
An object undergoing inelastic deformation will not return to its original shape once the forces being applied to it are removed.

Inertia
The tendency of an object to remain stationary or continue travelling at a constant velocity.

Inertial mass
The ratio between the resultant force acting on an accelerating object and its acceleration.

Infrared (IR) radiation
A type of electromagnetic wave continually emitted and absorbed by all objects.

Instantaneous velocity
The velocity of an object at a particular moment in time.

Insulator (electrical)
A material in which electrical charges cannot easily move.

Intensity
The power per unit area, i.e. how much energy is transferred to a given area in a certain amount of time. Its units are W/m^2.

Internal energy
The total energy that a system's particles have in their kinetic and potential energy stores.

Ion
An atom in which the number of electrons is different to the number of protons, giving it an overall charge.

Ionising radiation
Radiation that has enough energy to knock electrons off atoms.

Irradiation
Exposure to radiation.

Isotope
A different form of the same element, which has atoms with the same number of protons (atomic number), but a different number of neutrons (and so different mass number).

I-V characteristic
A graph of current against potential difference for a component.

K

Kinetic energy store
Anything that's moving has energy in its kinetic energy store.

L

Latent heat
The energy required to change the state of a substance without changing its temperature.

Law of reflection
The angle of reflection of a reflected ray is always equal to the angle of incidence.

Lever
A device that increases the distance between an applied force and a pivot. It transmits the rotational effect of a force and acts as a force multiplier.

Light-dependent resistor (LDR)
A resistor whose resistance is dependent on light intensity. The resistance decreases as light intensity increases.

Limit of proportionality
The point beyond which the force applied to an elastic object is no longer directly proportional to the extension of the object.

Line of action (of a force)
A straight line passing through the point at which the force is acting in the same direction as the force.

Linear graph
A straight line graph for which $y = mx + c$, where m = gradient and c = y-intercept.

Live wire
The brown wire in an electrical cable that carries an alternating potential difference from the mains.

Longitudinal wave
A wave in which the oscillations are along the same line as the direction of energy transfer.

Lubricant
A substance (usually a liquid) that can flow easily between two objects. Used to reduce friction between surfaces.

M

Magnetic field
A region where magnetic materials (like iron and steel) and current-carrying wires experience a force.

Magnetic flux density
The number of magnetic field lines per unit area. Its symbol is B and it is measured in tesla, T.

Magnetic material
A material (such as iron, steel, cobalt or nickel) which can become an induced magnet while it's inside another magnetic field.

Magnification
The ratio of the size of the image to the size of the object.

Main sequence star
A star in the main sequence of its life, which is stable because the nuclear fusion in the star provides an outward pressure that balances the inward pull of gravity.

Mass number
The number of neutrons and protons in the nucleus of an atom.

Mean (average)
A measure of average found by adding up all the data and dividing by the number of values there are.

Median (average)
A measure of average found by selecting the middle value from a data set arranged in ascending order.

Medical tracer
A radioactive isotope that can be injected into or swallowed by people. Their progress around the body can be followed using an external detector and can diagnose medical conditions.

Microwave
A type of electromagnetic wave that can be used for cooking and satellite communications.

Mode (average)
A measure of average found by selecting the most frequent value from a data set.

Model
Used to describe or display how an object or system behaves in reality.

Moment
The turning effect of a force.

Momentum
A property of a moving object that is the product of its mass and velocity.

Moon
A natural satellite which orbits a planet.

Motor effect
When a current-carrying wire in a magnetic field experiences a force.

N

National grid
The network of transformers and cables that distributes electrical power from power stations to consumers.

Nebula
A cloud of dust and gas in space.

Neutral wire
The blue wire in an electrical cable that carries away current from the appliance. It is around 0 V.

Neutron
A subatomic particle with a relative charge of 0 and a relative mass of 1.

Neutron star
The very dense core of a star that is left behind when a red super giant explodes in a supernova.

Newton's First Law
An object will remain at rest or travelling at a constant velocity unless it is acted on by a resultant force.

Newton's Second Law
The acceleration of an object is directly proportional to the resultant force acting on it, and inversely proportional to its mass.

Newton's Third Law
When two objects interact, they exert equal and opposite forces on each other.

Non-contact force
A force that can act between objects that are not touching.

Non-renewable energy resource
An energy resource that is non-renewable cannot be made at the same rate as it's being used, so it will run out one day.

Normal (at a boundary)
A line that's perpendicular (at 90°) to a surface at the point of incidence (where a wave hits the surface).

Nuclear fission
When an atomic nucleus splits up to form two smaller nuclei.

Nuclear fusion
When two nuclei join to create a heavier nucleus.

Nuclear model
A model of the atom that says that the atom has a small, central positively-charged nucleus with negatively-charged electrons moving around the nucleus, and that most of the atom is empty space. The nucleus is made up of protons and neutrons.

Nucleus (atom)
The centre of an atom, containing protons and neutrons.

O

Ohmic conductor
A conductor with resistance that is constant at a constant temperature. It has a linear *I-V* characteristic.

Orbit
The path on which one object moves around another.

P

Parallel circuit
A circuit in which every component is connected separately to the positive and negative ends of the battery.

Partial reflection
When waves are incident on a boundary, some are reflected and some are transmitted.

Peer-review
The process in which other scientists check the results and explanations of an investigation before they are published.

Period (of a wave)
The time taken for one full cycle of a wave to be completed.

Permanent magnet
A magnetic material that always has its own magnetic field around it.

Physical change
A change where you don't end up with a new substance — it's the same substance as before, just in a different form. (A change of state is a physical change.)

Planet
A natural object in space which orbits a star.

Potential difference
The driving force that pushes electric charge around a circuit, measured in volts (V). Also known as pd or voltage.

Power
The rate of transferring energy (or doing work). Normally measured in watts (W).

Precise result
When all the data is close to the mean.

Prediction
A statement that can be tested and is based on a hypothesis.

Pressure
The force per unit area exerted on a surface.

Principal focus of a concave lens
The point where rays hitting the lens parallel to the axis appear to have come from.

Principal focus of a convex lens
The point where rays hitting the lens parallel to the axis all meet.

Proton
A subatomic particle with a relative charge of +1 and a relative mass of 1.

Protostar
The earliest stage in the life cycle of a star. Protostars are formed when the force of gravity causes clouds of dust and gas to spiral together.

R

Radiation dose
A measure of the risk of harm to your body due to exposure to radiation.

Radio wave
A type of electromagnetic wave mainly used for radio and TV signals.

Radioactive decay
The random process of a radioactive substance giving out radiation from the nuclei of its atoms.

Radioactive substance
A substance that spontaneously gives out radiation from the nuclei of its atoms.

Radiotherapy
A treatment of cancer that uses ionising radiation (such as gamma rays and X-rays) to kill cancer cells.

Random error
A difference in the results of an experiment caused by unpredictable events, e.g. human error in measuring.

Range
The difference between the smallest and largest values in a set of data.

Range of human hearing
The range of frequencies of sound waves that humans can hear. It's 20 Hz to 20 kHz.

Ray
A straight line showing the direction of energy transfer of a wave, indicating the path along which the wave moves.

Ray diagram
A diagram that shows the path of light waves.

Reaction time
The time taken for a person to react after an event (e.g. seeing a hazard).

Real image
An image formed when light rays from a point on an object come together at another point — the light rays actually pass through that point.

Red giant
A type of star that is formed when a star around the same size as the Sun expands as it starts to run out of hydrogen.

Red super giant
A type of star that is formed when a large star (much bigger than the Sun) expands as it starts to run out of hydrogen.

Red-shift
The shift in observed wavelength of light from a source moving away from a stationary observer. The wavelength is shifted towards the red end of the electromagnetic spectrum.

Reflection
When a wave bounces back as it meets a boundary between two materials.

Refraction
When a wave changes direction as it passes across the boundary between two materials at an angle to the normal.

Reliable result
A result that is repeatable and reproducible.

Renewable energy resource
An energy resource that is renewable is one that is being, or can be, made at the same rate (or faster) than it's being used, and so will never run out.

Repeatable result
A result that will come out the same if the experiment is repeated by the same person using the same method and equipment.

Reproducible result
A result that will come out the same if someone different does the experiment, or a slightly different method or piece of equipment is used.

Resistance
Anything in a circuit that reduces the flow of current. Measured in ohms, Ω.

Resolution
The smallest change a measuring instrument can detect.

Resultant force
A single force that can replace all the forces acting on an object to give the same effect as the original forces acting altogether.

Right-hand thumb rule
The rule to work out the direction of the magnetic field around a current-carrying wire. Your thumb points in the direction of the current, and your fingers curl in the direction of the magnetic field.

Risk
The chance that a hazard will cause harm.

S

Satellite
An object which orbits a second more massive object.

Scalar
A quantity that has magnitude but no direction.

Scaling prefix
A word or symbol which goes before a unit to indicate a multiplying factor (e.g. 1 km = 1000 m).

Seismic wave
A wave which travels through (or over the surface of) the Earth when an earthquake occurs. Two important types are P-waves and S-waves.

Series circuit
A circuit in which every component is connected in a line, end to end.

S.I. unit
A unit recognised as standard by scientists all over the world.

Significant figure
The first significant figure of a number is the first non-zero digit. The second, third and fourth significant figures follow on immediately after it.

Solar cell
A device that generates electricity directly from the Sun's radiation.

Solenoid
A coil of wire often used in the construction of electromagnets.

Sound wave
A longitudinal wave caused by vibrating particles. When they reach the ears, they can cause vibrations that are converted into electrical signals, which cause the sensation of hearing.

Spark
The passage of electrons across a (usually) small gap between a static charge and an earthed conductor.

Specific heat capacity
The amount of energy (in joules) needed to raise the temperature of 1 kg of a material by 1°C.

Specific latent heat (SLH)
The amount of energy needed to change 1 kg of a substance from one state to another without changing its temperature. (For cooling, it is the energy released by a change in state.)

Specific latent heat of fusion
The specific latent heat for changing between a solid and a liquid (melting or freezing).

Specific latent heat of vaporisation
The specific latent heat for changing between a liquid and a gas (evaporating, boiling or condensing).

Specular reflection
When parallel waves are reflected in a single direction by a smooth surface.

Split-ring commutator
A ring with gaps in it that swaps the electrical contacts of a device every half-turn.

Standard form
A number written in the form $A \times 10^n$, where A is a number between 1 and 10.

State of matter
The form which a substance can take — e.g. solid, liquid or gas.

Static charge
An electric charge that cannot move. It often forms on electrical insulators, where charge cannot flow freely.

Stopping distance
The distance covered by a vehicle in the time between the driver spotting a hazard and the vehicle coming to a complete stop. It's the sum of the thinking distance and the braking distance.

Supernova
The explosion of a red super giant.

System
The object, or group of objects, that you're considering.

Systematic error
An error that is consistently made throughout an experiment.

T

Tangent
A straight line that touches a curve at a point but doesn't cross it.

Terminal velocity
The maximum velocity a falling object can reach without any added driving forces. It's the velocity at which the resistive forces (drag) acting on the object match the force due to gravity (weight).

Theory
A hypothesis which has been accepted by the scientific community because there is good evidence to back it up.

Thermal conductivity
A measure of how quickly an object transfers energy by heating through conduction.

Thermal insulator
A material with a low thermal conductivity.

Thermistor
A resistor whose resistance is dependent on the temperature. The resistance decreases as temperature increases.

Thinking distance
The distance a vehicle travels during the driver's reaction time (before the brakes have been applied).

Three-core cable
An electrical cable containing a live wire, a neutral wire and an earth wire.

Tidal barrage
A dam built across a river estuary with turbines connected to generators. When there's a difference in water height on either side, water flows through the dam, turning the turbines and generating electricity.

Transformer
A device which can change the potential difference of an ac supply.

Transmission (of a wave)
When a wave passes through a boundary from one material into another and continues travelling.

Transverse wave
A wave in which the oscillations are perpendicular (at 90°) to the direction of energy transfer.

Trial run
A quick version of an experiment that can be used to work out the range of variables and the interval between the variables that will be used in the proper experiment.

Ultrasound
Sound with a frequency higher than the range of human hearing (i.e. greater than 20 000 Hz).

Ultraviolet (UV) radiation
A type of electromagnetic wave, the main source of which is sunlight.

Uncertainty
The amount by which a given result may differ from the true value.

Upthrust
The resultant force acting upwards on an object submerged in a liquid, due to the pressure of the liquid being greater at the bottom of the object than at the top.

Valid result
A result that is repeatable, reproducible and answers the original question.

Vector
A quantity which has both magnitude (size) and a direction.

Velocity
The speed and direction of an object.

Velocity-time graph
A graph showing how the velocity of an object changes over a period of time.

Virtual image
An image that is formed when light rays appear to have come from one point, but have actually come from another — the light rays don't actually pass through that point.

Visible light
The part of the electromagnetic spectrum that we can see with our eyes.

Voltmeter
A component used to measure the potential difference across a component. Always connected in parallel with the component.

Wave
An oscillation that transfers energy without transferring any matter.

Wavefront
A line perpendicular to the direction of energy transfer of a wave. It is used to represent a crest (or trough) of a wave in a wavefront diagram.

Wavelength
The length of a full cycle of a wave, e.g. from a crest to the next crest.

Weight
The force acting on an object due to gravity.

White dwarf
The hot, dense core left behind when a red giant becomes unstable and ejects its outer layer of dust and gas.

Work done
The energy transferred when a force moves an object through a distance, or by a moving charge.

X-ray
A high-frequency, short-wavelength electromagnetic wave. It is mainly used in medical imaging and treatment.

Zero error
A type of systematic error caused by using a piece of equipment that isn't zeroed properly.

Acknowledgements

Cover photo **Chris Madeley**/Science Photo Library, p 5 **Alastair Philip Wiper**/Science Photo Library, p 6 **Frank Zullo**/Science Photo Library, p 7 **Belmonte**/Science Photo Library, p 8 **Trevor Clifford Photography**/Science Photo Library, p 9 **Philippe Plailly**/Science Photo Library, p 12 iStock.com/**tunart**, p 23 **Martyn F. Chillmaid**/Science Photo Library, p 31 **Tony McConnell**/Science Photo Library, p 36 **Martyn F. Chillmaid**/Science Photo Library, p 37 (top) Science Photo Library, p 37 (bottom) **Mark Sykes**/Science Photo Library, p 38 **Martyn F. Chillmaid**/Science Photo Library, p 41 **Sheila Terry**/Science Photo Library, p 42 **Mark Sykes**/Science Photo Library, p 48 **Ashley Cooper**/Science Photo Library, p 49 **Chris Hellier**/Science Photo Library, p 50 (top) **Paul Rapson**/Science Photo Library, p 50 (bottom) **Martin Bond**/Science Photo Library, p 51 **Martin Bond**/Science Photo Library, p 52 **David Parker**/Science Photo Library, p 53 **Martin Bond**/Science Photo Library, p 54 **Martin Bond**/Science Photo Library, p 55 **Matteis/Look At Sciences**/Science Photo Library, p 56 **U.S. Coast Guard**/Science Photo Library, p 57 **Martin Bond**/Science Photo Library, p 58 iStock.com/**Amr_Photos**, p 59 **David Woodfall Images**/Science Photo Library, p 67 **Trevor Clifford Photography**/Science Photo Library, p 68 (top) **Andrew Lambert Photography**/Science Photo Library, p 68 (bottom) **Lawrence Lawry**/Science Photo Library, p 71 **Cordelia Molloy**/Science Photo Library, p 75 **Philippe Psaila**/Science Photo Library, p 82 (top) **Martyn F. Chillmaid**/Science Photo Library, p 82 (bottom) **Martyn F. Chillmaid**/Science Photo Library, p 83 **Andrew Lambert Photography**/Science Photo Library, p 90 **Martyn F. Chillmaid**/Science Photo Library, p 92 **Martyn F. Chillmaid**/Science Photo Library, p 100 **Mike Hollingshead**/Science Photo Library, p 101 Science Photo Library, p 107 **Mehau Kulyk**/Science Photo Library, p 108 Science Photo Library, p 111 (top) **Charles D. Winters**/Science Photo Library, p 111 (bottom) **GiPhotoStock**/Science Photo Library, p 115 (top) **Ted Kinsman**/Science Photo Library, p 115 (bottom) **Ted Kinsman**/Science Photo Library, p 121 **Emilio Segre Visual Archives**/**American Institute of Physics**/Science Photo Library, p 132 **Trevor Clifford Photography**/Science Photo Library, p 135 **Lawrence Livermore Laboratory**/Science Photo Library, p 136 **Steve Allen**/Science Photo Library, p 138 **Oulette & Theroux, Publiphoto Diffusion**/Science Photo Library, p 139 **Dr P. Marazzi**/Science Photo Library, p 147 iStock.com/**Hallgerd**, p 149 **Martyn F. Chillmaid**/Science Photo Library, p 151 iStock.com/**Ideabug**, p 156 **Lee Powers**/Science Photo Library, p 160 **Sputnik**/Science Photo Library, p 166 (top) **Peter Menzel**/Science Photo Library, p 166 (bottom) **Trevor Clifford Photography**/Science Photo Library, p 167 **'Leonard Lessin, Fbpa'**/Science Photo Library, p 170 **Geoff Tompkinson**/Science Photo Library, p 173 (top) **Laurence Lawry**/Science Photo Library, p 173 (bottom) iStock.com/**Alxpin**, p 179 **Andy Williams**/Science Photo Library, p 181 **Alan and Sandy Carey**/Science Photo Library, p 190 iStock.com/**Bsnider**, p 191 **Gustoimages**/Science Photo Library, p 192 (top) **NASA**/Science Photo Library, p 192 (bottom) **Sputnik**/Science Photo Library, p 195 **NASA/Joel Kowsky**/Science Photo Library, p 197 **Ashley Cooper**/Science Photo Library, p 199 **Martyn F. Chillmaid**/Science Photo Library, p 201 **Sheila Terry**/Science Photo Library, p 209 (top) **Ton Kinsbergen**/Science Photo Library, p 209 (bottom) **David Woodfall Images**/Science Photo Library, p 209 data used to construct stopping distance diagram on page 209 from the Highway Code, contains public sector information licensed under the open government licence v3.0, **http://www.nationalarchives.gov.uk/doc/open-government-licence/version/3/**, p 213 **Sputnik**/Science Photo Library, p 217 **Sputnik**/Science Photo Library, p 220 (top) **CC Studio**/Science Photo Library, p 220 (bottom) **Gustoimages**/Science Photo Library, p 225 **David Weintraub**/Science Photo Library, p 227 **Tek Image**/Science Photo Library, p 232 iStock.com/**AlistairCotton**, p 235 **GiPhotoStock**/Science Photo Library, p 236 **GiPhotoStock**/Science Photo Library, p 237 **Andrew Lambert Photography**/Science Photo Library, p 243 **Ton Kinsbergen**/Science Photo Library, p 244 **David Parker**/Science Photo Library, p 246 **Alex Bartel**/Science Photo Library, p 247 **Chris Martin-Bahr**/Science Photo Library, p 248 (top left) **Sputnik**/Science Photo Library, p 248 (top right) **Tony McConnell**/Science Photo Library, p 248 (bottom) **Mark Sykes**/Science Photo Library, p 249 (top) **Alfred Pasieka**/Science Photo Library, p 249 (middle) Science Photo Library, p 249 (bottom) **Jim West**/Science Photo Library, p 250 **Du Cane Medical Imaging Ltd**/Science Photo Library, p 251 (top) **Sue Ford**/Science Photo Library, p 251 (bottom) **BSIP, Laurent**/Science Photo Library, p 252 **Public Health England**/Science Photo Library, p 254 **Martyn F. Chillmaid**/Science Photo Library, p 257 **Cordelia Molloy**/Science Photo Library, p 266 **David Parker**/Science Photo Library, p 267 **David Parker**/Science Photo Library, p 269 iStock.com/**Echo1**, p 272 (top) **Lea Paterson**/Science Photo Library, p 272 (bottom) iStock.com/**Jasantiso**, p 275 **Andrew Lambert Photography**/Science Photo Library, p 283 (top) **Cavallini James/BSIP**/Science Photo Library, p 283 (bottom) **NASA**/Science Photo Library, p 285 **Zephyr**/Science Photo Library, p 290 **Dorling Kindersley/UIG**/Science Photo Library, p 292 (top) **Andrew Lambert Photography**/Science Photo Library, p 292 (middle) **CERN**/Science Photo Library, p 292 (bottom) **Alex Bartel**/Science Photo Library, p 296 **Trevor Clifford Photography**/Science Photo Library, p 302 **Andrew Lambert Photography**/Science Photo Library, p 304 **Public Health England**/Science Photo Library, p 305 **Victor de Schwanberg**/Science Photo Library, p 306 Science Photo Library, p 307 **Dorling Kindersley/UIG**/Science Photo Library, p 318 **National Optical Astronomy Observatories**/Science Photo Library, p 319 **European Southern Observatory**/Science Photo Library, p 327 (top) **Andrew Lambert Photography**/Science Photo Library, p 327 (bottom) **GiPhotoStock**/Science Photo Library, p 328 **Charles D. Winters**/Science Photo Library, p 329 **GiPhotoStock**/Science Photo Library, p 330 (top) **Trevor Clifford Photography**/Science Photo Library, p 330 (bottom) **Crown Copyright**/Health & Safety Laboratory Science Photo Library, p 331 (top) Science Photo Library, p 331 (bottom) **Martyn F. Chillmaid**/Science Photo Library, p 332 Science Photo Library, p 337 iStock.com/**Wavebreak**, p 341 iStock.com/**Sorendls**

Index

Equations Page

In each paper you have to sit for your Physics GCSE, you'll be given an equations sheet listing some of the equations you might need to use. That means you don't have to learn them (hurrah), but you still need to be able to pick out the correct equations to use and be really confident using them. The equations sheet won't give you any units for the equation quantities — so make sure you know them inside out.

The equations you'll be given in the exam are all on this page. You can use this page as a reference when you're doing the exam-style questions at the end of each section.

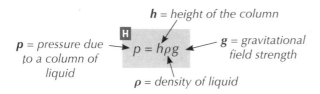

h = height of the column
p = pressure due to a column of liquid
$$p = h\rho g$$
g = gravitational field strength
ρ = density of liquid

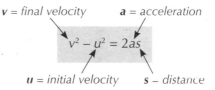

v = final velocity a = acceleration
$$v^2 - u^2 = 2as$$
u = initial velocity s – distance

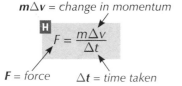

$m\Delta v$ = change in momentum
$$F = \frac{m\Delta v}{\Delta t}$$
F = force Δt = time taken

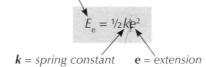

E_e = elastic potential energy
$$E_e = \tfrac{1}{2}ke^2$$
k = spring constant e = extension

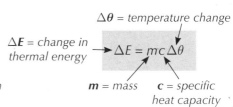

$\Delta\theta$ = temperature change
ΔE = change in thermal energy
$$\Delta E = mc\Delta\theta$$
m = mass c = specific heat capacity

$$\text{period} = \frac{1}{\text{frequency}}$$

$$\text{magnification} = \frac{\text{image height}}{\text{object height}}$$

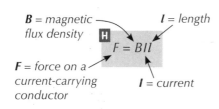

B = magnetic flux density l = length
$$F = BIl$$
F = force on a current-carrying conductor I = current

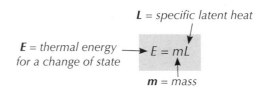

L = specific latent heat
E = thermal energy for a change of state
$$E = mL$$
m = mass

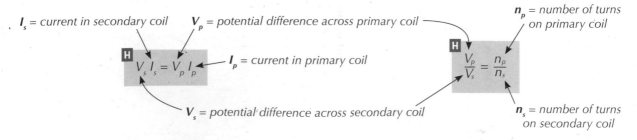

I_s = current in secondary coil V_p = potential difference across primary coil
$$V_s I_s = V_p I_p$$
I_p = current in primary coil
V_s = potential difference across secondary coil

n_p = number of turns on primary coil
$$\frac{V_p}{V_s} = \frac{n_p}{n_s}$$
n_s = number of turns on secondary coil

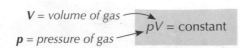

V = volume of gas
p = pressure of gas
$$pV = \text{constant}$$

PATB42